NOVEL NANOCOMPOSITE COATINGS

NOVEL NANOCOMPOSITE COATINGS

Advances and Industrial Applications

Rostislav Daniel | Jindřich Musil

Pan Stanford Publishing

Published by

Pan Stanford Publishing Pte. Ltd.
Penthouse Level, Suntec Tower 3
8 Temasek Boulevard
Singapore 038988

Email: editorial@panstanford.com
Web: www.panstanford.com

British Library Cataloguing-in-Publication Data
A catalogue record for this book is available from the British Library.

Novel Nanocomposite Coatings: Advances and Industrial Applications

ISBN 978-981-4411-17-2 (Hardcover)
ISBN 978-981-4411-18-9 (eBook)

Printed in the USA

To Jitka, Ellen, and Sophie

Contents

Preface

The field of plasma-based thin film processing has grown rapidly over the past two decades. The coating technology evolved and developed in many industries, including tooling industry and electronic or manufacturing industry. In all these fields, thin films have found their usage as decorative and metallurgical films, diffusion barriers in microelectronics, films for high-temperature applications, and cutting or forming applications. Indeed, plasma processing technology has a number of manifestations, from simple dc to pulsed glow discharges over microwave and rf plasmas to arc discharges of very specific characteristics. Due to high versatility in depositing a wide range of materials under a great range of conditions combined with high deposition rates, the plasma processing technologies have become the most preferred technologies in the last few years. However, very complex relations between the film growth conditions, forming structures and properties of thin films have to be fully understood in order to prepare new film systems with optimized properties.

This book is intended to provide a perspective look at a range of thin film plasma processing technologies and give an overview on principles of the film formation so that the complex structure–property relation can be more easily and intuitively understood. The mean part of the book comprises a review on Si-containing nanocomposite films based on transition metal nitrides with a wide range of compositions, focused especially on the novel amorphous-like nanocomposite Me-Si-N films with a high (≥25 at.%) Si content. Two selected high Si-containing nanocomposites are compared for their mechanical and high-temperature properties to demonstrate the importance of the structure and phase composition on their thermal stability and resistance against oxidation.

This book is presented in six chapters. Following the introductory Chapter 1, which gives an overview of materials development from traditional polycrystalline coarse-grained bulk materials to thin nanocrystalline and nanocomposite films with enhanced properties, Chapter 2 comprises fundamentals of thin film processing using plasma discharges as the working medium. Principles of plasma discharges and plasma chemistry are introduced for a better understanding of the complexity of the plasma processing technology. Further in this chapter, fundamentals of physical sputtering related to the processes involved during thin film deposition from solid targets are discussed. An overview of various deposition techniques helps the reader to understand the advantages and limitations in sputtering of thin films with specific structures and properties. Chapter 3 summarizes basic principles of condensation of sputtered adatoms on the substrate surfaces and film formation. This chapter also considers the manipulation of the plasma sputtering environment to influence the microstructure and properties of thin films by varying the deposition conditions. Here, the addition of alloying elements and the effect of their segregation to surfaces and grain boundaries on the film structure and morphology is discussed in details. Moreover, the structure–property relations in hard films are discussed in this chapter. Chapter 4 introduces the present status of the basic research on Me-Si-N films with low and intermediate Si content (<25 at.%). Here, the synergetic effect of the silicon and nitrogen content on the structure, morphology, and phase composition of the films is discussed with mechanical and other physical properties of the films. Development of the nanocomposite structure is also correlated with residual stresses and thermal stability of the Me-Si-N films with various crystallite-to-amorphous phase volume fraction ratio. Furthermore, oxidation mechanisms and oxidation kinetics through very broad Si content range in the films is also presented. Chapter 5 is dedicated to the investigation of the novel nanocomposite Zr-Si-N films with high Si content (≥25 at.%). This chapter summarizes the development of the structure, elemental composition, chemical bonding, and phase composition as a function of increasing partial pressure of nitrogen used during film deposition. The findings are further correlated with residual stresses and mechanical and physical properties of the films. Main scope of this chapter is focused on the thermal stability and resistance to oxidation of the Zr-Si-N films showing very close relations with the volume fraction ratio between crystalline and amorphous phases forming the films. Chapter 6 deals with another transition metal nitride material alloyed with a high amount of silicon. The W-Si-N films deposited from an alloyed WSi_2 target serve for comparison with the Zr-Si-N films with the same silicon content so that the difference in their properties can be analyzed and fully understood. Realization of the important role of phase composition on film properties helps to select the right material combination and consequently, to improve the film behavior for high-temperature applications by selecting the right material combination. Relationships between deposition conditions and film properties are also discussed in details in this chapter. The reader can also get acquainted with basic techniques for thin film characterization including methods for analyzing mechanical properties of thin films, their structure, residual stresses, elemental and phase composition, thermal stability and resistance against oxidation, which are summarized in the Appendix of this book.

Rostislav Daniel
Jindřich Musil
Autumn 2014

Nomenclature

a-	amorphous
a	radius of the contact circle
a_D	deposition rate
at.%	atomic percent
A	neutral atom in basic state
A^+	positively ionized atom
A^-	negative ion
A^*	atom in an excited state
A_e	work necessary to the elastic deformation of the film
A_t	total work done by the load applied to the film
AB	molecule consisting of two atoms A and B
b	magnitude of the Burgers vector
B	magnetic field
c-	crystalline
C	capacitance
$C_{1,..,x}$	fitting constants
d	displacement
d	average grain diameter
dc	direct current
$d_{s\text{-}t}$	substrate-to-target distance
e	lattice strain
E	intensity of electric field
E^*	effective (reduced) Young's modulus
E_{bi}	energy delivered to the film by ion bombardment
$E_{bi\ min}$	minimum energy delivered to the film
E_f	Young's modulus of the film

E_i	Young's modulus of the indenter
E_i	energy of an incident particle
E_s	Young's modulus of the substrate
E_t	energy of a target particle
fcc	faced centered cubic
f_r	repetition frequency
$F(h_c)$	area function of the real indenter shape
G	shear modulus
h-	hexagonal
h	penetration depth
h_a	depression of a sample around the indentation
h_b	Boltzmann's constant
h_c	contact depth
hcp	hexagonal closed packed
h_f	film thickness
h_{oxide}	thickness of the oxidized layer
h_r	depth of residual impression
h_s	substrate thickness
h_t	maximum penetration depth
H	hardness
H_{max}	maximum hardness
H_{post}	hardness of sample after annealing
i_s	substrate ion current density
I_d	magnetron discharge current
J	ion current density
J_{AC}	interfacial energy
J_C	grain boundary energy
k	Hall–Petch slope
K	relative dielectric constant
K	shape factor
l_c	critical spacing between dislocations

L	indenter load
m	Schmid factor
m	mass
M_A	atomic weight
Me	transition metal
M_i	mass of an incident particle
M_t	mass of a target particle
nc-	nanocrystalline
p_{Ar}	partial pressure of argon
p_{N_2}	partial pressure of nitrogen
p_{N_2min}	minimum value of partial pressure of nitrogen
p_T	total pressure of sputtering gas
P	indenter load
P_t	maximum load
r	radius of the film-substrate curvature
R	erosion rate
R_a	mean surface roughness
S	sputtering yield
S	material stiffness
t_1	length of the negative pulse
$t_{reverse}$	reverse time
T	temperature
T_a	annealing temperature
T_c	critical temperature
T_{cr}	crystallization temperature
T_e	temperature of electrons
T_i	temperature of ions
T_m	melting temperature
T_{ox}	onset temperature of oxidation
T_r	onset temperature of recovery
T_s	processing substrate temperature

U	surface binding energy
U_d	magnetron discharge voltage
U_{fl}	floating potential
U_s	substrate bias
v	wave length of photons
v_i	velocity of an incident particle
wt.%	weight percent
W_e	elastic recovery
x	stoichiometry

Units and Symbols

°C	degree Celsius
amu	atomic mass unit
A	ampere
Å	ångström
cm	centimeter
eV	electronvolt
F	farad
g	gram
GPa	gigapascal
h	hour
Hz	hertz
kW	kilowatt
kJ	kilojoule
kHz	kilohertz
K	kelvin
keV	kiloelectronvolt
l	liter
mg	milligram
mN	millinewton
min	minute
mol	mol
mW	milliwatt
mm	millimeter
mm^2	square millimeter
mm^3	cubic millimeter
mA	milliampere
MeV	megaelectronvolt

MJ	mega Joule
MPa	megapascal
nm	nanometer
pF	pikofarad
Pa	Pascal
s	second
V	volt
W	watt
μm	micrometer
μs	microsecond
α	thermal expansion coefficient
α_f	thermal expansion coefficient of the film
β	peak broadening
β_c	peak broadening due to crystallite size
β_e	peak broadening due to lattice strain
δ	density of material
Δ	average grain boundary thickness
ΔE	differential energy of sputtered atoms
ΔH_f	enthalpy of formation
ΔJ	energy difference
Δm	mass difference
ΔN	differential flux of sputtered atoms
ΔT	temperature difference
$\Delta \sigma$	macrostress difference
ε_0	permittivity of free space
ϕ	angle between the directions of the applied stress and the normal to slip plane
λ	wave length
λ	angle between the directions of the applied stress and the slip plane
Ω	ohm
ν	Poisson's ratio

ν_s	**Poisson's ratio of the substrate**
ν_f	**Poisson's ratio of the film**
ν_i	**Poisson's ratio of the indenter**
ψ	**sample tilt angle**
ρ	**electric resistivity**
σ_a	**applied stress**
σ_0	**friction stress**
σ	**macrostress**
σ_{th}	**thermal stress**
σ_i	**intrinsic stress**
σ_{RT}	**macrostress measured at room temperature**
τ	**charging time**
τ	**shear stress in a slip plane**
θ	**angle of diffraction**
θ_i	**angle of incidence**

Abbreviations

AFM	**atomic force microscope**
BSE	**backscattered electrons**
CVD	**chemical vapor deposition**
CCD	**charge-coupled device**
DSC	**differential scanning calorimetry**
EDX	**energy dispersive X-ray analysis**
FTIR	**Fourier transform infrared spectroscopy**
FWHM	**full width at half maximum**
HR	**high resolution**
LOM	**light optical microscope**
MeN	**transition metal nitride**
OP	**optical profilometry**
PA	**plasma assisted**
PVD	**physical vapor deposition**
RT	**room temperature**
SE	**secondary electrons**
SEM	**scanning electron microscope**
SZM	**structure zone model**
TEM	**transmission electron microscope**
TGA	**thermogravimetric analysis**
WDX	**wavelength dispersive X-ray analysis**
XRD	**X-ray diffraction**
XPS	**X-ray photoelectron spectroscopy**

Chapter 1

Synthesis of New Nanostructured Materials

Traditional materials widely used in industry have early met their limits as requirements especially of load-bearing and protective applications became extremely high. In such applications, improved surface properties of the base material enhancing its protecting capability are required. Ever increasing demands on materials performance are moreover coupled with the needs for weight and cost reduction and reduction of energy consumption. Over the years, plasma processing technologies have become to be widely used in the field of surface engineering due to a high potential to satisfy the actual requirements. Fast development in plasma-assisted diffusion treatments, hard and protective coatings, and composite surface layers has covered the most needs put on to new materials with enhanced properties. Above all, thin hard protective films have been successfully used for protection of materials since the 1970s with the aim to prolong the lifetime of working tools. The coating materials developed for wear protection can be classified according to the chemical bonding character as metallic, covalent, or ionic hard materials [1]. Especially, hard ceramic films have attracted considerable interest due to their potential applications in cutting and forming tools, engine parts, bearings, seals, bearing balls, gears, or drills. Protective coatings must, however, meet a set of various

Novel Nanocomposite Coatings: Advances and Industrial Applications
Rostislav Daniel and Jindřich Musil

ISBN 978-981-4411-17-2 (Hardcover), 978-981-4411-18-9 (eBook)
www.panstanford.com

demands depending on the specific applications. In spite of a variety of materials with a high hardness, the choice of possible candidates which would meet all requirements is extremely limited. As material specifications have become stricter, demands for high-performance structural materials have started to call for novel advanced hard and superhard coatings of a new generation having simultaneously a high hardness, stiffness, and a sufficient stability against environmental attack. Plasma processing has been shown to be a powerful tool in preparation of thin films with an ample range of properties that can be tailored to optimize the desired functionality. Recently, a special attention is focused on possibilities to meaningfully improve film properties by optimizing deposition conditions or by synthesizing a new generation of multiphase materials by addition of small amount of alloying elements in order to completely change the film structure and thus mechanical and other functional properties. Since there still exist several drawbacks, which limit the practical application of traditional single phase films of a first generation, such as high residual stresses and coefficient of friction, low thermal stability and oxidation resistance, these new ways might overcome these problems so that the films should be able to perform under extreme conditions.

Under various available physical vapor deposition (PVD) processes, sputtering is a deposition technique with the advantage of producing unique fine grained and metastable structures, including amorphous and high-temperature phases. By optimizing the energy and flux of impinging sputtered atoms and bombarding ions, it is possible to effectively alter the crystallographic orientation and size of grains in a large range from hundreds down to a few nanometers. The ever-growing need for superior materials to withstand severe operating conditions has driven the search for new nanocrystalline materials and thermodynamically metastable thin films exhibiting exceptional mechanical properties and enhanced strength and toughness characteristics. One of the design concepts is based on the formation of single phase films with ultra-fine grained microstructures, which consist of grains in the nanometer size level. Another approach utilizes an addition of alloying elements into the films in order to form completely new structures. Films composed of mixtures of various phases, metastable solid solutions, or composite structures in the nanometer range are very promising for variety of applications due to their new unique functional properties [2–6]. The aim of this chapter is to present a progress in development of thin protective films with superior properties over traditional materials and to show an impact of surface engineering technologies on the performance and lifetime of coated tools.

1.1 Nanocrystalline Materials

One of the very basic findings of the physics of solids is the insight that most properties of solids depend on their microstructure. In particular the arrangement of the atoms and the size of the building blocks of which the solid is consisted in one-, two-, or three-dimensions are the key factors determining functional properties of materials. A huge effort is therefore made by many researchers to prepare materials with exceptional properties through new unique structures. Nowadays, coating technologies attract a great attention since nonequilibrium processes used for the film preparation allow formation of such structures. Moreover, simply by changing the deposition conditions, a wide variety of film microstructures may be generated providing significantly enhanced materials properties. Nanocrystalline and nanocomposite materials are one of those replacing traditional coarse-grained materials in many applications.

1.1.1 Characteristics

In general, materials properties are strongly related to their structure and have been shown to vary considerably with the size of grains of which they are composed. A novel way of improving material properties has been found in reduction of grain sizes to the nanometer scale that significantly distinguishes these materials from conventional coarse-grained polycrystalline materials which have grains of several micrometers in size. Materials with a nanometer-scale microstructure where the atomic arrangement, crystallographic orientation, chemical composition, and the size of grains vary on a length scale of a few nanometers throughout the bulk are referred to as nanocrystalline materials. These materials are single or multiphase materials with the grain size typically below 100 nm in at least one-dimension and may be composed of crystalline, quasicrystalline, or amorphous phases. Unique properties of nanocrystalline materials are given mainly by a combination of rare compositions and novel microstructures. These presumptions ensure an exceptional potential of the nanocrystalline materials in many technological applications. However, a nanometer-sized microstructure is not the only feature responsible for improved mechanical, electrical, or magnetic properties of these materials. Another characteristic feature of nanometer-scale materials is the presence of incoherent or coherent interfaces formed between individual crystallites. The atomic structure, thickness, and chemical composition of the boundary regions are equally crucial for the material properties. It means that materials composed of grains with comparable size and composition may markedly differ from one another due to the various interfacial structures [7]. Figure 1.1 shows a schematic depiction of a typical nanocrystalline material composed of one kind of atoms forming equiaxed grains with long range order and grain boundaries exhibiting a broad spectrum of interatomic spacings.

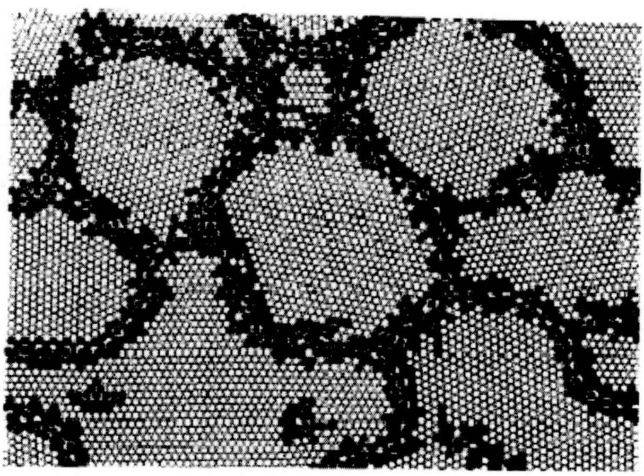

Figure 1.1 Two-dimensional model of a nanostructured material showing two types of atoms: crystal atoms corresponding to the lattice represented by open circles and atoms in boundary regions with variety of interatomic spacing (dark circles). Reprinted from Ref. [8], Copyright (1995), with permission from Elsevier.

While the atomic structure of all grains is generally identical and the grains only differ in their crystallographic orientation, completely different situation is observed in the grain boundary regions where the grains are joined together. In such network of intercrystalline regions, the proportion of atoms is equivalent to or higher than number of atoms inside the grains. A high number of interfaces with random orientation relationships are a main characteristic of the nanocrystalline materials. In the core of incoherent interfaces, the misfit between individual grains results in a local rearrangement of the atomic structure by reducing the atomic density and the coordination between the atoms relative to the perfect crystal [9]. As the grain size is decreased, the average atomic density in the grain boundaries dramatically increases together with deviation of the coordination between nearest neighbor atoms in comparison with the ones in the crystallites. The volume fraction of the grain boundaries in the nanocrystalline material may be calculated as

$$C = 3\Delta/d, \qquad (1.1)$$

where Δ is the average grain boundary thickness and d the average grain diameter [10]. Thus, the volume fraction of atoms in the grain boundaries is about 30% for 10 nm grains and can increase to 50% for 5 nm grains. The presence of these two various nanometer-scaled structural components—grains and grain boundaries—of comparable volume fractions is decisive for the properties of the nanocrystalline materials. Moreover, by a specific combination of grains with the same or different chemical composition, different or identical atomic structure, shape, and size, the properties of such materials may be varied in a wide range.

It was already mentioned that properties of nanocrystalline materials are greatly influenced by microstructural features. Mechanical and other physical properties strongly depend on the distribution and size of the grains, the nature and morphology of grain boundaries and interfaces, composition profiles across grains and interfaces, as well as on the amount and nature of defects developed in the material in the course of a preparation process. Extraordinary properties of nanocrystalline materials are exactly the result of a combination of equiaxed randomly oriented grains covered by disordered grain boundaries, both exhibiting no voids or dislocations. Only some local faceting between individual grains has been observed probably due to the natural tendency to bring the planes from both grains into conformity.

1.1.2 Grain Growth

The stability of the nanocrystalline structures is a key factor of their usage in technological applications. Nanocrystalline materials are in principal thermodynamically unstable due to the presence of a large fraction of interface boundaries. A high amount of energy stored in the grain or interphase boundaries acts as a driving force for the grain growth through which the system tends to reduce its interfacial and thus total energy. Although nanocrystalline materials exhibit a high energy state and hence a high driving force for grain growth, almost all nanocrystalline material revealed just very limited tendency of the grains to grow up to a reasonably high temperatures. It is due to the structural factors, such as equiaxed grain morphology, low energy grain boundary structures or narrow grain size distribution, which have been shown to stabilize structure of nanocrystalline materials through retarding the grain growth. However, since there is a strong tendency of nanocrystalline materials to convert to conventional coarser grained materials with less interfaces, further structure stabilization is needed to retain unique structure and properties of these materials. It has been found that the activation energy of the grain growth can be compared with that of grain boundary diffusion once the transport of atoms across or along the grain boundaries is involved. It has also been noted that with a decrease of the grain size, the starting temperature and the activation energy for the grain growth process increase significantly [11]. Both these findings are of high importance for the understanding of the grain growth in nanocrystalline materials and consequently for tailoring their structure to increase the stability. Basic approaches for preventing the grain growth suggest to slow down the growth kinetics by reducing the grain boundary free energy or grain boundary mobility or to increase the energy barrier which has to be overcome in order to start the grain growth. In first cases, the material remains in an unstable state where the total free energy is reduced through local re-arrangements of the grain boundary planes, while the second type of stabilization aims at achieving a truly metastable state. Here, any variation of the total grain boundary area causes an increase in the free energy of the material. When such states are achieved, much higher temperature is required for the grain growth to occur.

1.1.3 Properties of Nanocrystalline Materials

Nanocrystalline materials have been shown to exhibit a variety of properties that significantly differ from those of conventional coarse-grained polycrystalline materials. The reason is the very fine size of grains of the material. Grain size is known to have an outstanding effect on the mechanical behavior of materials, and it is also the reason for reduced density, enhanced diffusivity, higher electrical resistivity, increased specific heat, higher thermal expansion coefficient, lower thermal conductivity, and specific magnetic properties of nanocrystalline materials in comparison with traditional coarse-grained materials. In general, the unique properties of nanocrystalline materials are strongly influenced by the interfaces between the individual nanocrystals. A contribution of a very high number of atoms in the grain boundaries, and thus a high volume fraction of interfaces in a small volume region, to the materials properties is enormous, especially in the case of materials composed of grains with very small sizes lower than 10 nm. The explanations for enhanced materials properties of nanocrystalline materials can be found in the developed microstructure and type of phases of which the material is composed. For instance, the reduced elastic properties of nanocrystalline materials have been interpreted as an effect of increased average interatomic spacings in the boundary regions resulting in large free volume of the interfacial component [12]. Further, the Young's modulus has been observed to be significantly affected by the specific level of porosity and cracks in the microstructure [13]. However, the most significant effect of a reduced grain size to the nanometer level can be found in an enhancement of the hardness and toughness of nanocrystalline materials. Hardness of materials represents their resistance against plastic deformation which is mainly given by an efficiency of a material to hinder the dislocation movement. Structural arrangements of individual phases in nanocrystalline materials ensure that the dislocations are effectively stopped in their movement due to the fundamental existence of a high volume of grain boundaries. Hence, nanocrystalline structures are suitable in development of materials with enhanced hardness, reduced elastic modulus, and improved toughness. One possible explanation of enhanced mechanical properties of this kind of materials is given by the Hall–Petch relationship suggesting that the hardness of a material increases with a decreasing grain size according to the following equation [14]:

$$\sigma = \sigma_0 + kd^{-1/2}, \quad (1.2)$$

where d is the grain size, σ_0 the friction stress resisting the motion of gliding dislocation, and k is the Hall–Petch slope, which is associated with a measure of the resistance of the grain boundary to slip transfer. It is sometimes also called the Hall–Petch intensity parameter. In analogy, hardness H can be related to the grain size by

$$H = H_0 + k_H d^{-1/2}, \quad (1.3)$$

where H_0 and k_H are constants. The Hall–Petch relationship was derived on the basis of strengthening resulting from dislocation pile-ups at physical obstacles. The concept of pile-ups assumes that the number of dislocations piled up against a grain boundary decreases with decrease in grain size since it is a function of the applied stress and of the distance to the source. It means that a higher stress level is needed to generate the same amount of dislocations at the pile-up in fine-grained materials. In this way, the Hall–Petch relation can sufficiently elucidate behavior of polycrystalline materials; however, it shows certain limitations in explanation of hardness of nanocrystalline materials with the grain size below 2–3 nm. Although the relationship between the role of the grain boundaries, acting as efficient obstacles that inhibit the dislocation motion, and mechanical properties is in a good agreement with experimental results, once the grain size is reduced down below a critical value a reversed softening effect is observed. It is in principal a logical entailment of the fact that the strength value cannot increase indefinitely to beyond the theoretical strength limit of a material. The breakdown of the Hall–Petch relation is attributed to different deformation mechanisms in nanocrystalline materials that occur when a grain size decrease below a certain critical limit [15]. From the investigations, it appears that conventional dislocation mechanisms stop to function when the size of a Frank–Read source cannot exceed the grain size. This is a reason why pile-up of the dislocations against grain boundaries would not occur and the dominant deformation mechanism is associated mainly with the sliding of grain boundaries. It has predicted the grain size in various materials at which the Hall–Petch relationship will break down [16].

It is based on the assumption that at extremely fine grain sizes in the nanometer regime, the individual grains cannot support more than one dislocation. The critical spacing between dislocations l_c has been stated as

$$l_c = \frac{3Gb}{\pi(1-\nu)H}, \tag{1.4}$$

where G is the shear modulus, b is the magnitude of the Burgers vector, and H is the hardness. Thus, once the grain size d is smaller than l_c, no dislocation pile-ups occur and the Hall–Petch relation breaks down. Other explanations include suggestions that very fine grain size can naturally induce relaxation processes at the grain boundaries which could lead to a decrease in strength. The breakdown of the Hall–Petch relation may also occur when the triple junctions in the microstructure begin to comprise a significant volume fraction of the bulk specimen value [17]. Moreover, the hardening and softening of nanocrystalline materials may also depend on the way in which the grain size is varied. For instance, grain growth driven by annealing may result in hardnesses greater than those obtained directly during a preparation of a sample with similar grain sizes [18]. Annealing a sample may, however, also cause a softening through the relaxation of the interfacial structure. It depends on the microstructure and phase composition of the material.

Considering experimental findings published in the literature, a development of the material hardness as a function of the grain size can be summarized as follows (see Fig. 1.2), while a significant hardness enhancement has been observed with decreasing grain size, going from conventional coarse-grained polycrystalline materials to nanoscaled materials, a dramatic softening effect occurs below a critical grain size. This development reflects major processes operating in the material under applied loading. Thus, plastic deformation of coarse-grained materials driven by intergranular processes dominated by the dislocation activity proceeds further to the regime where the dislocation movement is effectively prevented by the grain boundaries which start to play a dominant role over that of the grains. In this region, hardness can be described by the Hall–Petch relation ($H \sim d^{-1/2}$) showing the maximum at a specific grain size, below which a softening effect is typically observed. A hardness decrease below the critical size of grain is attributed to a large amount of defects at the grain boundaries, which allow fast diffusion of atoms and vacancies. It results in small-scale grain boundary sliding that is the dominating deformation mechanism in this region. As such, further increase in hardness requires materials to involve qualitatively new processes in the grain boundary region which would result in higher complexity and strength of grain boundaries [20]. This idea has been realized through a new concept of microstructural design where a combination of different phases provides complex boundary interfaces to accommodate a coherent strain. In this way a formation of voids or flaws is prevented and the dislocation movement is even more effectively hindered through interfaces with improved cohesive strength. These multiphase structures are referred to as nanocomposite materials and will be discussed in Section 1.3.

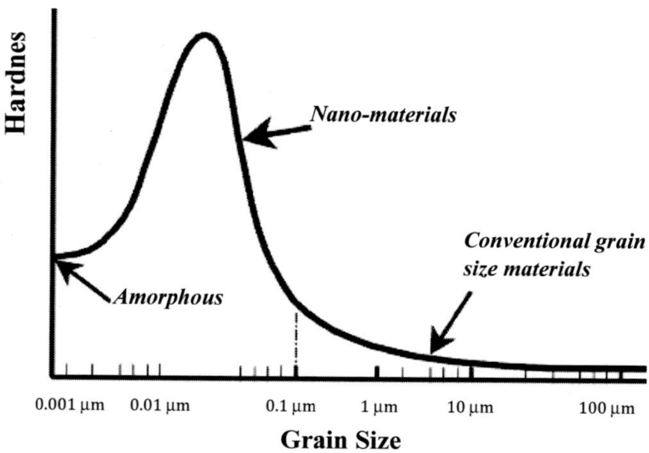

Figure 1.2 Hardness of a material as a function of the grain size. Reprinted from Ref. [19], Copyright (2003), with permission from Elsevier.

1.1.4 Transition Metal Nitride Films

Nanocrystalline nitride films of transition metals are widely used in many applications due to their unique microstructure and functional properties. Thin nitride films have replaced traditional materials since they exhibit a variety of new enhanced physical and chemical properties. The development of hard thin films for mechanical applications has made enormous progress in past years mainly in improvement of mechanical properties. Hence, thin nanocrystalline films of a new generation are known for their very high hardness and fracture toughness combined with low friction. Due to their good mechanical properties together with high resistance against wear, binary transition metal nitride films prepared by various deposition techniques have attracted considerable interest as protective coatings on cutting tools, bearings or forming tools. However, not only high hardness, toughness, and good tribological properties determine the durability and lifetime of protective coatings, but the thermal stability of the coated systems is also very important for practical usage. Though transition metal nitrides exhibit distinct properties at moderate temperature range, their performance is no longer exceptional at elevated temperatures. As an example, TiN films oxidize readily already at 500°C, leading to the formation of poorly adhered TiO_2 oxide phase on the sample surface [21]. It makes the films useless for high-temperature applications in air or other aggressive environments [22]. The film resistance against oxidation, however, strongly depends on the nature of the oxide layer formed on the sample surface. If a passive oxide layer is formed, further oxidation may be markedly restricted. It has been observed in the case of CrN and AlN films forming dense passivating Cr_2O_3 and Al_2O_3 oxides, respectively [23,24]. The wear protection efficiency of films is significantly deteriorated once thickness of the surface oxide layer exceeds certain critical value. Hence, intensive efforts have been made to develop new high-performance multicomponent materials with new structures and compositions where oxidation processes, commonly associated with early coating failure, will be efficiently prevented. Since improvements of thermal stability of the films through changes of their phase composition commonly result in a hardness reduction, synthesis of films with improved oxidation resistance and preserved enhanced mechanical properties for high-temperature applications is a subject of intensive research.

1.2 Multilayers and Superlattices

Another concept of nanoscaled structure has been proposed in the case of multilayered and superlattice films. In both cases, two or more layers of different materials are periodically alternated so that a film system with a thickness of several micrometers is finally produced. In general terms, a multilayer structure is consisted of many layers of different constituents stacking on top of one another with distinct interface between adjacent layers. Typical thickness of each layer is in the nanometer range, the composition within the single layer is commonly homogeneous.

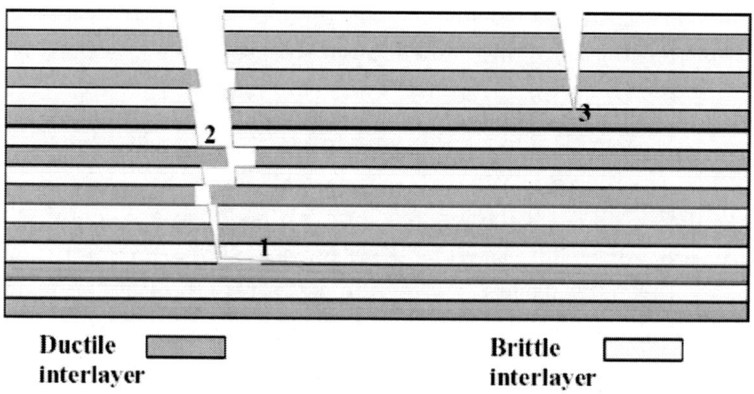

Figure 1.3 Schematic representation of (1) crack deflection, (2) ductile interlayer ligament bridging, and (3) crack tip blunting due to nanoplasticity at interlayer, as basic toughening mechanisms in multilayer films. Reprinted from Ref. [25], Copyright (2005), with permission from Elsevier.

Similar to nanocrystalline materials with strong boundary regions, here also a formation of dislocation and nanocracks and their propagation is strongly prohibited. In Fig. 1.3, typical toughening mechanisms observed in multilayer films are shown including crack deflection at interface between layers, ductile interlayer ligament bridging, and crack tip blunting due to nanoplasticity at the interface [26,27]. These effects can be even enhanced if materials with high and low elastic properties are used. Dislocations formed when material is exposed to mechanical loading cannot be transmitted across individual layers since they need to overcome very high repulsive image stress induced in the phase with higher modulus [28]. Hence, the generated dislocations or nanocracks are expressively suppressed already at the interface between individual layers. Many multilayers have been designed for various applications by optimizing the grain size, the number of interlayers, thickness, and composition of each interlayer, the thickness ratio of different interlayers and the structure of interfaces [29]. However, material selection is also an important part in multilayer film design [30]. While covalent bonding materials are characterized by high hardness and temperature strength, metallic bonding materials possess good adhesion and toughness. Ionic bonding materials are then used in order to improve stability and inertness of the system. Recent advances in coating technologies demonstrate that multilayers of different composition with superior mechanical properties may be synthesized using both PVD and chemical vapor deposition (CVD) techniques [31–34].

1.3 Nanocomposite Films

One of the fundamental concepts in searching for new materials with exceptional properties is based on the fact that the bulk modulus of a material can be varied by changes in structure and strength of interatomic bonds [35]. Following this concept, new superhard materials have been produced where the usual mechanisms leading to the fracture of crystalline (multiplication, movement, and pile up of dislocations) and amorphous (growth of microcracks) materials are avoided by combining of at least two different phases. Experimental results showed that this concept can be realized through the films composed of a nanocrystalline material embedded within a surrounding matrix forming a nanocomposite structure. Due to this unique structure, nanocomposite films exhibit properties that differ greatly from those of micro- or nanocrystalline materials. Recently, considerable attention has been focused on elaboration and characterization of nanocomposite films in order to synthesize extremely hard and stable structures for various applications. The idea finally involves the preparation mostly of nanocrystalline refractory nitrides of transition metals combined with a tough nanocrystalline or amorphous matrix, which substitutes the grain boundaries of the first phase. Qualitatively new processes, such as grain boundary interaction, grain boundary enhancement or sliding, which involve when material is loaded, result in completely new physical and functional properties of these materials.

1.3.1 Concept of Nanocomposite Materials

Nanocomposite coatings represent a new generation of materials exhibiting completely new properties with respect to the conventionally used materials. Nanocomposites are composed of at least two separated phases with a nanocrystalline and/or amorphous structure [36]. The formation of a nanocomposite structure is principally attributed to a segregation of one phase to the grain boundaries of the second phase. This effect is responsible for hindering of the grain growth and formation of the structure in the nanometer scale. New unique physical and functional properties of the nanocomposite materials results from this extraordinary microstructure where boundary regions surrounding individual grains play a dominant role. Due to very small grains (≤10 nm) separated by the second, mainly amorphous phase, nanocomposite materials behave in a different manner compared to the conventional materials with grains greater than 100 nm. Properties of two-phase nanocomposite materials, however, strongly depend on the volume fraction of the second phase. It has been observed that the superhardness, that is, hardness exceeding 40 GPa, occurs in the nanocomposite films just when hard nanocrystalline grains are separated by a tiny mesh of atoms of the second phase [37]. The optimum thickness of a tissue separating the nanocrystallites is between 1 and 3 nm that corresponds to only a few monolayers. The optimum size of grains forming the nanocomposite structure has been found in the range of 3–10 nm (see Fig. 1.4). A variety of boride, nitride, carbide, or oxide materials forming nanocomposite structure with ceramic, metal, or carbon amorphous matrix provides unique properties for each material combination. Such composition and property diversity of nanocomposite films enables their utilization in various fields of engineering applications.

Formation of films with nanocrystalline and even nanocomposite structure requires optimization of deposition conditions in order to reduce the size of grains and to ensure development of the segregating phases. Basic process parameters affecting the size and crystallographic orientation of grains include the discharge current, substrate temperature, bias voltage, substrate ion current density, and partial pressure of reactive gas. Especially low-energy ion bombardment has been shown as one of fundamental processes which allow deposition of films with controlled texture and grain size [39]. The grain growth can be effectively controlled through kinetic energy of bombarding particles transferred to very small surface region. The ion bombardment contributes to a minimum surface mobility of adatoms which is needed for formation of crystalline phases and a sufficient diffusion within the segregating system. However, the temperature is required to be controlled as well since by exceeding a critical value grain growth and grain boundary segregation is no longer prevented. Similar to low-energy bombardment, an addition of one or several alloying elements to a base material has the same effect on the formation of the nanocomposite structure. Moreover, no additional substrate bias or heating is needed to form films with nanocrystalline structure [40].

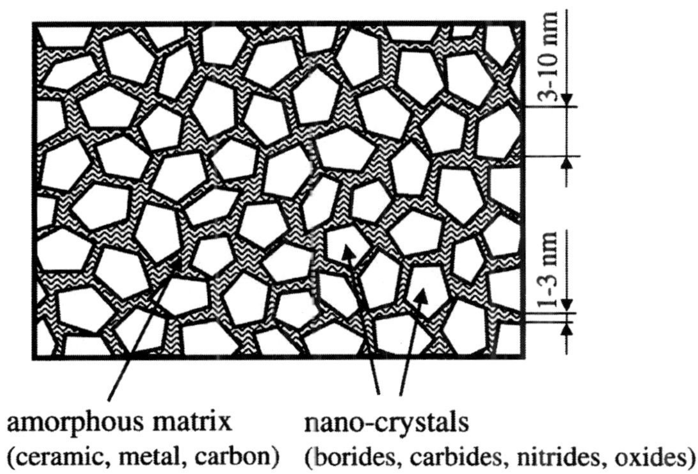

Figure 1.4 Typical structure of a nanocomposite film composed of crystallites in the nanometer range embedded within a thin amorphous matrix. Reprinted from Ref. [38], Copyright (2011), with permission from Elsevier.

1.3.1 Classification of Hard Nanocomposite Films

At present, two different groups of hard and superhard nanocomposite films are known. They are as follows:

(1) nc-MeN(C,B,O)/hard phase—nanocomposites composed of two hard phases and

(2) nc-MeN(C,B,O)/soft phase—nanocomposites composed of one hard and one soft phase (Me = Ti, Zr, W, V, Cr, Mo, and Ta are transition metals or Al).

In principal, all two-phase nanocomposite films contain one or two hard crystalline phases. The second phase is more complicated and has been found to be either amorphous or nanocrystalline, hard or soft. With respect to their composition, hard nanocomposite coatings can be divided into two main groups:

(1) Crystalline/amorphous nanocomposites
  (a) nc-MeN/a-nitride—nc-TiN/a-Si_3N_4 [41], nc-WN/a-Si_3N_4 [42], nc-ZrN/a-Si_3N_4 [43]
  (b) nc-MeN(C)/a-boron compounds—nc-Ti(B,O)/quasi-a-(TiB_2, TiB) [44], Ti-B-C [45]
  (c) nc-MeN/a-metal—nc-ZrN/Cu [46], nc-TiN/Ni [47], nc-TiN/Mo [48], nc-CrN/Cu [49], nc-CrN/Ni [50], nc-ZrN/Y [51], nc-ZrN/Ni [52], nc-AlN/Cu [53]
  (d) nc-MeC/a-C—nc-TiC/DLC [54]
  (e) nc-WC+nc-WS_2/DLC [55]
  (f) nc-MeC/a-C+a-nitride—nc-Mo_2C/a-C+a-Mo_2N [56]

(2) Crystalline/crystalline nanocomposites
  (a) nc-MeN/nc-nitride—nc-TiN/nc-BN [57], nc-TiN/nc-W_2N [58]
  (b) nc-MeN/nc-metal—nc-AlN/Ti [59], nc-TiN/Al [60].

Here nc- and a- denote the nanocrystalline and amorphous phases, respectively.

Various types of hard nanocomposite films synthesized so far illustrate extensive potentials and complexity of a plasma processing technology in preparation of thin films with enhanced properties. Although no definite and exact methodology has been proposed yet for selection and combination of elements to produce hard films with nanocrystalline or nanocomposite structure, based on experimental results in the field of plasma surface engineering some fundamentals can be summarized: typically, nanocomposite structure is formed if hard transition metal nitrides or borides are combined in the film with nitrides of binary alloys and other nitride or carbide systems [45]. Further, a miscibility gap between elements forming binary alloys and intermetallics is a necessary condition to prepare hard nanocomposite films. Typical examples are ZrYN and ZrN/Cu nanocomposite films, representing a combination of such immiscible compounds. However, nanocomposite structure may also be formed by nitrides of alloys forming a solid solution. $Ti_{1-x}Al_xN$ films represent such nanocomposites [61].

1.3.2 Hardening Effects

While individual grains in nanocrystalline materials are separated by grain boundaries, fine grains in nanocomposite materials are separated by phase boundaries with a various composition. This discrimination originates from a nature of nanocomposites as a material consisting of at least two phases. Enhanced properties of nanocomposite materials result from a large number of phase boundaries acting against the dislocation movement similar to the nanocrystalline materials. Also in this case, a large number of atoms in a boundary region dramatically changes interaction between individual grains. Phase boundaries in nanocomposite materials, however, hinder the generation of dislocations and prevent cracks propagation even more effectively. In contrast to nanocrystalline materials where plastic deformation is dominated by grain boundary sliding events controlled by grain boundary diffusion, the sliding triggered by atomic shuffling or by stress-assisted free volume migration is markedly suppressed in nanocomposites through strengthened interfaces [62,63]. This fact substantially increases the field of applicability of this new generation of materials. Enhanced properties of nanocomposite films, however, strongly depend on the size, amount and distribution of nanocrystals as well as on the volume of surrounding phase. Also the constitution of the interfaces and their relation to the phases developed in the film are essential to its properties. It has been shown that the volume of the surrounding phase needs to be optimized with respect to the volume of the nanocrystalline phase to obtain enhanced properties. The synergetic effect of the surrounding phase on the size of the nanocrystalline grains and mechanical properties of the system has been demonstrated by many researchers [64–66]. Only sufficiently high amount of the boundary phase can interrupt grain growth and forced newly arriving forming species to renucleate. With decreasing size of grains combined with a thin matrix separating the nanocrystals, indeed dislocation takes place but self-propagation of the cracks is limited.

In general, the distance between two nanocrystalline grains needs to be optimized to be within a few nanometers in order to block grain boundary sliding and achieve enhanced hardness. That is ensured in thin film which consists of hard nanocrystallites with grain sizes in the nanometer range encapsulated in a thin matrix. Thus, the hard nanocrystalline phase bears the high load, while the matrix phase provides structural flexibility. An optimum volume fraction of the matrix ensures that under mechanical load dislocation activities are dramatically limited and a material can react only by grain boundary sliding. In such nanocomposites, movement of single, undeformed nanocrystalline grains against each other requires very high energy due to the strengthened interfaces which lead to very hard and stable structure. When the nanocrystals are too close to each other, atomic planes in the neighboring grains may actively interact. It results in the formation and multiplication of dislocations decreasing the resistance of a material against plastic deformation. By exceeding a critical thickness of the surrounding phase, larger grain separation permits adaptation of incoherency strains and formation of nanocracks between the crystallites which can, in the extent, easily propagate in the matrix.

1.3.3 Design Concept for Nanocomposite Films

Although every nanocomposite system is composed of nanocrystalline grains embedded in a surrounding matrix forming phase boundaries, the films may completely differ in their physical properties. This is due to the different phase composition which determines resulting functional properties of the system. It has been found that a formation of interfaces having a high cohesive energy is not necessary for films to exhibit exceptional properties. To synthesize nanocomposite films with high hardness, stable structure, and high toughness, several prerequisites have to be satisfied [67]. First, the nanocrystalline grains should have random orientation. It ensures a minimum incoherent strain and facilitates grains to slide in a matrix to release strain. In this way, a high toughness may be obtained. However, the right selection of the surrounding matrix is also very important. Such a phase has to possess high structural flexibility in order to accommodate coherent strains without forming any defects. An amorphous phase in the grain boundaries has been shown to effectively prevent dislocation movement and crack formation. Moreover, it provides high fracture toughness of the system through the enhancement of grain boundary sliding and by deflecting, meandering, and terminating the nanocracks. However consequently, if plastic deformation is strongly prohibited through prevented dislocation movement and grain boundary sliding, film ductility can be partially lost. Therefore, it is assumed that a certain degree of grain boundary diffusion and grain boundary sliding are necessary in order to obtain superhard nanocomposite films with high toughness. Nanocomposite films with the most promising properties have been prepared by combining refractory and immiscible nanocrystalline phases of hard transition metal nitrides with a high strength amorphous phase as a matrix. These very complex systems are characterized by unique properties due to high grain boundary complexity and strength, optimized grain size and volume fraction of the surrounding phase, well-defined structure of grains and interfaces, and stable microstructure [68–71].

1.3.4 Stability of Nanocomposite Films

Nanocrystalline films have received much attention due to their improved mechanical, electronic, and magnetic properties compared to traditional coarse-grained materials. Even so, by reaching certain critical temperature the nanocrystalline single-phase films often succumb to a strong tendency to transform into a polycrystalline material. The mechanisms that ensure preventing of dislocation movement, grain boundary sliding, and formation and propagation of nanocracks and thus resistance against plastic deformation also strongly depend on temperature. Materials with this kind of instability are no longer of interest as hard protective coatings due to the loss of their improved properties at high temperatures. This phenomenon is a great problem in various applications where coated tools are heavily thermally loaded. The loss of their protecting capability usually leads to permanent damages of both the film and the tool. From this reason, hard films for high-temperature applications are often designed as nanocomposites which exhibit both improved hardness and increased thermal stability. Thermal stability of nanocomposite films is typically improved by including elements or phases of high thermal stability. In this way, the grain growth is hindered to high temperatures and the film structure remains stable. However, in some cases an improvement of thermal stability by addition of various phases may result in reduction of hardness or other mechanical properties. Thus, a finding of a nanocomposite film with an optimum set of properties for a wide field of applications is still an open task.

A novel concept for synthesis of a new generation of high-temperature stable nanocomposite coatings based on the formation of nanocomposite structure consisting of two thermally stable crystalline and amorphous phases with a specific volume fraction ratio has recently attracted a great research interest. By optimizing the volume of a silicon nitride matrix covering nanocrystalline grains of specific transition metal nitrides, it is possible to synthesize hard films, which are thermally stable and resistant against oxidation at temperatures exceeding 1000°C. Broadening the group of nanocomposite materials by a new generation of extremely stable hard nanocomposites should have a positive impact on the durability and lifetime of coated tools in high-temperature applications. These new perspective materials with unique properties representing a promising alternative to traditional coatings are in detail discussed in Chapters 5 and 6.

References

1. H. Holleck, V. Schier, Multilayer PVD coatings for wear protection, *Surf. Coat. Technol.* 76–77 (1995) 328–336.
2. C. Wiemer, F. Lévy, F. Bussy, Determination of chemical composition and its relationship with optical properties of Ti-N and Ti-V-N sputtered thin films, *Surf. Coat. Technol.* 68–69 (1994) 181–187.
3. M. Zhou, Y. Makino, M. Nose, K. Nogi, Phase transition and properties of Ti-Al-N thin films prepared by r.f.-plasma assisted magnetron sputtering, *Thin Solid Films* 339 (1999) 203–208.
4. S.J. Bull, D.G. Bhat, M.H. Staia, Properties and performance of commercial TiCN coatings. Part 1: Coating architecture and hardness modeling, *Surf. Coat. Technol.* 163–164 (2003) 499–506.
5. J.G. Han, H.S. Myung, H.M. Lee, L.R. Shaginyan, Microstructure and mechanical properties of Ti-Ag-N and Ti-Cr-N superhard

nanostructured coatings, *Surf. Coat. Technol.* 174–175 (2003) 738–743.

6. H. Holzschuh, Deposition of Ti-B-N (single and multilayer) and Zr-B-N coatings by chemical vapor deposition techniques on cutting tools, *Thin Solid Films* 469–470 (2004) 92–98.
7. C. Suryanarayana, Nanocrystalline materials, *Int. Mat. Rev.* 40 (1995) 41–64.
8. H. Gleiter, Nanostructured materials: State of the art and perspectives, *Nanostruct. Mat.* 6 (1995) 3–14.
9. H. Gleiter, Nanostructured materials: basic concepts and microstructure, *Acta Mater.* 48 (2000) 1–29
10. M.A. Meyers, A. Mishra, D.J. Benson, Mechanical properties of nanocrystalline materials, *Prog. Mat. Sci.* 51 (2006) 427–556.
11. K. Lu, The thermal instability of nanocrystalline Ni–P materials with different grain sizes, *Nanostruct. Mater.* 2 (1993) 643–652.
12. H. Gleiter, Nanocrystalline materials, *Prog. Mater. Sci.* 33 (1989) 223–315.
13. G.W. Nieman, J.R. Weertman, R.W. Siegel, Mechanical-behavior of nanocrystalline Cu and Pal, *J. Mater. Res.* 6 (1991) 1012–1027.
14. E.O. Hall, The deformation and ageing of mild steel: III Discussion of results, *Proc. Phys. Soc.* B64 (1951) 747–753.
15. C.C. Koch, Optimization of strength and ductility in nanocrystalline and ultrafine grained metals *Scripta Mater.* 49 (2003) 657–662.
16. T.G. Nieh, J. Wadsworth, Hall-Petch relation in nanocrystalline solids, *Scripta Metall. Mater.* 25 (1991) 955–958.
17. G. Palumbo, U. Erb, K.T. Aust, Triple line disclination effects on the mechanical behaviour of materials, *Scr. Metall. Mater.* 24 (1990) 2347–2350.
18. G.E. Fougere, J.R. Weertman, R.W. Siegel, On the hardening and softening of nanocrystalline materials, *Nanostructured Mater.* 3 (1993) 379–384.
19. S. Zhang, D. Sun, Y. Fu, H. Du, Recent advances of superhard nanocomposite coatings: a review, *Surf. Coat. Technol.* 167 (2003) 113–119.
20. V.V. Brazhkin, A.G. Lyapin, R.J. Hemley, Harder than diamond: Dreams and reality, *Philo. Mag.* A82 (2002) 231–253.
21. P. Panjan, B. Navinšek, A. Cvelbar, I. Milošev, A. Zalar, Oxidation of TiN, ZrN, TiZrN, CrN, TiCrN and TiN/CrN multilayer hard coatings reactively

sputtered at low temperature, *Thin Solid Films* 281–282 (1996) 298–301.

22. C. Subramanian, K.N. Strafford, Review of multicomponent and multilayer coatings for tribological applications, *Wear* 165 (1993) 85–95.
23. H. Jensen, U.M. Jensen, G. Sorensen, Reactively sputtered Cr nitride coatings studied using the acoustic emission scratch test technique, *Surf. Coat. Technol.* 74–75 (1995) 297–305.
24. L. Hultman, Thermal stability of nitride thin films, *Vacuum* 57 (2000) 1–30.
25. S. Zhang, D. Sun, Y. Fu, H. Du, Toughening of hard nanostructural thin films: A critical review, *Surf. Coat. Technol.* 198 (2005) 2–8.
26. S. Hogmark, S. Jacobson, M. Larsson, Design and evaluation of tribological coatings, *Wear* 246 (2000) 20–33.
27. H. Holleck, H. Schulz, Preparation and behaviour of wear-resistant TiC/TiB_2, TiN/TiB_2 and TiC/TiN coatings with high amounts of phase boundaries, *Surf. Coat. Technol.* 36 (1988) 707–714.
28. J.S. Koehler, Attempt to design a strong solid, *Phys. Rev.* B2 (1970) 547–551.
29. S.C. Tjong, H. Chen, Nanocrystalline materials and coatings, *Mat. Sci. Eng.* R45 (2004) 1–88.
30. H. Holleck, Material selection for hard coatings, *J. Vac. Sci. Technol.* A4 (1986) 2661–2669.
31. M. Berger, U. Wiklund, M. Eriksson, H. Engqvist, S. Jacobson, The multilayer effect in abrasion—Optimising the combination of hard and tough phases, *Surf. Coat. Technol.* 116–119 (1999) 1138–1144.
32. Y. Zhou, R. Asaki, W.-H. Soe, R. Yamamoto, R. Chen, A. Iwabuchi, Hardness anomaly, plastic deformation work and fretting wear properties of polycrystalline TiN/CrN multilayers, *Wear* 236 (1999) 159–164.
33. P.C. Yashar, W.D. Sproul, Nanometer scale multilayered hard coatings, *Vacuum* 55 (1999) 179–190.
34. M.T. Vieira, A.S. Ramos, The influence of ductile interlayers on the mechanical performance of tungsten nitride coatings, *J. Mat. Proc. Technol.* 92–93 (1999) 156–161.
35. M.L. Cohen, Predicting properties and new materials, *Solid State Comm.* 92 (1994) 45–52.
36. J. Musil, J. Vlček, F. Regent, F. Kunc, H. Zeman, Hard nanocomposite coatings prepared by magnetron sputtering, *Key Eng Mater.* 230–232 (2002) 613–622.

37. J. Musil, Hard and superhard nanocomposite coatings, *Surf. Coat. Technol.* 125 (2000) 322–330

38. A.A. Voevodin, J.S. Zabinski, Supertough wear-resistant coatings with “chameleon” surface adaptation, *Thin Solid Films* 370 (2000) 223–231.

39. J. Musil, J. Vlček, Magnetron sputtering of films with controlled texture and grain size, *Mat. Chem. Phys.* 54 (1998) 116–122.

40. J. Musil, J. Vlček, Magnetron sputtering of hard nanocomposite coatings and their properties, *Surf. Coat. Technol.* 142–144 (2001) 557–566.

41. S. Veprek, S. Reiprich, A concept for the design of novel superhard coatings, *Thin Solid Films* 268 (1995) 64–71.

42. A. Cavaleiro, C. Louro, Nanocrystalline structure and hardness of thin films, *Vacuum* 64 (2002) 211–218.

43. M. Nose, M. Zhou, T. Nagae, T. Mae, M. Yokota, S. Saji, Properties of Zr-Si-N coatings prepared by RF reactive sputtering, *Surf. Coat. Technol.* 132 (2000) 163–168.

44. C. Mitterer, P. Losbichler, F. Hofer, P. Warbichler, W. Gissler, Nanocrystalline hard coatings within the quasi-binary system TiN-TiB_2, *Vacuum* 50 (1998) 313–318.

45. C. Mitterer, P.H. Mayrhofer, M. Beschliesser, P. Losbichler, P. Warbichler, F. Hofer, P.N. Gibson, W. Gissler, H. Hrubý, J. Musil, J. Vlček, Microstructure and properties of nanocomposite Ti-B-N and Ti-B-C coatings, *Surf. Coat. Technol.* 120–121 (1999) 405–411.

46. J. Musil, P. Zeman, H. Hrubý, P.H. Mayrhofer, ZrN/Cu nanocomposite film—A novel superhard material, *Surf. Coat. Technol.* 120–121 (1999) 179–183.

47. M. Mišina, J. Musil, S. Kadlec, Composite TiN-Ni thin films deposited by reactive magnetron sputter ion-plating, *Surf. Coat. Technol.* 110 (1998) 168–172.

48. F. Regent, J. Musil, Magnetron sputtered Cr-Ni-N and Ti-Mo-N films: Comparison of mechanical properties, *Surf. Coat. Technol.* 142–144 (2001) 146–151.

49. J. Musil, I. Leipner, M. Kolega, Nanocrystalline and nanocomposite CrCu and CrCu-N films prepared by magnetron sputtering, *Surf. Coat. Technol.* 115 (1999) 32–37.

50. J. Musil, F. Regent, Formation of nanocrystalline NiCr-N films by reactive dc magnetron sputtering, *J. Vac. Sci. Technol.* A16 (1998) 3301–3304.

51. J. Musil, H. Poláková, Hard nanocomposite Zr-Y-N coatings: correlation between hardness and structure, *Surf. Coat. Technol.* 127 (2000) 99–106.
52. J. Musil, P. Karvánková, J. Kasl, Hard and superhard Zr-Ni-N nanocomposite films, *Surf. Coat. Technol.* 139 (2001) 101–109.
53. J. Musil , H. Hrubý, P. Zeman, H. Zeman, R. Čerstvý, P.H. Mayrhofer, C. Mitterer, Hard and superhard nanocomposite Al-Cu-N films prepared by magnetron sputtering, *Surf. Coat. Technol.* 142–144 (2001) 603–609.
54. A.A. Voevodin, J.S. Zabinski, Superhard, functionally gradient, nanolayered and nanocomposite diamond-like carbon coatings for wear protection, *Diamond Relat. Mater.* 7 (1998) 463–467.
55. A.A. Voevodin, J.P. O'Neill, J.S. Zabinski, Tribological performance and tribochemistry of nanocrystalline WC/amorphous diamond-like carbon composites, *Thin Solid Films* 342 (1999) 194–200.
56. M. Benda, J. Musil, Plasma nitriding enhanced by hollow cathode discharge—A new method for formation of superhard nanocomposite coatings on steel surfaces, *Vacuum* 55 (1999) 171–175.
57. S. Veprek, P. Nesládek, A. Niederhofer, F. Glatz, M. Jílek, M. Šíma, Recent progress in the superhard nanocrystalline composites: Towards their industrialization and understanding of the origin of the superhardness, *Surf. Coat. Technol.* 108–109 (1998) 138–147.
58. L.R. Shaginyan , M. Mišina, J. Zemek , J. Musil , F. Regent , V.F. Britun, Composition, structure, microhardness and residual stress of W-Ti-N films deposited by reactive magnetron sputtering, *Thin Solid Films* 408 (2002) 136–147.
59. J. Musil, H. Hrubý, Superhard nanocomposite $Ti_{1-x}Al_xN$ films prepared by magnetron sputtering, *Thin Solid Films* 365 (2000) 104–109.
60. M. Zhou, Y. Makino, M. Nose, K. Nogi, Phase transition and properties of Ti-Al-N thin films prepared by r.f.-plasma assisted magnetron sputtering, *Thin Solid Films* 339 (1999) 203–208.
61. J. Musil, H. Hrubý, Superhard nanocomposite $Ti_{1-x}Al_xN$ films prepared by magnetron sputtering, *Thin Solid Films* 365 (2000) 104–109.
62. H. Van Swygenhoven, M. Spaczer, A. Caro, D. Farkas, Competing plastic deformation mechanisms in nanophase metals, *Phys. Rev.* B60 (1999) 22–25.
63. H. Van Swygenhoven, A. Caro, Molecular dynamics computer simulation of nanophase Ni: Structure and mechanical properties, *Nanostruct. Mater.* 9 (1999) 669–672.

64. R. Birringer, Nanocrystalline materials, *Mater. Sci. Eng.* A117 (1989) 33–43.

65. R.W. Siegel, Cluster-assembled nanophase materials, *Annu. Rev. Mater. Sci.* 21 (1991) 559–579.

66. H.S. Kim, A composite model for mechanical properties of nanocrystalline materials, *Scripta Mater.* 39 (1998) 1057–1061.

67. J. Patscheider, T. Zehnder, M. Diserens, Structure-performance relations in nanocomposite coatings, *Surf. Coat. Technol.* 146–147 (2001) 201–208.

68. N. Nose, W.A. Chiou, M. Zhou, T. Mae, M. Meshii, Microstructure and mechanical properties of Zr-Si-N films prepared by rf sputtering, *J. Vac. Sci. Technol.* A20 (2002) 823–828.

69. S. Veprek, A.S. Argon, Towards the understanding of mechanical properties of super- and ultrahard nanocomposites. *J. Vac. Sci. Technol.* B20 (2002) 650–664.

70. P. Zeman, R. Čerstvý, P.H. Mayrhofer, C. Mitterer, J. Musil, Structure and properties of hard and superhard Zr-Cu-N nanocomposite coatings, *Mater. Sci. Eng.* A289 (2000) 189–197.

71. J. Musil, R. Daniel, O. Takai, Structure and mechanical properties of magnetron sputtered Zr-Si-N films with a high (≥25 at.%) Si content, *Thin Solid Films* 478 (2005) 238–247.

Chapter 2

Thin-Film Processing

Plasma science and technology have fundamental importance in the modern electronics, aerospace, metallurgical, biomedical, and waste treatment industries. Many of the plasma processing applications were developed in recent decades in order to fulfill ever-growing demands of technology applications. A huge variability of plasma processing has led to new discoveries that can only be achieved using plasma-related methods. This fact is also attributed to the fast development of plasma science. It has, in turn, opened up new avenues of basic science, in particular the nonlinear dynamics or science of chaos. Understanding the complex behavior of confined plasmas has led researchers to formulate the fundamental equations of plasma physics that resulted lately in important advances in fields as diverse as computers, lighting, waste handling, space physics, switches, relays, and lasers. Studies of high-intensity laser interactions, new highly efficient lighting systems, and plasma–surface interactions in materials engineering have become important for various applications where the usage of traditional technologies were obviously limited. Plasma applications have also attracted an interest due to their potential in reduction of energy consumption and waste productions in major industrial processes. Most of the research work mentioned in this book is based on low-pressure plasma technology as an environmental friendly and cost-efficient technique for modification of surfaces of traditional materials on

Novel Nanocomposite Coatings: Advances and Industrial Applications
Rostislav Daniel and Jindřich Musil

ISBN 978-981-4411-17-2 (Hardcover), 978-981-4411-18-9 (eBook)
www.panstanford.com

a microscopic level. Surface modification and deposition of thin films have gained an enormous interest in the last years to enhance functional properties of traditional bulk materials. As an alternative technology to chemistry and metallurgy, it is widely used for preparation of new materials with new possibilities. In addition, plasma technologies have opened new perspectives in material production and development of completely new applications. Indeed, thin-film processing represents a highly versatile method of manufacturing new advanced materials of a highly metastable nature exhibiting exceptional properties.

2.1 Principles of Plasma Discharges

A plasma is a collection of free charged particles, such as electrons, positive and negative ions, atoms or molecules, moving in random directions (see Fig. 2.1a). On the average, a plasma is electrically neutral. It is caused by electric fields that eliminate any charge imbalances in the plasma discharge by the charge rearrangement. As a result, the density of electrons together with negative ions in a plasma is equal to the density of positively charged ions. Plasma is also called a fourth state of matter. While a solid substance is in thermal equilibrium, it generally passes into a liquid state with increasing temperature, further into a gas and at a sufficiently high temperature the molecules in the gas decompose to form a gas of atoms that move freely in random directions. If the temperature is further increased, the substance enters the plasma state consisted of freely moving charged particles. It similarly happens in the plasma discharge. However, low-pressure plasma discharges are far away from thermal equilibrium since they are electrically driven and the applied power preferentially heats the mobile electrons with the mass much lower than the mass of ions. Hence, $T_e \gg T_i$ in plasma discharges. Consequently, the electrons promptly accumulate sufficient kinetic energy to have a high probability of producing excitation or ionization during collisions with heavy particles. On the other hand, the heavy ions exchange their energy mostly by collisions with the background gas. The degree of ionization is an important parameter of a plasma discharge, which represents the fraction of the original neutral atoms and molecules which have become ionized. A weakly ionized plasma is that one which has only a small fraction of the gaseous species ionized with the rest being neutrals. Then, the high number of neutrals determines the behavior of this type of plasma commonly used for film deposition.

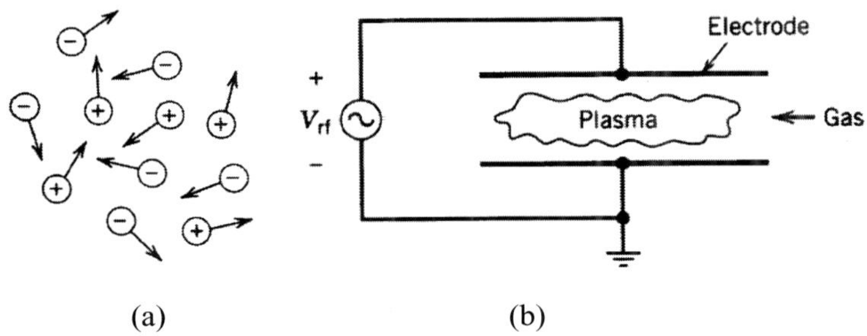

Figure 2.1 Schematic view of (a) a plasma and (b) a discharge. Reprinted from Ref. [1], copyright (2005), with permission from John Wiley and Sons.

A simple discharge is shown schematically in Fig. 2.1b. It consists of a voltage source that drives current through a gas between two conducting electrodes. In deposition processes, a glow discharge plasma is typically used which is characterized as a partially ionized low-pressure gas with approximately equal numbers of positive and negative particles. The production of ionized and excited species in glow discharge plasmas through collision processes and their interactions with surfaces and growing films is one of the reasons of increasing interest of plasma technology in materials processing. Its potential for the production of advanced materials is highly recognized.

2.1.1 Plasma Chemistry

A plasma is generally an energetic environment where many collisions between electrons and all other species, charged or neutral, typically occur. Rich diversity of elastic and inelastic collisions and chemical processes that occur in the plasma discharge make the plasma processing very complex. Many of the chemical processes, occurring as a consequence of electron–atom collisions, are very important in plasma materials processing for sputtering deposition,

ion implantation, and other processes. In principle, collisions within the plasma dominate the properties of the glow discharge. Collisions between individual species can be elastic or inelastic, depending on whether the internal energy of the colliding specie is preserved. An elastic collision is characterized by the kinetic energy interchange, where momentum and kinetic energy of translational motion is conserved. An inelastic collision is a collision in which momentum is conserved, but kinetic energy is not. Since the potential energy resides within the electronic structure of the colliding species, almost any change in potential energy is reflected by excitation or ionization processes. In elastic collisions, potential energy is conserved and no atomic excitation occurs. On the other hand, practically all of the electron's kinetic energy can be transferred to the heavier species in inelastic collisions. Therefore, electrons accelerated in an electric field play an essential role to sustain the plasma discharge. Some important collisional and chemical processes are summarized in the following paragraph and schematically shown in Fig. 2.2.

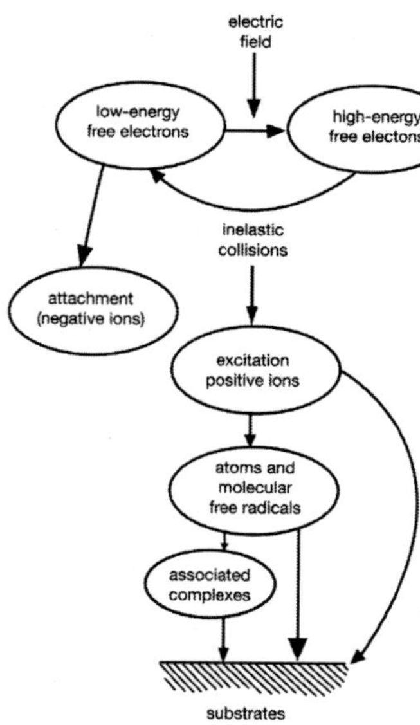

Figure 2.2 Production of active species in a plasma through collisional processes.

One of very important processes is electronic excitation $e + \mathrm{A} \rightarrow \mathrm{A}^* + e$ where outer shell electrons of the atom A are elevated to a higher energy state as a consequence of the inelastic collision between an atom and an electron. The lifetime of the excited states may be very short, so that electrons return spontaneously to the ground energy state by radiative decay $A^* \rightarrow A + h\nu$, where a photon of specific frequency is emitted. The de-excitation emission spectrum from the plasma is characteristic of the species in the plasma and can be used for plasma diagnostics and to monitor the density of species in the plasma. Excitation may, however, be very stable where some additional collision process is necessary to de-excite the atom. These long-lived states are called metastable states and can have lifetimes of 1 ms or longer. Since the metastable states have high energies, by collision of metastable species A^* with ground state neutrals B, they can be excited or even ionized. These processes are called the Penning excitation $A^* + B \rightarrow A + B^*$ and the Penning ionization $A^* + B \rightarrow A + B^+ + e$ and significantly contribute to the ionization and excitation rates in plasmas. However, these specific plasma processes occur only if the excitation energy from a metastable atom is greater than the ionization or excitation energy of the first colliding atom. The ionization of the atom by inelastic collision with energetic electrons $e + A \rightarrow A^+ + 2e$ is a process, where positive ions and additional electrons are formed. This process strongly influences the degree of ionization of the plasma and is given mainly by the electron density and energy distribution in the gas. In electron attachment ionization, negative ions are formed by electron attachment in the gas making plasmas very electronegative. The inverse processes, where electrons are lost by recombination, are also very important. Three body recombination $A^+ + 2e \rightarrow A + e$, where two electrons are necessary to conserve momentum, and radiative recombination $A^+ + e \rightarrow A + h_b\nu$, where energy and momentum are balanced by emission of a photon after recombination, are often observed in plasma discharges. Electron-ion recombination $A^+ + e \rightarrow$ surface $\rightarrow$ A, also called neutralization, occurs when ions and electron combine to form a neutral species. This process is typical for the interaction of the species on surfaces followed by an energy release corresponding to the energy taken up in the ionization process. An important process for electron loss that occurs in a plasma discharge is electron capture $A + e \rightarrow A^-$. This usually happens if the neutral species have a high electron affinity. If electrons elastically collide with neutrals, only very little energy is transferred to neutral due to a high mass difference between both colliding species. However, the electron will experience a large change in the direction of its velocity, and hence its momentum is changed. Unlike an electron–neutral collision, the electrons exchange energy very effectively in electron–electron collisions since their masses are equal. This kind of collision becomes important even at low degrees of plasma ionization.

Since atoms other than rare gases present in plasmas have lower lying levels for excitation than molecules, the excitation processes dominate generally in atomic systems. In the case of molecules, rotational and vibrational states require much less energy than excited electronic states that results in much greater electron energy losses compared to nonrare gas plasmas. Thus, when an electron collides with a molecule AB, a number of processes typically occur. Dissociation $e + \mathrm{AB} \rightarrow \mathrm{A} + \mathrm{B} + e$ is the electron impact fragmentation of molecules which results to form charged chemically reactive radicals or smaller atomic or molecular uncharged fragments of the original colliding molecule. This collision process is very important for plasma chemistry in discharges since the products are generally much more chemically active than the parent gas molecule and serve to accelerate reactions. During dissociation $e + \mathrm{AB} \rightarrow \mathrm{A}^+ + \mathrm{B} + 2e$, one of the excited species may become ionized and thus simultaneously ions and radicals as dissociation products of molecules may be formed. These chemically active species can consequently undergo surface reactions to form additional species, as during deposition. This collisional process is called dissociative ionization. A high number of electrons are lost after collision with molecules also due to associative recombination $e + \mathrm{A}^+\mathrm{B} \rightarrow \mathrm{AB}$ and dissociative attachment $e + \mathrm{AB} \rightarrow \mathrm{A}^- + \mathrm{B}$. In order to sustain the discharge, the electron losses have to be recompensed by newly formed electrons. They originate mainly through secondary electron emission from surfaces bombarded by energetic ions; however, electrons are also often formed directly in the plasma discharge through some processes listed above. Excitation and ionization processes in a plasma discharge also have been observed to occur in molecules when the photon radiation is absorbed. These processes are called photoionization $\mathrm{AB} + h\nu \rightarrow \mathrm{A} + \mathrm{B}^+ + e$ and photoexcitation $\mathrm{AB} + h\nu \rightarrow \mathrm{A} + \mathrm{B}^*$ processes and take place if the $h\nu$ energy of a photon is high enough to emit or excite an electron present in a molecule. Another important ion process is charge transfer $\mathrm{A}^+ + \mathrm{B} \rightarrow \mathrm{A} + \mathrm{B}^+$. It occurs when an energetic ion passes close to a thermal neutral. This process gives rise to a spectrum of energies of the ions and neutrals in a plasma since after a charge transfer an energetic neutral and a thermal ion are formed. It is also an important loss mechanism for energetic ions in the sheath region of discharges. It is worthy to point out that many of these plasma processes contribute to make atoms and molecules more chemically active so they readily react on the substrate surface and thus enhance reactive deposition processes. Moreover, if the species are ionized, they deliver additional energy to the growing film while they are accelerated to the biased substrate.

2.1.2 Discharge Regimes

After a dc voltage is applied between two electrodes in a vacuum system that is filled with a homogeneous neutral gas, a discharge is formed after breakdown at certain voltage. It is seen that characteristic regions are established between both electrodes as a consequence of natural tendency of the discharge to effectively set energy input to balance particle generation and losses. Electrons emitted from gas atoms by natural radiation are accelerated toward electrodes generating new ion–electron pairs by inelastic electron–atom collisions. While electrons are accelerated toward the anode, heavy ions impinge on the cathode surface, which typically serves as a target for thin-film deposition. Besides its role as a source of material for film formation, it is also a source of secondary electrons generated from the ion impact. An amount of both secondary electrons and sputtered particles from the target surface strongly depends on energy and mass of impinging ions.

Very important for breaking down and sustaining a plasma discharge is the degree of ionization. It depends on a balance between the rate at which ionization is produced by energetic electrons and the rate at which particles are lost by volume recombination and by passage to the walls of the apparatus. Since wall losses generally dominate over volume recombination and electron collisions with gas atoms may become so frequent that the electrons cannot accumulate sufficient energy to overcome the ionization potential, apparatus scaling, and adjusting the operating conditions within a given device is very important in order to produce and sustain the plasma discharge. Breakdown of a plasma discharge is described by an experimentally determined Paschen's law relating the breakdown voltage as a function of the gas pressure, distance between two electrodes, and the type of gas. Other variables determining conditions for establishing a steady-state discharge, where enough ions are created due to the collisional processes of the secondary electrons with gas atoms, are the electric field strength and the surface-to-volume ratio of the plasma.

The voltage–current characteristic of a dc discharge reveals several regimes, which significantly differ in their behavior depending on an amount and energy of electrons and ions formed (see Fig. 2.3). Just at breakdown, the current increases with only little increase in the voltage. Since there is a continuous loss of electrons, there must be an equal degree of ongoing ionization to maintain the steady state. The energy is being continuously transferred out of the discharge and hence the energy balance must be satisfied as well. With increasing voltage, the glow gradually builds up until the Townsend discharge is established. At this stage, the number of electrons generated and lost via inelastic collisions becomes equal reaching the steady state. The glow discharge operates at near constant voltage in this regime while most of the potential drop occurs near the cathode. The luminous glow is produced because the electrons have enough energy to generate visible light by excitation collisions. The electrode current density is independent of the total current in this regime since the plasma is in contact with only a small part of the cathode surface at low currents. Although the cathode surface is not completely covered by the glow at lower currents, its coverage gradually increases with increasing power. It is due to a higher number of secondary electrons emitted from the cathode surface by ion impact which results in an avalanche of ion production through enhanced electron-atom collisions. This regime precedes a normal glow discharge, which builds up at higher current. As the power increases further, the glow expands to completely cover the cathode surface which is followed by an increase in both voltage and current. By entering the abnormal discharge, typical conditions for sputtering are reached. A further increase in current results in heating the cathode surface and an arc discharge is consequently formed due to the thermionic electron emission. That is reflected by a discharge voltage drop as the glow evolves into an arc.

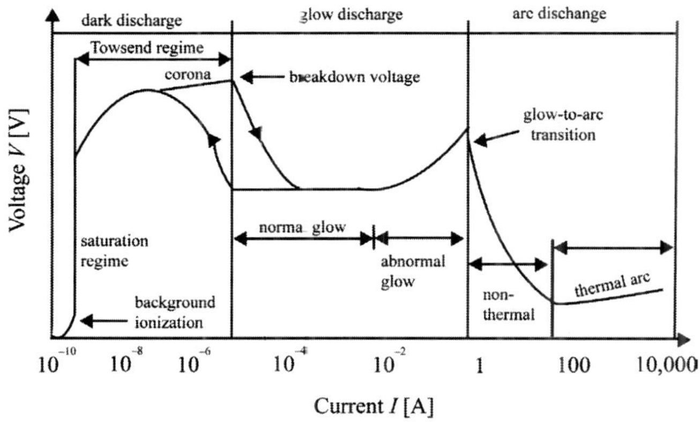

Figure 2.3 Typical voltage–current characteristic of a dc glow discharge.

2.1.3 Structure of Glow Discharge

The potential drops rapidly close to the cathode, varies slowly in the plasma, and changes again close to the anode. The electric fields in the system are restricted to sheaths at each of the electrodes. The sheath fields are such as to repel electrons trying to reach the electrode. Electrons originating at the cathode absorb energy from the field, are accelerated, collide, transfer energy, leave by diffusion and recombination, slow down by the anode, and get transferred into the outside circuit. This process route between both electrodes becomes continuous once a sustainable discharge is established due to the production of high number of ion–electron pairs in the course of the collisional processes. Then, the plasma discharge fills the region between the electrodes with a striking appearance of alternating light and dark spaces. The structure of a typical low-pressure glow discharge is shown in Fig. 2.4. All of the regions are gas; pressure and voltage are dependent on their size and intensity. The size and intensity of all regions are gas-, pressure- and voltage-dependent.

A thin region next to the cathode with a strong electric field is the Aston dark space. The electrons are accelerated through this space away from the cathode. This region has a negative space charge, meaning that stray initial electrons together with the secondary electrons from the cathode outnumber the ions in this region. The electrons are also of low density and energy to excite the gas, so it appears dark. The next structure to the Aston dark space is the cathode glow. Here, the electrons are energetic enough to excite and ionize the neutral atoms they collide with and hence the cathode glow is luminous exhibiting a relatively high ion density. The cathode glow sometimes clings to the cathode and masks the Aston dark space. Most of the voltage drop across the discharge occurs between the cathode and the negative glow with simultaneous dissipation of most of the power in this region. The electrons are accelerated during transition between both columns to energies high enough to induce ionization and avalanching in the negative glow. This column exhibits the brightest intensity of the entire discharge and is characteristic with relatively low electric field which is the most intense on the cathode side. The exponentially increasing density of high-velocity electrons near the cathode leads rapidly to a bright cathode glow in which intense ionization and excitation occurs. The electric field must decrease rapidly at the end of this region, where the transition to the positive column occurs. Electrons carry almost the entire current in the negative glow region. Electrons that have been accelerated in the cathode region to high speeds produce ionization and slower electrons that have already experienced inelastic collisions produce excitations. These slower electrons are responsible for the negative glow. Actually, the high electron velocities must be dissipated by elastic and inelastic collisions before the equilibrium conditions of the positive column can be established. This is done in a rather complicated process in which the electrons first lose almost all of their energy and then are reaccelerated in a weak field over approximately a mean free path. The electron number density in the negative glow is about 10^{16} electrons/m^3. As these electrons slow down, energy for excitation is no longer available and this is the point where Faraday dark space begins. The electron energy is very low in this region. The electron number density decreases by recombination and diffusion to the walls, the net space charge is very low and the axial electric field is small. On the contrary, the positive column is quasi-neutral with small electric field just large enough to maintain the degree of ionization at its cathode end. The positive column is a long, uniform glow, maintained by the $J.E$ power, which balances the loss of energy per electron–ion pair created. The electron number density is about

10^{15} to 10^{16} electrons/m^3 in the positive column. The next glow to the positive column is the anode glow which is slightly brighter than the positive column, but not always present. It is the boundary of the anode sheath. Anode dark space is the region between the anode glow and the anode itself. It has negative space charge due to electrons traveling from the positive column to the anode. There is a higher electric field than in the positive column. The anode pulls electrons out of the positive column and acts like a Langmuir probe in electron saturation. The drift velocity of the electrons in the weak electric field of the positive column is typically less than their thermal velocity. This required a retarding electric field in the neighborhood of the anode to prevent the full thermal electron current from reaching the anode. It must be positive with respect to the positive column to maintain the current.

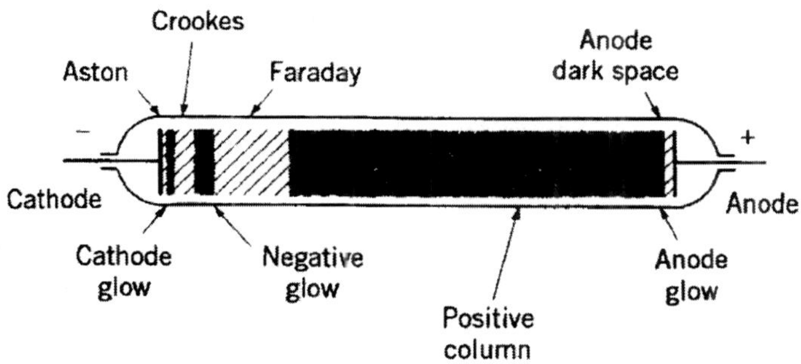

Figure 2.4 Structure of glow discharge in the discharge tube. Reprinted from Ref. [1], copyright (2005), with permission from John Wiley and Sons.

A phenomenon very important for applications is cathode sputtering. It is a consequence of collisional processes occurred within a cathode sheath with the potential drop of typically several hundred volts. A detail of the discharge near the cathode is depicted in Fig. 2.5. Ions from the negative glow fall through the cathode sheath causing secondary electron emission. The secondaries are accelerated back through the sheath and comprise the main energy input to the negative glow which supports the ionization process, both through direct ionization and through ionization by the high energy tail of the electron distribution. These regions comprise a self-sustaining combination. High ion-bombarding voltages in the cathode sheath lead to severe sputtering of the cathode surface and consequently deposit material on other surfaces. The principles of physical sputtering are described and explained in the next section.

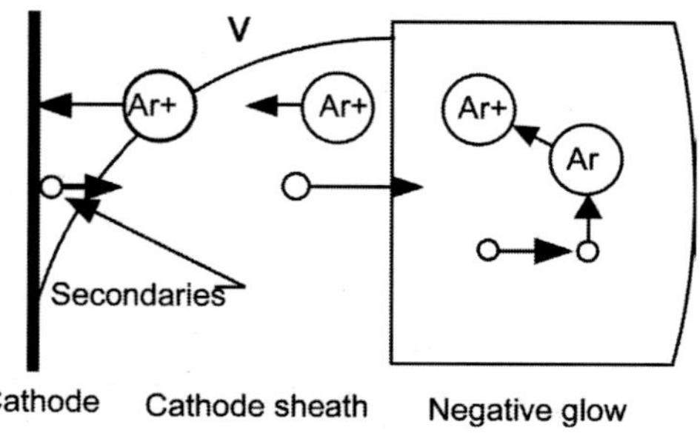

Figure 2.5 Detail of the discharge behavior in the cathode sheath-negative glow regions.

2.2 Physical Sputtering and Transport of Sputtered Material

The physical sputtering process involves the physical vaporization of atoms from a target surface by momentum transfer from bombarding energetic atomic-sized particles. In principal, it is a statistical process that occurs as a result of a collisional cascade process initiated near the target surface by incident energetic projectile particles. The energetic particles are usually ions of a gaseous material accelerated in an electric field. Sputter deposition of thin films as a term then characterizes the process where a surface being sputtered serves as the source of the deposited material. Sputter deposition is used to deposit pure metal films or compound materials exploiting a low pressure gas environment. The sputtered particles are transported from the target to the substrate typically without collisions in a plasma discharge serving as the source of ions.

Sputtering can occur whenever any particle strikes a target surface with energy high enough to eject an atom from the surface. However, physical sputtering relies on the transfer of physical momentum and kinetic energy from the incident particle to the surface atoms. It depends on the energy and mass of colliding particles and can be expressed by

$$\frac{E_t}{E_i} = \frac{4M_t M_i \cos^2 \theta_i}{(M_i + M_t)^2}, \tag{2.1}$$

characterizing the energy transferred to the target particle by the physical collision. Here, M_i, E_i, and v_i are the mass, energy, and velocity of a particle which impact on a line of centers with a target particle of mass M_t and energy E_t that is initially at rest. θ_i is the angle of incidence (see Fig. 2.6).

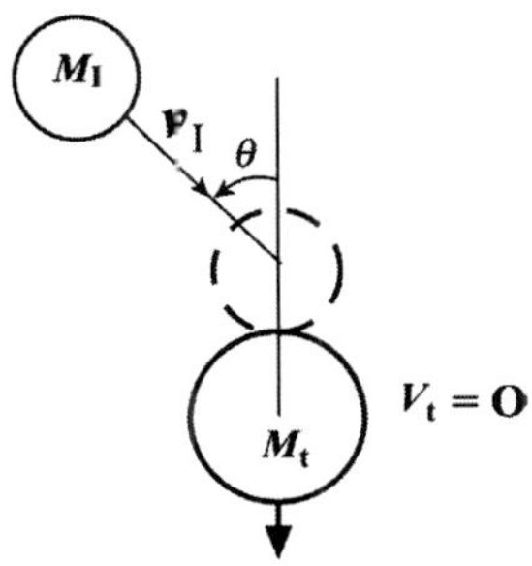

Figure 2.6 Schematic diagram showing collision of particles and momentum exchange that occur during sputtering.

From this equation, it is seen that the probability to eject an atom from the target surface depends on both the energy of the incident particle and its mass. It directly influences the sputtering efficiency of the target surface. However, both experiments and calculations show that the sputtering efficiency is more dependent on the incident-ion/target-atom mass ratio, expressed as a monotonically increasing function, compared to the effect of the incident ion energy. The maximum energy is transferred when cos θ_i = 1 (zero degrees) and $M_i = M_t$. Therefore, matching the atomic mass of the bombarding ion to the target atom is important to the sputtering yield. The sputtering yield is defined as the ratio of the number of emitted particles per incident particle and characterizes the sputtering efficiency. The flux of sputtered particles leaving the cathode depends on the elemental composition of the target, particularly on their mass and chemical bonding and the nature, energy, and angle of incidence of the bombarding species. Note that for typical ion acceleration energies, the sputtering yields of most metals are near unity and within an order of magnitude of one another. It dramatically decreases if the target is composed of strongly bonded compounds or if a reactive product is formed on the target surface in the reactive mode of sputtering. This phenomenon will be discussed further. Although the sputtering yield has been shown independent of the charge of the incident particle, the efficiency to eject the target particles is given by the mass of incident particle showing the highest efficiency of heavy ions accelerated in the field of the biased substrate. The mass of the bombarding particles is important to the energy and momentum transferred to the film atom during the collision. It is also relatively insensitive to the target temperature. Moreover, the sputtering yield is also independent of whether or not the bombarding particles are ionized. In fact, incident ions have a high probability of being neutralized by a field-emitted electron prior to impact [2]. Higher effect to the sputtering yield has been found in roughening and feature formation of the target surface. While the target becomes rough after removing several monolayers of target atoms, the sputtering yield typically decreases. Note that the sputtering yields are generally less than one at bombarding energies of several hundred electron volts, indicating the large amount of energy input necessary to eject one atom. Sputtering is much less energy efficient and the vaporization rates are much lower than can be attained by thermal vaporization. However, the physical sputtering is shown much easier to control and thin films with new exceptional properties can be prepared. It is also ascribed to the fact that sputter-ejected particles have kinetic energies considerably greater than thermal.

Sputtering yields of individual elements or compounds are mostly determined experimentally and may serve for estimating the amount of material removed from the target during deposition. It is characterized by the so-called erosion rate, which is given by

$$R = 6.23\frac{J.S.M_{\mathrm{A}}}{\delta}\ (\mathrm{nm/min}), \tag{2.2}$$

where J is the ion current density in mA/cm^2, S is the sputtering yield in atoms/ion, M_{A} is the atomic weight in grams, and δ is the density of the target material in g/cm^3.

2.2.1 Mechanisms of Sputtering

Statistical analyses show that sputter ejection very rarely occurs due to collision cascades initiated more than five atomic layers below the surface [3]. The mechanism involved in the course of sputtering depends mainly on the energy of the impinging ions. However, most of the energy transferred to the target surface region during collisions of ions in the lattice is lost as heat. The sputtering process is schematically shown in Fig. 2.7. The incident ions with specific energy impact the surface or near-surface atoms of the solid. If this energy is high enough to break bonds and dislodge atoms of the target, one or more atoms are removed from the solid. The energy of ejected particles depends on the energy and mass of impinging ions and bonding energies of atoms in the target. It influences the sputtering mechanism involved in the near-surface region. The atoms removed from the target are considered to be sputtered atoms.

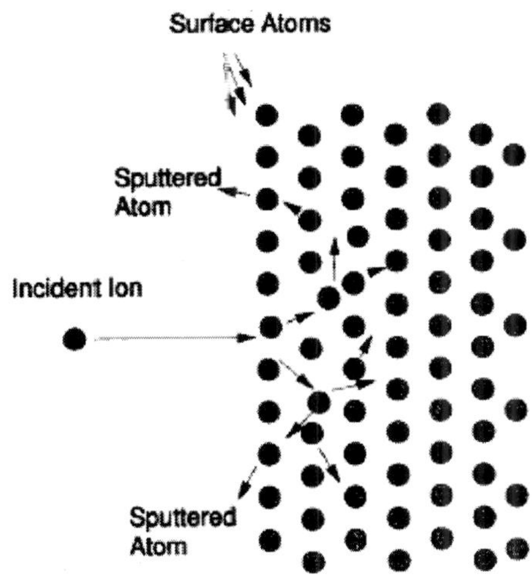

Figure 2.7 Schematic diagram of a physical sputtering process.

In dependence on the energy of incident ions, four different regimes of the physical sputtering can be defined. It comprises energies from near the binding energy to many MeV.

It has been shown that at very low energies of a few eV up to several tens of eV, the sputtering yield is very low and atoms ejected from the target surface exhibit low kinetic energy. Moreover, the presence of a sputter threshold within this energy regime has been observed. Below the threshold energy no sputtering occurs irrespective of the bombarding flux. In a polycrystalline material, some crystallographic planes are sputtered faster than the others.

This effect is called preferential sputter etching and is very strong in some materials. On the other hand, there is no sputtering observed by electrons even at very high temperatures.

At incident energies of up to 1000 eV, the incoming ion has energy sufficiently high to dislodge tens to hundreds of atoms. Within this knock-on energy regime the collision sequence is erratic and strongly depends on the impact site. In this case, the sputtering yield is also very sensitive to the angle-of-incidence of the bombarding ion. After the first collision, the incident ion and the impacted one move on into the material causing more and more collisions.

At even higher energies of impact ions of nearly 1 keV to 50 keV, the sputtering yield significantly increases. The incident ion has sufficient energy to break all the bonds between atoms in a spherical region around the impact site, which causes emitting of higher amount of the target atoms. Although the sputtering yield in this energy range regime called collision-cascade sputtering is much higher than that in the knock-on regime, the higher energies associated to high voltages make it impractical to use for most industrial-scale sputter deposition applications. Mainly within this energy regime, sputtered atoms have kinetic energies much higher than those of thermally evaporated atoms.

At ion energies above 50 keV, the incident ions can penetrate into the bulk of the solid before transferring its energy. This regime is called high energy implantation and is associated with a dramatic decrease of the sputtering yield. Although high-energetic ions can cause damages in a micrometer range below the surface, much of their energy is lost far away from the surface resulting in almost no sputtering of the target atoms. In addition, the incident particle is often trapped or implanted within the target.

Some of the above-mentioned effects of the momentum-transfer theory are very important for the film growth from condensing atoms on the substrate surface. It has been observed that those effects are more pronounced at low pressures since during the low-pressure sputtering the energetic sputtered atoms and reflected high energy neutrals are not thermalized by collision between the sputtering target and the substrate. Thus, the time associated with a particular collision event is short with respect to the ion time of flight between collisions. The same effects are observed at the biased substrates, which subsequently allows an effective modification of surfaces of growing films.

2.2.2 Practical Implementation of Sputtering

Sputter deposition of thin films is principally a vacuum process. The source of a coating material exits in the form of a target mounted opposite to the substrate in a vacuum deposition system. A negative potential is applied to the target to attract positive ions and initiate the ion bombardment. Since the sputtering yield tends to be greatest when the mass of the bombarding particle is of the same order of magnitude or larger than that of the target atoms, the selection of suitable working gas for the plasma discharge is essential. In order to avoid chemical reactions of species of working gas on the target or substrate surfaces or directly in the discharge, inert gases are commonly used for plasma processing technologies. Typically argon is used for inert gas sputtering because of its relatively high mass (40 amu) and low cost. On the contrary, mixtures of argon with nitrogen, oxygen, methane, or acetylene are used for reactive sputter depositions. The discharge in a deposition system is commonly operated in the abnormal negative glow mode [4]. The ions of working gas approach the target at certain energies in a direction normal to the target surface. The accelerated ions are commonly neutralized prior to impact and are reflected as energetic neutrals. Thus, they are not influenced by the electrical field over the target surface and together with sputtered particles are transported to the substrate. Sputtered atoms leave the target surface at each point with a cosine distribution. The distribution of the sputtered particles dramatically alters in a consequence of the gas scattering. Collisions of sputtered particles with working gas species are markedly pronounced at higher pressures. At high pressures, a portion of the sputtered material is even scattered back to the target [5]. The reduction of the energy of sputtered particles by their collisions in the discharge during their transport to the substrate is called thermalization. Generally in low-pressure deposition, energetic sputtered atoms approach the surface of the substrate at relatively high energies since they do not lose much of the kinetic energy by collisions with working gas species. The pressure and distance for thermalization depend on the relative masses of the particles and the collision probability. In some applications, scattering is used to advantage to improve the surface coverage by randomizing the flux direction. If reflected high-energy neutrals reach the substrate without thermalization, they can significantly contribute to substrate heating [6]. For the same reason, the sputtered target is actively cooled, which minimizes the amount of radiant heat in a sputtering system and prevents the diffusion of atoms within the target. Diffusion processes may markedly change the elemental composition in the surface region of alloy targets and thus the composition of the deposited films.

2.2.3 Sputtering of Alloys and Mixtures

Sputter deposition technique has found a widespread use in the early 1980s due to the possibility to deposit films of elemental materials. An important advantage of the sputtering process is, however, a fact that metal alloy films can be easily prepared as well. In contrast to chemical vapor deposition, where the deposited films exhibit different composition than the target source, the composition of a sputter-deposited film tends to be the same as that of the target. Targets are usually formed by casting or by hot pressing of powders of different compositions. It allows sputter deposition of variable compound materials. Sputtering of solid surfaces proceeds under ideal conditions so that each layer of atoms is removed from the surface before the next layer is subject to sputtering. However, targets are typically composed of species having different individual sputtering yields or masses. It leads to preferential sputtering of the component with higher yield and the sputtered flux is rich in this constituent. Consequently, the surface layer of the target will be enriched with the material having the lower sputtering yield [7]. The composition of this altered surface layer continues to change until a steady state is reached. The steady-state altered layer ensures the sputtered flux to have composition equal to the bulk target composition. This process is schematically illustrated in Fig. 2.8.

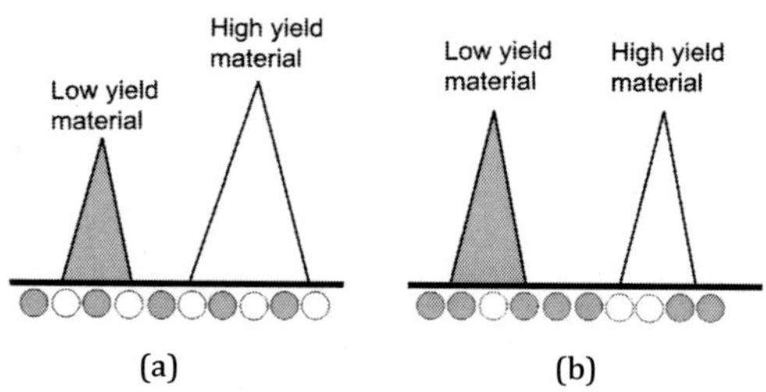

Figure 2.8 Schematic illustration of the surface composition modification which occurs during sputtering of a single-phase alloy: (a) before and (b) after surface equilibration.

In the case of chemical vapor deposition, each alloy component also evaporates with a different vapor pressure, however, the melts readily homogenize due to rapid atomic diffusion and convection effects in the liquid phase. This effect is negligible during sputtering since cooled target enables just minimal solid-state diffusion and thus the target composition alters minimally. Moreover, during sputtering the material is removed from the target layer-by-layer that allows to reach a steady state after removing a few monolayers of the target atoms. Hence, the composition of the target is commonly reached in the sputter deposited film. It is also the main reason why sputter deposition has attracted a huge interest for many practical applications. However, deposition of films having the same composition as the target source requires several conditions to be fulfilled. The target must be sufficiently cooled to avoid bulk diffusion of individual components and so to alter the target composition during the sputtering process. The target should remain stable since its decomposition may result in changes of composition of the sputtered flux as well. Furthermore, the gas-phase transport of the components and their sticking coefficients on the substrate must be the same. If composite targets are made from discs, wafers, or wires of one material mounted over a target of another material, the situation during sputtering becomes more complex. In that case, the composition of the sputtered flux may differ from that reached at a steady-state sputtering of alloyed targets, typically due to significant difference in sputtering yields of individual constituents. If the second phase, or any included impurity particles, has very low sputtering yield, the surface may develop into an irregular surface topography (see Fig. 2.9). It has also been observed in sputtered two-phase alloys in which the phases have significantly different sputtering yields. However, also in that case a steady-state surface topography is finally developed.

Sputtering initiated from compound targets, where strong chemical bonds are involved, also results in a loss of stoichiometry in the deposited film compared to the target material. Especially if the target is composed of elements with different mass, the lighter and more volatile species may get lost in the transport between the target and the substrate.

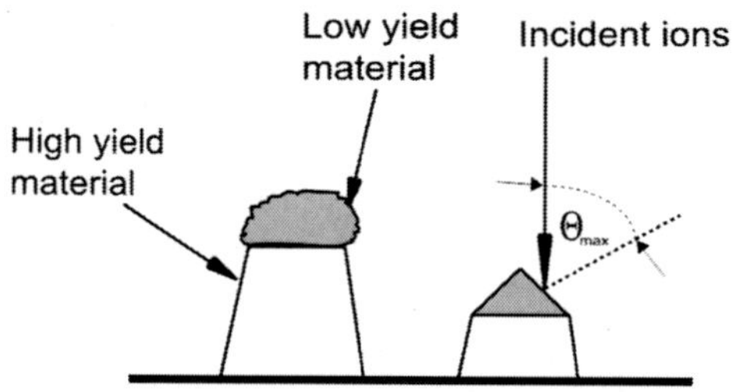

Figure 2.9 Effect of a material dependent sputtering yield on the surface topography development.

2.2.4 Distribution of Sputtered Atoms

To understand the film forming phenomena, it is required to fully describe and understand the transport of sputtered atoms from the target to the substrate. In the case of a low-pressure sputtering process in a collisionless plasma, the distribution of sputtered atoms from the target surface is a meaningful parameter. It has been shown that atoms emitted from a flat surface by sputtering, using near-normal high energy incidence particle bombardment, are commonly described by a cosine emission distribution. In other words, an amount of atoms ejected at any particular angle can be compared to the amount sputtered at normal incidence times the cosine of the angle from normal incidence. In general, the overall distribution can be drawn as a sphere centered on the impact point which represents an amount of atoms sputtered at any particular angle. A schematic illustration of the angular emission distribution for sputtered atoms is given in Fig. 2.10.

Although some deviations from the ideal case have been observed if the incident ions approach the target surface in off-normal direction or if the target is not ideally flat, the ejected flux still have approximately a cosine distribution. The real distribution, however, strongly depends on the energy of the incident ions. It becomes typically an over-cosine distribution skewed in a forward direction with less emission out to the side if the incident particle energy is low. On the other hand, high energies result in under-cosine emission. An interaction distribution is just the opposite. In the knock-on regime, low incident ion energy leads to an under-cosine distribution [2].

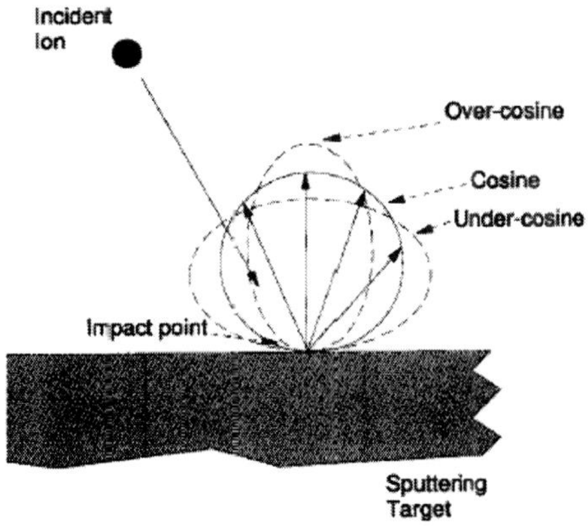

Figure 2.10 Cosine emission distribution for sputtered atoms from a flat target. Reprinted from Ref. [8], Copyright (2001), with permission from Elsevier.

2.2.5 Energy Distribution

Among other things, physical sputtering differs from other vapor-phase deposition techniques in kinetic energies of sputtered atoms. The average ejection energy of atoms during high-energy sputtering may achieve values one to two orders of magnitude higher compared to those obtained by evaporation. This significant difference is given by different dynamics of both processes. Moreover, due to the low-pressure deposition regime and typical target-to-substrate distance less than a few mean free paths, the sputtered atoms approach the substrate with the energies comparable to the ejection energies. From collision cascade theory, the energy distribution of sputtered atoms is given by

$$\frac{\Delta N}{\Delta E} \propto \frac{E}{\left(E+U\right)^3}, \tag{2.3}$$

where $\Delta N/\Delta E$ is the differential flux of sputtered atoms with energy E, and U is the surface binding energy. The most probable ejection energy corresponds approximately to one half the surface binding energy, stated at a few eV to 10 eV (see Fig. 2.11). A decrease from the maximum value of $\Delta N/\Delta E$ to the high-energy tail of sputtered-atom energy distribution follows an E^{-2} dependence. The atomic number of the target material also plays an important role. The difference in the energy distribution for sputtered atoms, particularly in their average energy, results in qualitatively different mechanism of the film growth during physical sputtering and evaporation. This is also the reason for exceptional properties of thin films grown under highly non-equilibrium conditions typical for sputtering processes.

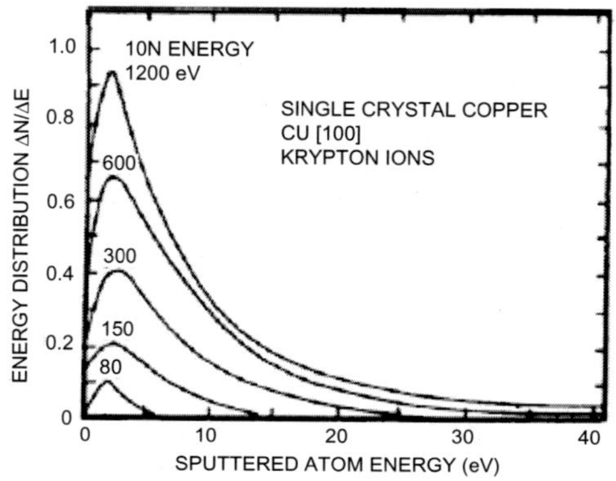

Figure 2.11 Energy distributions of sputtered Cu atoms ejected by Kr^+ ions at various bombarding energies. Reprinted with permission from Ref. [9]. Copyright [1969] American Institute of Physics.

2.2.6 Sputter Efficiency

Very important for practical usage of sputter deposition in industrial applications is the deposition rate. It is given by the sputter efficiency of the target which significantly varies in dependence of gas used as the working medium in the plasma discharge, the material of the target, and the deposition conditions. The deposition rate represents a general relation between atoms emitted from the target and atoms deposited on the substrate. It is also sometimes called the deposition probability. The higher number of atoms are sputtered from the target and transported without any loss to the substrate, the higher deposition rate will be achieved. The main parameters influencing the deposition rate are the discharge current, total gas pressure, and sputtering yield of the target material, which can be expressively lowered in the course of the reactive sputtering. Experimental results suggest that the highest deposition rates are achieved at low pressures, small target-to-substrate distances, and by using a gas species which are slightly lighter than the sputtered species.

Although sputtering with a heavy gas results in higher sputter yields, the probability of atoms to condensate on the substrate is reduced significantly. Deposition of films on the substrate can also be dramatically influenced when the bias voltage is applied on the substrate and resputtering of the growing film occurs.

2.3 Sputter Deposition Techniques

2.3.1 DC Plasma Discharges

The planar direct current diode is the simplest sputtering configuration. A low-pressure dc diode discharge operates in the abnormal negative glow discharge region and is typically maintained between two electrodes. In the dc diode discharge, the cathode, simultaneously serving as a source for deposited films, attracts ions from near the edge of the plasma region due to its high negative potential. The ions are accelerated in strong electric fields across the cathode fall region and after passing the distance from the plasma volume to the cathode sheet region, they impact the cathode surface. The impinging ions and energetic neutrals, produced by charge-exchange collisions in the dark space of the discharge, cause first the emission of atoms from the cathode surface (target) and second the ejection of small number of secondary electrons, approximately one for every ten ions in the case of Ar^+ ions impacting on a metal cathode [10]. Secondary electrons are very important to sustain the discharge since they create new ions during their motion back across the cathode fall region [11]. In the dc diode discharge, the cathode current density depends mainly on the applied voltage [1,4]. The dc diode discharge configuration is characterized by the consumption of most of the electrical potential that is applied between the anode and cathode in a cathode dark space or sheath region. It is caused by relatively low mobility of the ions in the discharge compared to the electrons. The thickness of the dark space is commonly several centimeters, depending on the pressure in the deposition system. It strongly affects the sputtering efficiency through the collisions of the accelerated ions in the discharge. If the number of physical collisions is higher, the higher pressure is established in the deposition system, which consequently influences the energy distribution of ions impacting the target surface. The ions accelerated in the cathode sheet region mostly experience physical collisions with the loss of their kinetic energy or become neutralized by charge-exchange processes. Relatively high probability of ions to neutralize results in a bombardment of the target mostly by fast neutrals and low energetic ions instead of being bombarded by ions having an energy equal to the potential drop across the cathode dark space. The energetic neutrals are not affected by the electric field and reach the surface of the target having energy typically less than 10–20% of the applied potential. The sputtering rate can be effectively increased at a given voltage by increasing the pressure of working gas. However, by reaching certain critical limit, the motion of the ions across the dark space is dramatically disrupted by collisions with gas atoms so that the deposition rate decreases instead.

The dc diode configuration schematically depicted in Fig. 2.12 requires many prerequisites to deposit films with exceptional properties. The cathode-to-anode spacing should be minimally several centimeters, otherwise insufficient number of the secondary electrons is produced to sustain the discharge. To ignite and establish a discharge in the dc diode configuration, the gas pressure must be greater than 1 Pa and the cathode must be of an electrically conductive material. All above-mentioned conditions, which must be necessarily fulfilled, demand the configuration system and the deposition conditions to be optimized. Only by operating at optimum pressure, discharge voltage, and cathode current density, the highest sputtering rates for preparation of films with desired properties can be achieved.

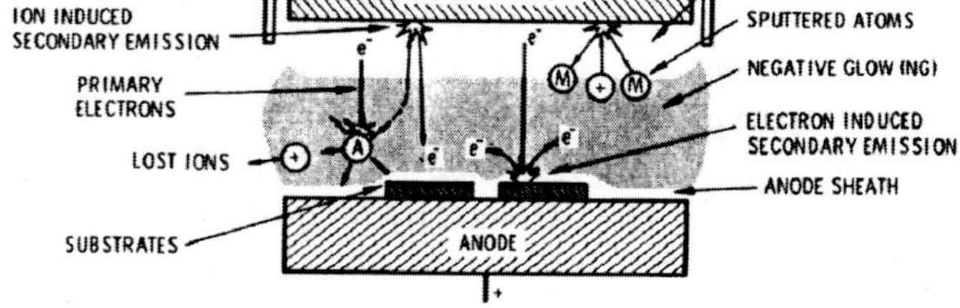

Figure 2.12 Schematic representation of a dc planar diode sputtering system. Reprinted from Ref. [12], Copyright (2010), with permission from Elsevier.

The reasons why the dc diode discharges have not attracted great concern in industrial applications for thin-film sputtering is given by many drawbacks of this configuration. In particular, the deposition rate is dramatically reduced by gas-scattering of sputtered atoms during transport between the target and the substrate. For the same reason, the kinetic energy of sputtered atoms approaching the substrate surface markedly decreases, which leads to low surface diffusion of the adatoms and consequently formation of non-compact film structure [13]. In addition, in this configuration the substrates are in contact with the plasma discharge and so subjected to the bombardment of the secondary electrons and the ions from the edge of the discharge. Thus, the substrate is heated which may result to permanent damages of heat-sensitive substrate materials.

2.3.2 Magnetically Confined Plasmas

It was shown in the previous section that the electrons in the dc diode sputtering participate highly ineffective on sustaining the discharge. Furthermore, the configuration of the system does not allow depositing films at very low pressures and without irradiating and thus heating the substrate surface. By the suitable application of a magnetic field, the secondary electrons, emitted from the target due to ion bombardment, can be constrained to stay near the target surface. In general, an electron motion follows an orientation of the electric E and magnetic B fields. The secondary electrons emitted from the cathode surface spiral around the magnetic field lines if the magnetic field is parallel and the electric field is normal to the cathode surface. The electrons are then accelerated away from the surface with the frequency and the radius of the spiraling motion depending on the strength of the magnetic field. A combination of both the electric E and magnetic B fields in an appropriate arrangement, so the electrons are constrained to move in a direction perpendicular to E and B fields ($E \times B$ drift), causes a formation of a current loop of drifting secondary electrons. Their intensive interaction with working gas in the drift region creates very dense plasma confined near the cathode surface. In this arrangement, the produced ions have a high probability to approach the cathode surface with sufficiently high energy to sputter the target atoms. The secondary electrons emitted from the target as a consequence of sputtering process are trapped in the discharge by the magnetic field which results in an electron having more ionizing collisions before being lost from the discharge. Thus, the sputtering efficiency in a magnetron sputtering configuration is enormously high allowing to sputter thin films at high deposition rates [14]. On the other hand, ions are relatively unaffected by the magnetic field.

2.3.2.1 Balanced magnetron configuration

Magnetron sputtering is presently the most widely used sputtering configuration. The simplest magnetron source is the planar magnetron with the sputter erosion track of a closed or elongated circle (Fig. 2.13a) [15]. However, a closed circulating path of electrons confined near the target surface using a magnetic field can be principally generated on any surface of revolution. Commonly used configurations are a hollow cylinder [16,17], post or spool [18] and magnetrons of a conical [19] or a hemispherical section [15]. Magnetic fields for the magnetron sputtering are generated using permanent magnets or electromagnets or a combination of the two. Permanent magnets have the advantage that they are relatively easy to be placed so that the magnetic field lines are set in a desirable manner. This is quite complicated in the case of electromagnets. On the other hand, by varying the strength of the magnetic field, the position and discharge characteristics can be easily optimized for sputtering certain materials. For that purpose, a two-coil Helmholtz arrangement is commonly used to produce a space with nearly parallel magnetic field lines.

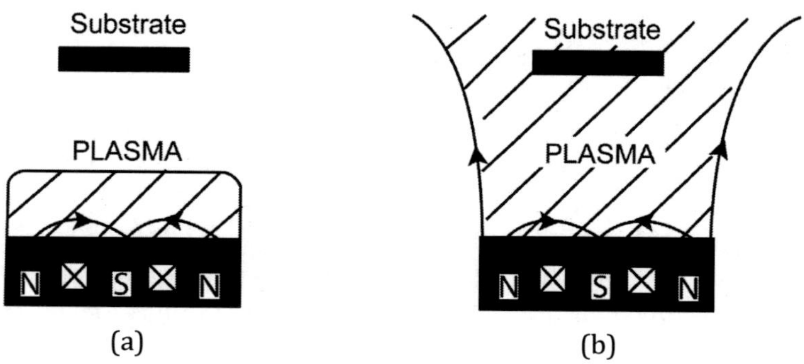

Figure 2.13 The plasma discharge and magnetic field configuration in (a) balanced and (b) unbalanced magnetron system.

2.3.2.2 Unbalanced magnetron configuration

The principal advantage of a magnetron sputtering configuration is given by a formation of a dense plasma discharge near the target surface. The ions bombarding a target exhibit high kinetic energy which is unaffected by any energy loss occurring typically in the dc diode sputtering at high pressures due to physical and charge-exchange collisions. Hence, this configuration allows the sputtering at high sputtering rates, low and high pressures, and with or without thermalization of sputtered atoms. One of the disadvantages of the balanced magnetron sputtering configuration is the fact that the plasma is typically confined so near the target substrate that neither ions nor electrons can be extracted from the plasma discharge to bombard the growing film. The activation of reactive gases near the substrate or the ion bombardment of thin film is required in many applications in order to deposit films with very dense and compact morphology. This problem can be solved by generating an additional plasma discharge near the substrate surface or by re-arranging the magnetic fields. An unbalanced magnetron configuration is designed to allow some of the electrons to escape from the cathode region towards the substrate surface (see Fig. 2.13b) [20]. The additional magnetic fields are deliberately arranged to expand the plasma away from the target so that it can consequently provide the ions for bombardment of the substrate and activate a reactive gas during reactive film deposition. The magnetic field for unbalancing the magnetron configuration can also be supplied either by permanent magnets or by electromagnets.

A major problem of the unbalanced magnetron sputtering using magnetic fields is the difficulty in generating a uniform magnetic field all over a target surface. Thus, the flux of emitted secondary electrons is not uniform and so is the plasma. This non-uniformity in the plasma discharge results in a non-uniform ion bombardment of the target and thus the erosion zone on the target surface covers only a specific part. In some cases, only 10–30% of the target material is being used before the target is scrapped. This effect also causes a formation of the compound material on the edges of the erosion zone and other surface areas during reactive sputter deposition. Once the compound is accumulated in these areas, a surface charge can be built up and cause arcing over the target surface. This problem is commonly solved by applying an rf potential to the target often combined with the dc potential. Another problem related to the nonuniform target sputtering is the position of the substrate with respect to the target in order to obtain a film homogeneously covering the substrate surface. It can be partially overcome by utilization of the target material and by moving the magnetic field over the target surface to ensure a uniform sputtering of the target. In this way, the target wear can be improved up to 40–50% and large surfaces can be homogeneously coated.

2.4 Reactive Sputter Deposition

Sputter deposition is commonly used to deposit thin films of compound materials in two principal ways: by sputtering from a compound target or from an elemental target in a partial pressure of a reactive gas. The sputter deposition performed in the presence of reactive gas is called reactive sputtering process. Reactive sputtering is commonly used for the deposition of nitride thin films. However, a universality of the reactive sputtering allows synthesizing many other diverse compounds using various reactive gases: oxides using O_2 or H_2O, nitrides using N_2 or NH_3, oxynitrides using a mixture of $O_2 + N_2$, carbides using C_2H_2, sulfides using H_2S, etc. A major advantage of the reactive sputtering process is that many various compound films can be deposited using pure metal targets. Reactive sputtering from metal targets has attracted the most attention mainly due to the possibility of high rate fully controlled film sputtering using dc rather than rf power. It has been shown that using dc power supplies insulating films can also be prepared [21]. The sputter deposition from the compound targets can increase the deposition rate and avoid unfavorable effects occurring typically during the reactive sputtering, although in this case the reactive gas is commonly used. It is due to the loss of a portion of the reactive gas in the transport to the substrate and during condensation and reaction processes. However, the sputtering is operated at much lower partial pressures of the reactive gas compared to conventional reactive deposition.

Reactive gases used for the reactive sputtering are in the molecular state but are also activated through collisions in the plasma discharge to form more chemically reactive or more adsorbed species. Since the reactive gases have typically low atomic masses (N = 14, O = 16), they are highly ineffective in sputtering. Hence, usually an inert-reactive gas mixture is preferred to aid in sputtering using heavier inert gas ions as well. By mixing inert gas with the reactive one also aids in activating the reactive gas by the ionization and excitation processes. Since sputtering is an atomistic process, particles are mostly emitted from a target in the atomic form than as molecules or clusters. Although these atoms are exposed to many collisions with the reactive gas, chemical reactions during their transport from the target to the substrate are rare. Reactions occur with a higher probability on the surfaces of the targets and substrates. Here, deposited adatoms form a compound film with gas atoms that impinge and get stuck on the surface. It can cause troubles if the target surface is fully covered by a compound layer. If the compound layer is formed on the target surface at higher rate than it is removed, the deposition of the film expressively slows down. It is caused by lower sputtering yields of compounds formed on the target surface compared to pure metal materials [22]. A decrease in the deposition rate also contributes higher secondary electron emission yield of the sputtered compounds. Once the reactive gas is introduced into the deposition system, the discharge voltage is markedly reduced as well as an ion component of the target current at constant voltage. Thus, most of the ion energy is consumed to produce and accelerate secondary electrons during sputtering. This poisoning effect causes several practical problems. Besides low deposition rates, charge buildup on the target surface can prevent sputtering of the cathode if a dielectric compound is formed in the course of the reactive sputtering. Arcing over the insulating target surface often results in poor film quality and also the process itself becomes more difficult to be controlled. Every change in the target composition caused by reactions of reactive gases with the target surface leads to significant changes of the plasma parameters and thus conditions for the film growth. Another problem is given by the transition behavior where the deposited material suddenly changes from metal to a nearly stoichiometric compound in a very narrow range of the partial pressure of the reactive gas. This phenomenon is shown in Fig. 2.14.

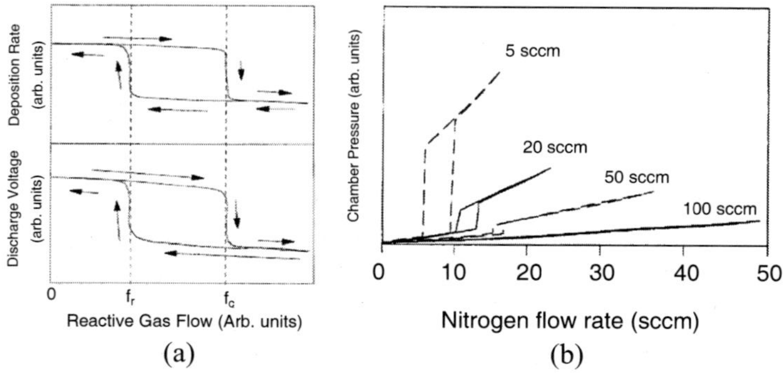

Figure 2.14 Hysteresis curves in reactive sputter deposition as a function of gas flow and pumping speed. Reprinted from Ref. [8], Copyright (2001), with permission from Elsevier, and reprinted from Ref. [23], copyright (1994), with permission from IEEE.

By keeping a partial pressure of the inert gas constant, it is able to investigate the effect of a partial pressure of the reactive gas on the plasma parameters and the deposition rate. If the discharge voltage and the deposition rate are plotted as a function of the partial pressure of the reactive gas, the highest rates and voltage can be seen at the lowest or no reactive gas flow. The target is sputtered mainly with Ar ions and the sputtered film is metallic. It is characteristic for this mode of reactive sputtering that an introduction of the reactive gas at its low flow results just in a very slight change of deposition parameters. The reactive gas is being absorbed on the substrate surface where it reacts with sputtered target material. If the partial pressure of nitrogen or oxygen is increased further, no more of the reactive gas can be absorbed on the substrate and thus a compound layer starts to be formed on the target surface consuming the reactive gas from the deposition system. At a certain pressure, a slight decrease in both the deposition rate and the discharge voltage merge into a sudden drop and at this point almost fully stoichiometric product is formed. As long as the target surface is covered by the reactive compound with the sputtering yield typically up to an order of magnitude lower compared to pure metals, the supply of metallic atoms is dramatically reduced. It leads to a higher amount of unused reactive gas thus reacting with the target even more intensively. Similar transition from a metallic to a transition and further to a reactive mode of sputtering is observed using any of the reactive gases. The transition behavior is, however, most abrupt during deposition in the gas mixture of argon and oxygen and more gradual during sputtering of nitrides. The consumption of the reactive gas is very strong in the metallic mode and decreases at higher reactive gas flows. This is, however, just one-way transition and occurs abruptly at two different critical values of the reactive gas flow depending on the direction of the transition [24]. Conversion back to the metal mode occurs only when the flow of the reactive gas is reduced to the point where it can no longer maintain the compound layer [25]. In other words, the residual compound layer has to be removed from the target surface by sputtering that however takes longer time due to its lower sputtering yield. This phenomenon is known as a hysteresis effect of the reactive sputtering and presents a real problem for the deposition of compound films at reasonable rates.

In order to prevent the target poisoning and to control the deposition process at that point, several approaches have been established. By increasing the sputtering rate enough target atoms are available to react with the reactive gas which shifts the critical point to higher flows. The amount of unused reactive gas can be decreased if the pumping speed of the deposition system is sufficiently high so that the excess reactive gas can no longer react with the target surface. This effect significantly flattens out the hysteresis curve [26,27]. Another approach is based on the feedback control of the reactive gas flow. In this case, the discharge voltage is monitored and the partial pressure of the reactive gas is controlled to operate in the metallic mode of sputtering. This feedback control is, however, applicable to small systems but tends to fail in the industrial scale. Further, reactive gas pulsing has been demonstrated to effectively control the reactive sputtering on the substrate surface. Here, the flow of the reactive gas is varied in order to operate in the metallic mode without poisoning of the cathode. Inserting a positively biased electrode near the substrate together with an incorporation of a baffle over the target can also limit the impingement of reactive gas molecules on the target as well as a formation of the plasma near the target. Both have been shown to effectively intensify the sputtering process [28,29]. The poisoning of the target can also be reduced if the inert gas is injected around the target surface and the reactive gas is injected near the substrate surface. So formed area of the inert gas over the target prevents reactions on the surface and shifts the transition between the metallic and reactive mode of sputtering into higher partial pressures of the reactive gas.

References

1. M.A. Lieberman, A.J. Lichtenberg, *Principles of Plasma Discharges and Materials Processing*, Wiley, New York, 1994.
2. G.K. Wehner, G.S. Anderson, in L. Maissel, R. Glang (Eds.), *Handbook of Thin Film Technology*, McGraw-Hill, New York, 1970.
3. K. Macak, P. Macak, U. Helmersson, Monte Carlo simulations of the transport of sputtered particles, *Comp. Phys. Comm.* 120 (1999) 238–254.
4. B. Chapman, *Glow Discharge Processes*, Wiley, New York, 1980.
5. I. Abril, A. Gras-Marti, J.A. Valles-Abarca, The influence of pressure on the operation of glow-discharge sputtering systems, *Vacuum* 37 (1987) 391–394.
6. J.A. Thornton, Substrate heating in cylindrical magnetron sputtering sources, *Thin Solid Films* 54 (1978) 23–31.
7. G. Betz, G.K. Wehner, Sputtering of multicomponent materials, in R. Behrisch (Ed.), *Sputtering by Particle Bombardment II*, Springer-Verlag, Berlin, 1983.
8. S. Rossnagel, Sputtering and sputter depositions, in K. Seshan (Ed.), *Handbook of Thin Film Deposition Processes and Techniques*, Noyes Publications, William Andrew Publishing, New York, 2002.
9. R.V. Stuart, G.K. Wehner, Energy distribution of sputtered Cu atoms, *J. Appl. Phys.* 35 (1964) 1819.
10. E.S. McDaniel, *Collision Phenomena in Ionized Gases*, Wiley, New York, 1964.
11. J.A. Thornton, Diagnostic methods for sputtering plasmas, *J. Vac. Sci. Technol.* 15 (1978) 188–192.
12. J.A. Thornton, J.E. Greene, Sputter deposition processes, in R.F. Bunshah (Ed.), *Handbook of Deposition Technologies For Films and Coatings*, Noyes Publications, New Jersey, 1994.
13. P.B. Barna, M. Adamik, Fundamental structure forming phenomena of polycrystalline films and the structure zone models, *Thin Solid Films* 317 (1998) 27–33.
14. R.D. Arnell, P.J. Kelly, Recent advances in magnetron sputtering, *Surf. Coat. Technol.* 112 (1999) 170–176.
15. J. Musil, Recent advances in magnetron sputtering technology, *Surf. Coat. Technol.* 100–101 (1998) 280–286.

16. K.I. Kirov, N.A. Ivanov, E.D. Atanasova, G.M. Minchev, Dc magnetron system for cathode sputtering, *Vacuum* 26 (1976) 237–241.

17. J.A. Thornton, A.S Penfold, Cylindrical magnetron sputtering, in J.L. Vossen, W. Kern (Eds.), *Thin Film Processes*, Academic Press, New York, 1978.

18. D.M. Mattox, R.E. Cuthrell, C.R. Peeples, P.L. Dreike, Design and performance of a moveable-post-cathode magnetron sputtering system for making PBFA II accelerator sources, *Surf. Coat. Technol.* 33 (1987) 425–432.

19. D.B. Fraser, H.D. Cook, Film deposition with the sputter gun, *J. Vac. Sci. Technol.* 14 (1977) 147–151.

20. B. Window, N. Savvides, Unbalanced dc magnetrons as sources of high ion fluxes, *J. Vac. Sci. Technol.* A4 (1986) 453–456.

21. M.S. Wong, W.J. Chia, P. Yashar, J.M. Schneider, W.D. Sproul, S.A. Barnett, High-rate reactive d.c. magnetron sputtering of ZrO_x coatings, *Surf. Coat. Technol.* 86–87 (1996) 381–387.

22. J.-E. Sundgren, B.O. Johansson, S.E. Karlsson, Kinetics of nitride formation on titanium targets during reactive sputtering, *Surf. Sci.* 128 (1983) 265–270.

23. K. Ngan, R. Mosely, Z. Xu and I. Raaijmakers, *Proc. VMIC Conf.* (1994) 452–455.

24. W.D. Sproul, High rate reactive sputtering process control, *Surf. Coat. Technol.* 33 (1987) 73–81.

25. S. Schiller, U. Heisig, C. Korndörfer, G. Beister, J. Reschke, K. Steinfelder, J. Strümpfel, Reactive dc high-rate sputtering as production technology, *Surf. Coat. Technol.* 33 (1987) 405–423.

26. P. Baroch, J. Musil, J. Vlček, K.H. Nam, J.G. Han, Reactive magnetron sputtering of TiO_x films, *Surf. Coat. Technol.* 193 (2005) 107–111.

27. K. Koski, J. Hölsä, P. Juliet, Voltage controlled reactive sputtering process for aluminium oxide thin films, *Thin Solid Films* 326 (1998) 189–193.

28. S. Logothetidis, I. Alexandrou, S. Kokkou, Optimization of TiN thin film growth with in situ monitoring: The effect of bias voltage and nitrogen flow rate, *Surf. Coat. Technol.* 80 (1996) 66–71.

29. W.D. Sproul, P.J. Rudnik, M.E. Graham, The effect of nitrogen partial pressures, deposition rate and substrate bias potential on the hardness and texture of reactively sputtered TiN coatings, *Surf. Coat. Technol.* 39–40 (1989) 355–363.

Chapter 3

Film Formation and Structure

Thin film technology is commonly used in many applications including microelectronics, optics, hard, and corrosion resistant coatings or micromechanics due to a wide variety of properties of thin films. Since the macroscopic properties of a material strongly depend on the film structure characterized in terms of grain size and crystallographic orientation, lattice defects, phase composition, and surface morphology, it is necessary to control the film growth at the atomic level. Indeed, progress in thin film technology depends upon the ability to selectively and controllably deposit the films with required physical properties. Very promising way how to engineer the properties of thin films has been found in manipulating the real structures of materials. Hence, understanding of the structure–property relationships is recently among the most important issues in material science. It is even more important for thin film processing since thin films are typically formed far away from equilibrium conditions. This is also a reason why thin films in most cases contain a much higher content of defects than conventionally prepared bulk materials. Due to the highly nonequilibrium state of the deposition process, besides vacancies, dislocations, grain boundaries, or precipitates, thin films are also characteristic for their anisotropic properties. Although it is very difficult to predict the structure and properties of deposited films due to a high complexity of the deposition process, it is simultaneously a reason for a high variety of films with different elemental

Novel Nanocomposite Coatings: Advances and Industrial Applications
Rostislav Daniel and Jindřich Musil

ISBN 978-981-4411-17-2 (Hardcover), 978-981-4411-18-9 (eBook)
www.panstanford.com

and phase compositions and microstructures that can be synthesized simply by varying the deposition conditions. Thin films can be prepared using a large number of deposition methods developed recently to cover all the demands of industry applications. They combine various physical principles for the film deposition so that the whole spectra of microstructures and physical properties of thin films are obtained. Due to specific limitations of every deposition process, there does not exist a universal technique for preparation films for any application. Especially the needs for low processing temperatures require very specific deposition conditions since in that case films are formed far away from thermodynamic equilibrium. Indeed, it is always necessary to optimize deposition conditions in order to enhance surface and bulk diffusion as the determinant atomic processes controlling structure evolution during the film growth. For this purpose, energetic particle bombardment has been shown very useful as an essential tool for enhancing adatom mobility and manipulating nucleation rates. On the other side, a competitive fashion and the kinetic limitations induced by low-temperature growth allows the controlled synthesis of unique metastable structures and structures of a high disorder showing completely new properties. The restricted surface and bulk diffusion results in limited migration of adatoms from the random points of impingement onto energetically favorable sites and thus films are formed in a different manner compared to equilibrium conditions.

The film formation is reviewed as a process starting with condensation and nucleation of the adatoms on the surface followed by the nuclei growth, coalescence, interface formation, and subsequent thickness growth by the continued nucleation of depositing atoms on previously deposited material. All the above given stages of film growth can be effectively influenced by deposition parameters. The nucleation and growth kinetics, microstructural evolution, and consequently physical properties of films are determined mainly by the composition and energy distribution of the incident sputtered flux, energy of species bombarding the growing film surface, and the deposition temperature. Also the substrate material and its condition play a very important role. Surface roughness, crystallinity, and orientation significantly influence the growth and the final film structure. In principal, these control variables can be used to tailor the film structure and physical properties of the as-deposited material. Another approach to vary film properties has been shown by addition of alloying elements into the film that tends to modify the structure evolution already at their extremely low concentrations. Hence, various elements are now generally applied as additives into thin films in order to form high sophisticated structures in polycrystalline thin films with unique physical and chemical properties. It, however, requires extensive studies of the correlation between film structure and deposition parameters. Only after the basic film formation processes are fully understood new materials can be designed for specific technological applications.

3.1 Condensation of Sputtered Material

A product of the sputtering process is a thin film formed on the substrate surface. The composition of the films strongly depends on the sputtering process and thus may be near to that of the target or may completely differ if the reactive gas is introduced into the deposition system during sputtering. The sputtered atoms, however, must transit first the region between the target and the substrate before they condensate on the substrate surface. Since this region is composed of a large number of sputtered and gas atoms or molecules, which are ionized or in a neutral or excited state, all the sputtered atoms commonly experience many collisions. Their amount depends on the pressure and the target-to-substrate distance and may vary from zero to many hundreds. Each collision changes the velocity and direction of the sputtered atom and results in their thermalization (cooling) at simultaneous heating of the background gas. In addition, not all of the atoms which impinge on a surface always condense. The atoms, which do not immediately react with the surface, move along the surface before they get stuck. During this stage, the so-called adatoms may be immediately reflected, re-evaporate after a residence time, or re-sputtered away from the substrate surface. The re-evaporation and re-sputtering are given by the bonding energy between the condensated atoms and the surface, the surface energy, and the flux and energy of the impinging atoms. The probability of the condensing atoms to stick on surfaces varies depending on the deposition conditions and is given by the sticking coefficient. It is the ratio of the condensing atoms to the impinging atoms. The probability of the atom deposition can be then expressed as a ratio of atoms which are actually deposited on the substrate divided by the total number of atoms emitted from the target. Besides the deposition on the substrate surface, the sputtered atoms commonly also condensate on the walls of the deposition system, shutters or even back onto the target. Although atoms stuck within the erosion zone of the target are immediately resputtered, they may get stuck in the areas of the target where the probability of being resputtered is relatively low. Thick deposits formed in the central regions of the targets may, in extreme cases, cause discharge instabilities or arcing.

The formation of a film on a substrate surface is typically enormously influenced by the energy of condensing particles. After approaching the substrate, condensing particles transferred their energy to other particles in the near region. It gives rise to a marked heating of the substrate which consequently increases the surface mobility of adatoms. Substrate heating arises, however, not only from the condensation energy of the depositing adatoms but also from the high kinetic energy of the depositing particles. This effect is very strong at low pressures where the particles are not significantly thermalized through the gas scattering. Substrate heating can also arise from plasma effects, such as radiation and surface recombination [1].

3.2 Interface Formation

Very important for a good performance of sputtered films in many applications is a good adherence at the substrate-layer boundary. The adhesion of the film to the substrate is usually a real problem which has to be solved. One of the possible solutions is the right material selection of both the film and the substrate combination with respect to their characteristics. Especially for multiphase films, it is necessary to analyze the constitution of interfaces. Based upon the results on bulk materials, three different kinds of interfaces can be stated (Fig. 3.1). Metallic hard materials typically form coherent or semicoherent interfaces with metals or other metallic hard materials, where low energy interfaces generally results in a good adherence. Interfaces between metallic and ionic hard materials usually exhibit intermediate regions with variable composition. The properties of the system strongly depend on the constitution and structure of this boundary phase. The right materials selection to form strong interfaces is important for all tribological applications where the films are heavily mechanically loaded.

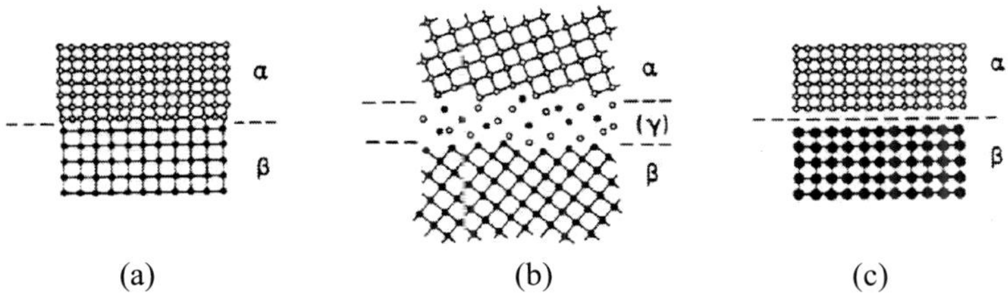

Figure 3.1 Various substrate-layer interfaces between different materials: (a) partially coherent boundaries, (b) phase-boundary-(mixed) phase, and (c) interaction-free phase boundaries [2]. Reprinted from Ref. [2], with kind permission by Prof. Dr. Helmut Holleck KIT (Karlsruhe Institute of Technology). Copyright (1986) AIP Publishing LLC.

3.3 Nucleation and Growth

The process of the films formation begins with a clean surface of the substrate material, which is exposed to a flux of sputtered atoms. After the film material approaches the substrate, it has to adhere to it and settle into possible equilibrium positions. Once the atoms come into contact with the substrate surface, they form chemical bonds with the atoms of the substrate. Nucleation is the earliest stage of film formation when incoming adatoms condense and establish a permanent resistance on the substrate. After the substrate is exposed to the flux of sputtered atoms, a uniform distribution of small but highly mobile clusters or islands is observed. Since the temperature of the sputtered atoms differs from that of the substrate, the deposited material must first be thermally accommodated. It is believed that stuck particles are thermalized quickly while they take the temperature of the substrate. Partial supersaturation of the deposited atoms on the substrate surface occurs during this process. The prior nuclei formed in this very early stage grow in size owing to the surface diffusion of impinging adatoms and clusters, which interact with other adatoms and nuclei. Although many studies have been performed in order to understand this phenomenon, it is still unclear if there a clear demarcation exists between the end of nucleation and the onset of nucleus growth.

At higher amount of deposited adatoms, merging of the formed islands occurs. This stage is called coalescence and is similar to a coalescence of liquid phases especially at high substrate temperatures. As it is seen from Fig. 3.2, the coalescence results in a decrease in the island density which results in local denuding of the substrate. It enables to form new nucleation sites on uncovered substrate surface which continue to grow in size and finally lead to a formation of a compact film. However, during the early stages of deposition, unfilled channels between individual islands develop and are filled in and shrinked first during next stages of the coalescence process. Once isolated voids are also filled in completely, the film is said to be continuous. It happens usually at first few tens to hundreds of Angströms of the film thickness. It has been observed that whether amorphous or crystalline film is formed depends on the deposition rate and temperature. It strongly influences the adatom surface diffusivity and the nucleation site density. If the adatoms do not have enough time or energy to diffuse along the surface and find low energy sites before their movement is prevented by subsequently deposited adatoms, amorphous films may be formed. Mostly covalently and ionically bonded materials are deposited in the amorphous state due to their low packing densities and strong bonding directionality. On the contrary, materials exhibiting higher surface mobility easily form crystalline structures. It is frequently the case of metals.

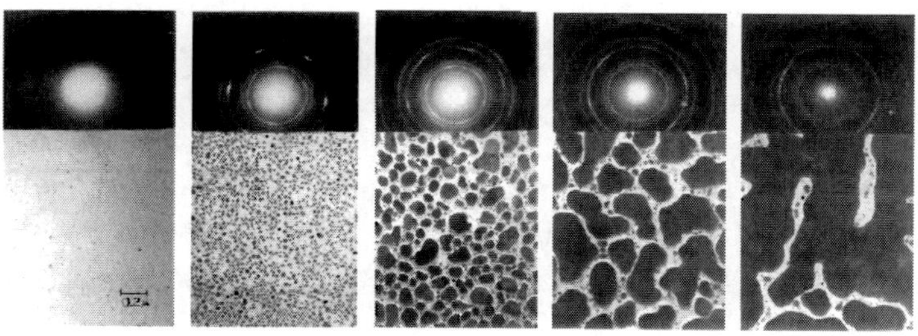

Figure 3.2 Nucleation, growth, and coalescence of the Ag film investigated by transmission electron microscopy (TEM). All stages are characterized by corresponding diffraction patterns. Micrographs courtesy P. Dixit, and reprinted from Ref. [3], copyright (1982), with permission from W. S. Maney & Son Ltd. http://www.maneyonline.com/doi/abs/10.1179/imr.1982.27.1.209

Binding energies between individual adatoms and adatoms and the substrate atoms strongly influence the film growth mechanisms. Based upon results obtained by systematic studies of film formation performed in last decades, three different growth modes have been stated. They are schematically shown in Fig. 3.3.

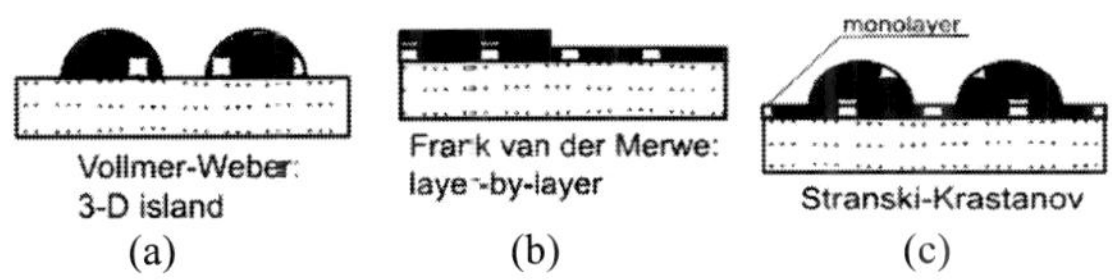

Figure 3.3 Basic modes of thin film growth: (a) Volmer-Weber, (b) Frank-van der Merwe and (c) Stranski-Krastanov [4].

If the film atoms are more strongly bound to each other than to the substrate, the three-dimensional island (Volmer-Weber) film growth occurs. In this case, small stable clusters nucleate directly on the substrate surface and grow in three dimensions to form separated islands. These islands grow further and coalesce to form a continuous film when higher number of adatoms is deposited on the substrate surface. It is characteristic for metal films on insulators or contaminated substrates. The opposite behavior is observed if the sputtered atoms are more strongly bound to the substrate than to each other. Here, the stable nuclei grow in two dimensions and form planar sheets. Once the first complete monolayer of the deposit covers the substrate surface, it is then covered with a somewhat less tightly bound second layer. Such two-dimensional layer-by-layer (Frank-van der Merwe) film growth is observed mainly on clean substrates in metal-metal and semiconductor-semiconductor systems. The third growth mode (Stranski-Krastanov) is a combination of the first two. In this case, after forming one or more monolayers, subsequent layer growth becomes unfavorable and islands form. It seems that the reason is given by the release of elastic energy stored in the film due to film-substrate lattice mismatch, which disturbs the monotonic decrease in binding energy characteristic of layer growth. Then, the high energy at the deposit-intermediate-layer interface may trigger island formation. This growth mode occurs in metal–metal and metal–semiconductor systems. In addition, it has been observed that crystallographic facets are frequently preserved on islands and at interfaces between initially disoriented, coalesced particles [3].

Although during the initial stages of the film growth, the density of stable nuclei dramatically increases with the deposition time, this increasing tendency is early followed by a gradual decrease. It is caused by a coalescence of the formed islands which leads to the formation of compact film. The kinetic theories of nucleation show that the main driving force for coalescence is the effort of the system to reduce the surface energy. However, before a compact film is formed, several characteristic processes affect the completed coalescence. Above all, nuclei decrease their projected area by increasing in height. In fact, larger islands grow to detriment of the smaller islands and clusters by capture of mobile adatoms. The coalescence also occurs with nearby smaller islands, occasionally with larger ones, whereby the smaller islands assume the crystallographic orientation of the larger island. It has been observed that during coalescence the clusters migrate over the substrate surface in a process described as cluster-mobility coalescence. After a while, the film becomes semicontinuous with a network of channels and holes, which are gradually filled in so that the compact film is formed. This film forming process is shown in Fig. 3.4a. The coalescence is driven by several mass-transform mechanisms: Ostwald ripening, sintering, or cluster migration. Ostwald ripening is observed if diffusion of adatoms proceeds from the smaller to larger island until the former disappears entirely. The former is driven to minimize the surface free energy of the island structure. On the contrary, sintering is a coalescence mechanism involving islands in contact. In this case, a neck is formed between two islands serving as a diffusion path for other adatoms resulting in a rapid formation of semicontinuous film. The third mechanism for coalescence is characteristic with migration of clusters on the substrate surface. They interact and collide with other separated islands and clusters to form larger system. The coalescence efficiency in this case is mainly given by the surface mobility of the islands.

Binding energy of adatoms, with respect to the substrate atoms and other adatoms, as well as the substrate diffusivity plays an essential role in the film formation mainly during the initial stages of deposition (Fig. 3.4b). While films forming a dense population of small islands become continuous at relatively low average film thicknesses, films consisting of only a few large islands are characterized by an island structure up to relatively large film thicknesses. Which growth mode will dominate depends on the film and substrate materials, energy of impinging adatoms, and the temperature of deposition. These deposition conditions can also strongly influence an amount of defects formed in the film and thus the stress level since the mechanical defects, such as dislocations, twins, stacking faults, and other line and volume defects, are generated even by interaction of individual islands during coalescence. The early stages of coalescence have been shown to initiate the island rotation and recrystallization to eliminate misorientation during coalescence while the islands are in contact. The high stresses in the very thin films due to film-substrate lattice constant and thermal expansion mismatches are frequently relieved by the generation of climbing dislocations while the film thickness increases. Contamination also affects the film growth, resulting in preventing island re-orientation and recrystallization during coalescence.

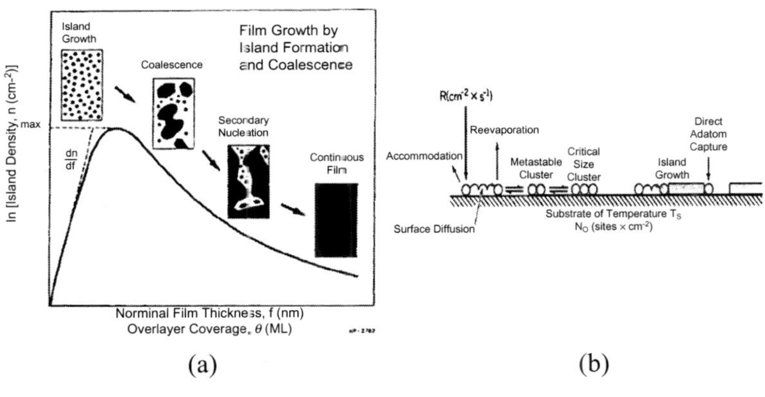

Figure 3.4 (a) Schematic illustration of the nucleation, nuclei and island growth, coalescence, and secondary nucleation from the very early stage of the film formation to the continuous film. (b) The surface processes commonly occurring during film deposition. Reprinted from Ref. [5], Copyright (2010), with permission from Elsevier.

3.3.1 Particle Bombardment Effects

Similar to the physical sputtering of targets, substrates, and growing films may also be significantly affected by a bombardment of high energetic particles. Some of the processes occurring at the surface region and in the near-surface region of the bombarded surface are summarized in Fig. 3.5. The ion bombardment of growing films is required in many applications in order to improve the structure and properties of the sputtered films. Negatively biased substrates are permanently exposed to the bombarding of particles with various masses and energies. This concurrent energetic particle bombardment may consequently cause a physical penetration of the bombarding particles into the surface region, where a collision cascade is typically initiated. High energetic particles can cause even an ejection of the deposited target atoms which is known as a re-sputtering effect. However, in the energy range of several tens of eV, the sputtering yield is relatively low and most of the energy is transferred into the surface region as heat. Some of the impinging particles are reflected as high energy neutrals, some are implanted into the surface, and some causes the emission of secondary electrons. Through the energetic ion bombardment, the surface mobility of adatoms is typically enhanced as well as chemical reactions on the substrate surface. All those effects strongly affect the film formation, especially if the sputtering is performed at low pressures.

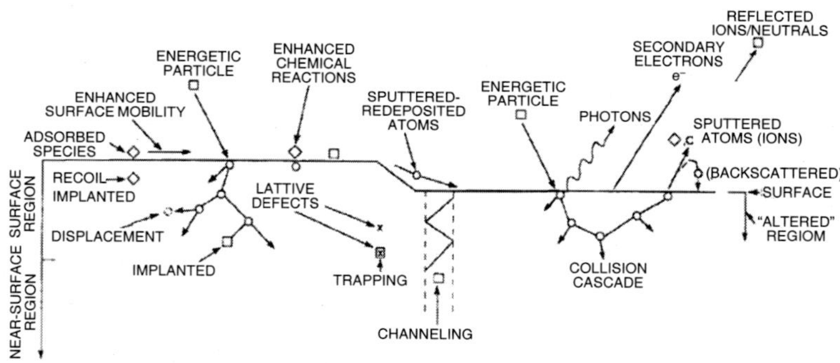

Figure 3.5 Schematic depiction of energetic particle bombardment effects on a target surface and growing films—events that occur on a surface being bombarded with energetic atomic-sized particles. Reprinted with permission from Ref. [6]. Copyright [1969], American Vacuum Society.

The condition of the substrate is another very important factor significantly affecting the properties of formed films. The surface morphology, in particular the roughness, may significantly influence the film growth through the preferential nucleation sites, inclusions or particulate contaminations. In general, the rough surfaces initiate formation of more complex film morphology and less dense films. Geometrical shadowing is for instance known to generate porosity in the coatings. If the surface mobility of the depositing adatoms and their reactions with the substrate are affected by the composition and properties of the substrate, the nucleation density as well as the stability of the interface between the formed film and the substrate may be dramatically altered. Homogeneity of the substrate is referred to affect the uniformity of the film properties over the surface. Moreover, other effects, such as angle of incidence of the sputtered flux, energy delivered into the growing film through the substrate heating or concurrent bombardment, gas entrapment, and other deposition conditions, may dramatically influence the nucleation and growth of films on the substrate. Hence, in order to deposit films with demand structure and properties, suitable deposition conditions must be selected.

In summary, the film structure together with the amount and distribution of growth defects, voids, or pores have a profound effect on the engineering properties of produced films, such as magnetic, optical, electrical, and mechanical properties. It has to be taken into account while new coating systems are developed.

3.4 Microstructure of Thin Films: Structure Zone Models

Thin films generally strongly differ in properties compared to bulk materials. The differences are mainly given by the conditions under which they are prepared. The contamination by impurities, density variations with thickness, short-range density fluctuations, or internal surfaces and interfaces are all responsible for the unique properties of the films. Especially fundamental aspects, such as substrate temperature and energetic particle bombardment, strongly influence the conditions for the film growth. The advantage of thin films is based upon the fact that they can be prepared so they cover a broad continuum of energy states, macro- and microscopic structures. The variability of thin films with a wide range of characteristics and properties make them very attractive for many scientific and technological applications. However, since resulting film properties can be significantly influenced by the deposition conditions, the effects on the film growth have to be fully understood. This requirement has led to the development of structure zone models (SZMs), which predict the morphology and microstructure of a film, independently of its material, as a function of the adatom mobility. Here, the effect of the deposition temperature [7], energetic particle bombardment [8,9], and the flux and energy distribution of the condensing particles are taken into account [10]. Especially at low pressures, the increased energetic particle bombardment of the growing film plays an essential role on the structure, grain size, and stress and impurity incorporation levels. The extended Thornton's version [8] of Movchan and Demchishin's structure zone diagram [7] is depicted in Fig. 3.6 [9].

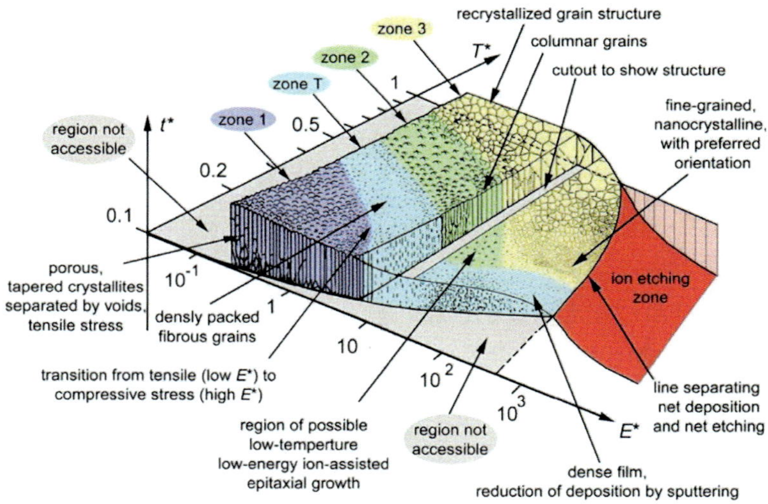

Figure 3.6 Structure zone model for film growth proposed by Movchan and Demchishin [7] and extended by Thornton [8] and Anders [9] to include the bombardment-induced mobility process in sputtering. Reprinted from Ref. [7], Copyright (2012), with permission from Elsevier.

Nucleation and growth of films strongly depend on the sublimation energy and activation energies for surface and bulk diffusion. Since for many metals these energies are proportional to their melting temperature T_m, the T_s/T_m ratio has been taken as a basic parameter to express the formed film morphology. Here, T_s is the substrate temperature.

Films sputtered at low substrate temperatures ($T_s/T_m < 0.2$–0.3—zone 1) are characterized by an extremely high nucleation density in the early stage of the film formation. It is because the impinging adatoms have no or extremely low surface mobility on the substrate surface. In a consequence of the high overcooling, the adatoms stick in the near vicinity where they approach the substrate surface. Under such conditions neither coalescence nor grain boundary migration can occur and thus, because the development of large grains is inhibited, a fine-grained structure consisting of textured and fibrous grains is formed. Films with a fiber like porous structure exhibit typically high surface area. Since the main structure forming mechanism is nucleation, insufficient surface diffusion results in a high density of lattice imperfections and open boundaries that are formed between individual columns. The structures related to zone 1 are frozen in structures which means that no segregation, even of the impurity species, can take place and the adatoms are incorporated into the growing fibers practically at their impinging sites [11]. An intensive restructuration is commonly observed at annealing temperatures higher than the deposition temperature with the increase in the crystallite width. At higher gas pressures, zone 1 is extended to higher temperatures due to gas scattering process and decreased surface adatom mobility due to gas adsorption and collisions on the surface.

At higher substrate temperatures ($0.3 \leq T_s/T_m < 0.5$—zone 2), the formation of the film structure is mainly driven by the grain growth. The growth process is dominated by adatom surface diffusion which allows columnar structure to develop with columns which proceed from the substrate to the free surface of the film. The same effect provides the densification of the intercolumnar boundaries and an increase of the grain size so that the surface features tend to be faceted. In addition, the grain boundaries are usually perpendicular to the substrate surface. Due to enhanced grain boundary migration the initially formed small grained structure is dissolved during coalescence. The driving force of grain boundary migration is the grain boundary energy and the surface energy difference of the neighboring crystal faces due to their orientational relationship.

Average grain width, typically less than the film thickness, increases with the increase in the substrate temperature, and the grains may extend through the film thickness at high T_s/T_m. In zone 2, grains tend to have highly faceted top surfaces giving rise to a smooth, but mat surface appearance.

In zone 3 ($T_s/T_m \leq 0.5$), the film formation is strongly affected by bulk diffusion which leads to recrystallization, grain growth, and densification. The columnar grains increases in width significantly, so the film morphology in some cases reveals highly modified columns being single crystals of sputtered material. The formed microstructure consisted of equiaxed grains often provides smooth surface of the films.

Zone T represents a transition zone between zones 1 and 2 where the film growth is affected by the energetic bombardment by high energetic ions from the plasma discharge and/or neutrals from the sputtered flux that is very effective mainly at low pressures [8]. It influences the surface diffusion which sets in during the film growth making the crystal growth the dominant structure forming phenomenon. While in the early stages the film starts to develop with a small grained structure reflecting the nucleation density, when it becomes continuous, typical cone-shaped columnar structure develops. It is given by transition from the film growth of immobile grain boundaries to the growth driven by surface diffusion which makes the migration of adatoms possible between neighboring grains. The difference in the surface energy of individual crystallites then causes the crystallites having lower surface energy incorporate more material and grow over the neighboring crystals having higher surface energy [12]. This competitive growth also influences the texture of the films by selecting the faster growing directions as texture components. The upper zone boundary reflects the onset of grain boundary migration. Thus, compared to films of the two surrounding zones, zone T films exhibit a denser fibrous morphology with smoother surfaces. The columnar grain structure of films within zone T are characterized by a low-density material surrounding an array of parallel rod-shaped columns of higher density. It has been found that the intrinsic structure inside the columns is either polycrystalline or amorphous depending on the sputtered material and conditions of the film growth.

3.4.1 Effect of Substrate Temperature, Pressure, and Ion Bombardment

Surface adatom mobility is affected by many factors, especially by heating through condensed particles since most of their energy is transferred into the surface region as heat. Thornton has investigated the effect of the pressure of the sputtering gas on the film growth and has observed that increasing argon pressure causes an increase of the normalized temperature at which the zone boundaries occur [8]. In fact, the pressure affects the film microstructure through several indirect mechanisms. Due to the low operating pressure of the magnetron, and thus a low number of collisions of sputtered atoms during their transport to the substrate, the kinetic energy of sputtered particles remains relatively high and is effectively transferred to the growing film. Another obvious contribution to the energy delivered to the forming film is registered especially from gas neutrals reflected from the target. These energetic neutrals are principally Auger neutralized ions accelerated to high energy to the target surface and elastically reflected to the substrate with a high kinetic energy. It happens usually if the reflection coefficients are larger than the sputter yields. A role of a gas pressure can be seen, for instance, in more opened zone 1 at higher pressures resulting from an increase of the oblique component of the deposition flux. It commonly occurs if the mean-free path for elastic collisions between sputtered atoms and working gas decreases to a magnitude corresponding to the target-to-substrate distance. Thus, for high pressures, a higher T_s/T_m is needed to obtain a zone T or zone 2 structures [13]. On the contrary, the film densification and reduction in film morphology with decreasing gas pressure is related to increased energetic-particle bombardment from elastically backscattered sputtering gas atoms. Finally, optical emission from the plasma as well as bombardment with hot gas atoms can also significantly heat the substrate surface. Moreover, if the energy of adatoms is still insufficient, various external ohmic heating sources are typically used in order to enhance their surface mobility.

The film formation can be dramatically affected by another important variable in sputter deposition of thin films—by a bombardment of the growing film by energetic ions. An ion bombardment is realized by having the surface at a negative potential so that ions from the plasma discharge are accelerated towards the substrate surface. This concurrent bombardment enhances the surface mobility of adatoms and their chemical reactivity with the reactive gas during the reactive sputtering process. Bombardment by energetic ions causes typically an atomic peening where surface atoms are struck and recoil into voids and interstitial sites in the lattice of the surface region. This causes incorporation of weakly adsorbed adatoms, densification of the material, reduces the grain size and introduces compressive stresses into the film. Strong densification of the formed film results in compact film morphology and changes a number of properties of the deposited films. Ion-surface interactions play an essential role in deposition of thin films for many applications since very specific materials with new metastable structures or compositions can be obtained at lower deposition temperatures and as a result a wide and controllable range of film properties can be attained. From the technological point of view, a very important finding is that the zone 1-T boundary is nonlinear and strongly depends on the gas pressure and the applied substrate bias potential. The extension of the boundary zone to lower T_s/T_m values as the bombardment energy increases is given by the thermal and bombardment-induced adatom surface mobility, see Fig. 3.7 [10,14].

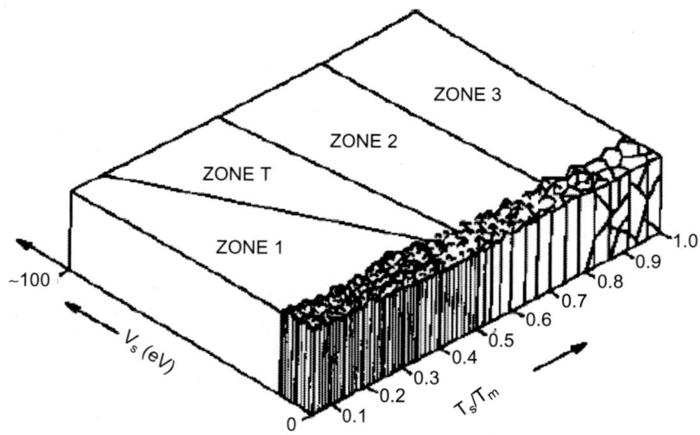

Figure 3.7 Modified structure zone model proposed by Messier including an effect of bombardment-induced mobility. Reprinted with permission from Ref. [10]. Copyright [1984], American Vacuum Society.

3.4.2 Effect of Impurities

Contamination of sputtered films is a phenomenon of a great importance in surface engineering. Even very small concentrations of incorporated impurities have dramatic effects on nucleation and growth kinetics. For instance, oxygen as a chemically reactive element originated from the residual gas in deposition systems plays a significant role in determining the film microstructure. In terms of SZMs, contamination generally reduces adatom mobility and therefore acts to promote zone 1 structures. However, also purposely incorporated additives, such as metal atoms or other elements, act in similar way which is often used in the development of thin films with unique structures. Since the Thornton's SZM is strongly affected by impurity content in thin films, it was extended by Grovenor et al. [15] and lately by Barna and Adamik [16] to include this effect as well. The development of the film structure with increasing impurity content is schematically illustrated in Fig. 3.8.

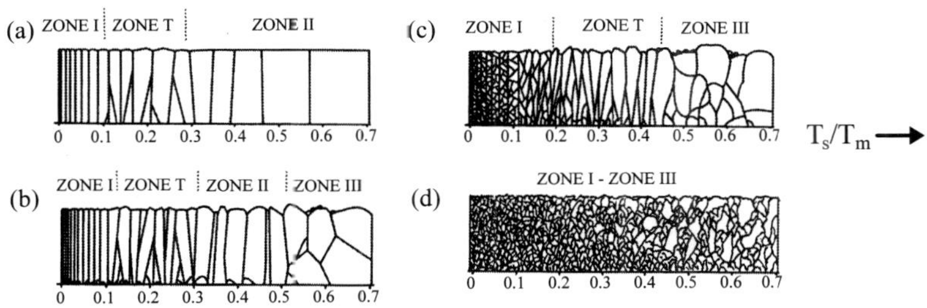

Figure 3.8 Effect of impurities/additives on the film structure from (a), (b) low to (c) intermediate and (d) high content. Reprinted from Ref. [17], with permission from World Scientific Publishing.

Zone 1 represents fiber-like structure which has been shown to develop due to the minimum surface mobility of impinging adatoms. Since the impurity species cannot be segregated due to their extremely low mobility, they are incorporated into the growing fibers practically at their impinging sites [11]. Thus, very fine grained structure is formed especially at the high contamination level. In zone 2, the grain boundary mobility is dramatically limited if films contain impurities. Hence, the abnormal grain growth cannot be completed and the films exhibit bimodal grain size distribution. Low grain boundary migration consequently results in a shift of the lower zone boundary to higher temperatures, where zone 3 appears. At increased contamination level, grain boundaries become immobile and zone 2 disappears from the SZM. Because of the same reason, the lower and upper zone boundaries also shift to higher temperatures in contaminated films within the zone T. Moreover, impurities may have an unusual effect on the texture evolution in case of competitive growth. At low or medium contamination level, the growth of certain crystal faces can be blocked while others can grow which typically leads to the development of specific preferred orientations. This effect is relatively strong especially if anisotropy exists in the surface chemical interaction of impurity species with various crystal faces [18,19]. In highly contaminated films, impurity phase can cover crystallites and thus hinder their growth. Therefore, the competitive growth cannot be evolved and zone T does not appear. Instead, globular structure with grain boundaries covered and stabilized by the impurity phase will develop. Zone 3 is typical zone structure of films containing impurities or additives. The structure forming phenomenon is the process-induced segregation of impurities. At high temperatures, the impurity species are segregated on grain boundaries where they form a covering layer which effectively stops the grain growth. The growth can continue just by repeated nucleation leading to development of the equiaxed structure at high deposition temperatures. Very effectively hindered grain growth results only in weak texture evolution even at high temperatures. By increasing the contamination level, the lower boundary of zone 3 is shifted to lower temperatures while at first zone 2 and later zone T will disappear gradually until the lower boundary of zone 3 reaches the upper boundary of zone 1. The nanocrystalline structure is thus formed mainly at lower temperatures.

3.5 Advantages and Limitations of Sputter Deposition Processes

Physical vapor deposition (PVD) processes become very popular for thin film synthesis in past decades due to several advantages. Very important for industry applications is mainly an extreme versatility in composition of sputtered films. Magnetron sputtering is used to deposit metal, alloy, refractory, or intermetallic compounds as well as polymeric materials. Herewith, the sputtering target represents a stable and long lived vaporization source. Due to the ability to easily vary deposition conditions, films with unusual microstructures and unique properties can be produced. Furthermore, sputtering conditions are highly reproducible so the films with the same properties can be deposited at high deposition rates. Sputter deposited films commonly exhibit very high purity and a good adherence to the substrate through a strong bonding. Compared to other deposition techniques, during the deposition pollutants and effluents are eliminated from the process which is a very important ecological factor.

On the other hand, some limitations make a preparation of thin films with required properties relatively difficult. For instance, the fact that the reactive species are activated in the plasma discharge making the reactive sputtering process very effective can be simultaneously a disadvantage since activated gaseous contaminants may influence the film growth and resulting film properties. Also higher degree of sophistication of the processing equipment and source design is needed to sputter homogeneous films with uniform properties. In the same way, substrate heating from electron bombardment, by short wavelength radiation or high energy particle bombardment, has to be avoided since it is mostly detrimental to their performance. Although bombardment by high energy reflected neutrals in low-pressure sputtering can be very useful for development of films with enhanced mechanical properties, it is often uncontrolled process variable. However, most of disadvantages and limitations of the PVD processes have been already overcome by a special design of the deposition systems and using special sputter sources and power supplies.

3.6 Structure–Property Relation in Hard Films

Microstructure of thin films is known to play a determinant role for the film properties. Especially hardness is strongly influenced by microstructural features that can meaningfully affect the involved deformation mechanisms. In order to understand properties of a hard material, its microstructure and in addition the electronic and atomic structure has to be known since it contributes to the so-called intrinsic hardness. The intrinsic hardness of a material is determined both by the strength of the interatomic forces and by the crystal structure. Mainly materials characterized by a high cohesive energy, short bond length, and a high degree of covalent bonding exhibit a high intrinsic hardness [20]. Hard coating materials with extremely high hardness are mainly the ionic hard oxides of Al, Zt, Ti, etc., the covalent hard borides, carbides, and nitrides materials of Al, Si, and B and the metallic hard compounds consisting of the transition metal borides, carbides, and nitrides. However, not only the hardness is determined by the strength of the interatomic forces or the bonding length. The special nature of bonding in such hard compounds also ensures very high melting points and decomposition temperatures. Moreover, the linear thermal expansion coefficient generally increases from the covalent to metallic to ionic hard compounds. Compared to these materials, metals exhibit thermal expansion coefficients higher typically of a factor of two or more. In the same way, the elastic properties of materials are also strongly affected [21,22].

Hard protective coatings are expected to be extremely resistant to plastic deformation. It requires a high resistance to dislocation generation, propagation, and multiplication within the film that is markedly limited in materials with high bond strength [23]. It has been shown that mainly highly directional bonds, such as covalent bonds, will restrict dislocation propagation [24]. Since film hardness is usually measured using nanoindentation techniques, the deformation mechanisms occurring underneath the diamond indenter have to be understood to be able to enhance film hardness by optimizing the structure of thin films. Dislocation slip, structural densification, and microcrack propagation intensively contribute to the plastic deformation of mechanically loaded materials. It has been proved by many researchers that any microstructural features which can hinder dislocation propagation can also effectively increase the hardness of the material. Grain boundaries, precipitates, dislocations, impurity atoms, and vacancies are effective inhibitors for dislocation generation and movement. However, existing imperfections in the film structure can act counterproductively as well. For instance, a high amount of imperfections of the grain boundaries, such as porosity and microcracks, can lower strength and hardness of thin films. Therefore, the optimization of the deposition conditions or the post-deposited annealing is often needed to reduce the grain-boundary defect structure. Both procedures effectively eliminate void networks and promote the strengthening of grain boundaries and thus enhance the hardness. It is typical for thin films that they exhibit smaller grain size and a higher concentration of defects compared to bulk materials, which make them much harder. It can be explained by a hardening of the material composed of grains smaller than 20 nm, which is commonly observed in metal and alloy films. However, the same effect has been found in multilayers and nanocomposite films [25,26]. The hardening of thin films due to a small grain size is in agreement with the Hall–Petch relation which originally was developed for explanation of yield strength of bulk materials. It is based on dislocation pile-ups at grain boundaries [27,28]. Another effect, which can lead to the hardness enhancement, is the effect of incorporated impurities. Although hard thin films are typically prepared at low pressures in pre-pumped deposition systems, the incorporation of noble gases from residual atmosphere and impurities from chamber walls into the growing films is relatively common phenomenon. Impurities, such as C, N, and O from residual gases, are located in substitutional or interstitial sites and are also usually found in grain-boundary positions. Their present involves the electrostatic attachment to charged dislocations in ionic materials and to dangling dislocation bonds in covalent compounds. Such interactions result in a pronounced hardening effect even at low concentrations through the limited dislocation mobility by pinning effects. If C or N is located in interstitial lattice sites, metastable structures can be formed. It usually happens if the films are prepared at high deposition rates, low substrate temperatures, or under high ion bombardment or at combination of all. Thus, the incorporated interstitials tend to distort the lattice which causes difficulty in initiating dislocation motion. In this way, formed supersaturated solid solutions are mostly observed in PVD films exhibiting an increased hardness.

The properties of hard thin films show large variations in dependence of their microstructure which strongly varies with the growth conditions. Besides the grain size, the film texture and defect density also have to be considered while the hardness-structure relationship of thin films is determined. Hardness values often vary with orientation of grains exhibiting maximum values at certain preferred orientation [29]. It is characteristic mainly for ceramic materials and thin films with other than cubic structures. On the other hand, hardness of thin films with a high density of defects, such as dislocations, point defects, or trapped gases, is strongly affected through the stress state of the films. Especially in films sputter deposited under an intensive ion bombardment, a high compressive level is induced. It results in altered lattice parameters and thus in a different elastic response of the material [30]. This fact also significantly affects the film hardness which is higher in compressive stressed materials. To this contributes a high defect concentration in films restricting the plastic flow.

The film stress phenomenon is very important in the surface engineering since by exceeding the critical values, film failure typically occurs. Films grown by PVD techniques are generally highly stressed and the stresses can be either tensile or compressive. While compressive stress can lead to the hardness enhancement, tensile stresses often results in the film damages [31]. Stresses in thin films are generally attributed to the thermal and intrinsic stresses, which both can be tensile or compressive depending on the substrate and film material and the deposition temperature. Thermal stresses are generated by the difference in thermal expansion coefficients between the film and the substrate after cooling the coated substrate down to room temperature. The higher is the stress state, the greater is the misfit in thermal expansion coefficients. The intrinsic stresses, usually called growth stresses, are attributed to a nonequilibrium state of the deposition process resulting in nonequilibrium microstructures formed during sputtering. Tensile stresses are generated through the grain boundaries, present small voids and pores in the films. Especially voids located in grain or column boundaries contribute significantly to the tensile stress state in the films. It is due to the attractive atomic forces acting across the voids, which however decreases with distance and so the tensile stress component is typically reduced as the voids become large. The tensile stresses are also generated by quenching the adatoms at spacings greater than they would be at the surface temperature. On the contrary, the origin of the compressive stresses is associated with energetic particle bombardment, typical for PVD processes. The main contribution to compressive stresses has been shown from backscattered energetic neutrals impinging on the growing film surface mainly at low sputtering pressures [32,33]. In order to generate the compressive stress in thin films, a negative substrate bias is usually applied during film deposition. Thus, the energetic ions commonly relocate the adatoms from their impact positions and make them closer together than they would be under equilibrium conditions. Deposited atoms may also be situated in interstitial or substitutional positions in the lattice by an atomic peening at higher bombarding ion energies, which induces relatively high compressive stress level in the films [34]. If the energy of bombarding particles is high enough, they can be even implanted in the growing films. It increases dramatically the defect concentration in the films and so the values of the compressive stress through a nonconventional lattice distortion. A controlled ion bombardment is also used to shift the unfavorable tensile stress in the film to the compressive state and in this way to avoid the possible film failure. The stresses generated in thin films during their growth are very important in a particular application. When the coated tools are subjected to a high loading intensity, the additional stresses contribute to those concentrated at the interface between the film and the substrate during deposition, which can result in film cracking and delamination if the critical values are exceeded. The values of stresses commonly approach the yield or fracture strength of the material if the mismatches between the elastic moduli of the film and the substrate are very high [35]. Also thermal stresses may signify a problem of coating integrity. The films should remain intact and adherent to the substrate even if they are mechanically or thermally loaded. However, it is not always that case. The tensile stress caused by the larger thermal expansion coefficient of the substrate compared to the film can result to initiate crack formation especially if coated tools are operated at elevated temperatures. On the other hand, compressive stress acts positively and has been shown to increase the resistance against crack propagation [36,37].

References

1. J.A. Thornton, J.L. Lamb, Substrate heating rates for planar and cylindrical-post magnetron sputtering sources, *Thin Solid Films* 119 (1984) 87–95.
2. H. Holleck, Material selection for hard coatings, *J. Vac. Sci. Technol.* A4 (1986) 2661–2669.

3. R.W. Vook, Structure and growth of thin films, *Int. Metals Rev.* 27 (1982) 209–245.
4. J.E. Greene, in D.T.J. Hurle (Ed.), *Handbook of Crystal Growth*, Elsevier Science Publishers, Amsterdam, 1993.
5. M.T. Taschuk, M.M. Hawkeye, M.J. Brett, Glancing angle deposition, in P.M. Martin (ed.) *Handbook of Deposition Technologies for Films and Coatings: Science, Applications and Technology*, Elsevier, Massachusetts, 2010.
6. D.M. Mattox, Particle bombardment effects on thin-film deposition: A review, *J. Vac. Sci. Technol.* A7(3) (1989) 1105–1114.
7. B.A. Movchan, A.V. Demchishin, Study of the structure and properties of thick vacuum condensates of nickel, titanium, tungsten, aluminium oxide and zirconium dioxide, *Phys. Metals Metall.* 28 (1969) 83–90.
8. J.A. Thornton, High rate thick film growth, *Ann. Rev. Mater. Sci.* 7 (1977) 239–260.
9. A. Anders, A structure zone diagram including plasma-based deposition and ion etching, *Thin Solid Films* 518 (2010) 4087–4090.
10. R. Messier, A.P. Giri, R.A. Roy, Revised structure zone model for thin film physical structure, *J. Vac. Sci. Technol.* A2 (1984) 500–503.
11. A.G. Dirks, H.J. Leamy, Columnar microstructure in vapor-deposited thin films, *Thin Solid Films* 47 (1977) 219–233.
12. G. Knuyt, C. Quaeyhaegens, J. D'Haen, L.M. Stals, A quantitative model for the evolution from random orientation to a unique texture in PVD thin film growth, *Thin Solid Films* 258 (1995) 159–169.
13. D.W. Hoffman, J.A. Thornton, The compressive stress transition in Al, V, Zr, Nb and W metal films sputtered at low working pressures, *Thin Solid Films* 45 (1977) 387–396.
14. R. Messier, Toward quantification of thin film morphology, *J. Vac. Sci. Technol.* A4 (1986) 490–495.
15. C.R.M. Grovenor, H.T.G. Hentzell, D.A. Smith, The development of grain structure during growth of metallic films, *Acta Metall.* 32(5) (1984) 773–781.
16. P.B. Barna, M. Adamik, Formation and characterization of the structure of surface coatings, in Y. Paleau, P.B. Barna (Eds.), *Protective Coatings and Thin Films*, Kluwer Academic Publisher, 1997.
17. P.B. Barna, M. Adamik, in F.C. Matacotta, G. Ottaviani (Eds.), *Science and Technology of Thin Films*, World Scientific Publishing, Singapore, 1995.

18. C.W.B. Martinson, S.A. Flodström, J. Rundgren, P. Westrin, Oxygen chemisorption on aluminum single crystals: Site determination by LEED studies, *Surf. Sci.* 89 (1979) 102–113.
19. R. Michel, J. Gastaldi, C. Allasia, C. Jourdan, J. Derrien, Initial interaction of oxygen with aluminium single crystal faces: A LEED, AES and work function study, *Surf. Sci.* 95 (1980) 309–320.
20. M.L. Cohen, Novel materials from theory, *Nature* 338 (1989) 291–292.
21. M.L. Cohen, Calculation of bulk moduli of diamond and zinc-blende solids, *Phys. Rev.* B32 (1985) 7988–7991.
22. M.L. Cohen, Theory of bulk moduli of hard solids, *Mat. Sci. Eng.* A105–A106 (1988) 11–18.
23. E. Knittle, R.M. Wentzcovitch, R. Jeanloz, M.L. Cohen, Experimental and theoretical equation of state of cubic boron nitride, *Nature* 337 (1989) 349–352.
24. A.Y. Liu, M.L. Cohen, Prediction of new low compressibility solids, *Science* 245 (1989) 841–842.
25. S. Zhang, D. Sun, Y. Fu, H. Du, Toughening of hard nanostructural thin films: A critical review, *Surf. Coat. Technol.* 198 (2005) 2–8.
26. S. Veprek, A.S. Argon, Towards the understanding of mechanical properties of super- and ultrahard nanocomposites. *J. Vac. Sci. Technol.* B20 (2002) 650–664.
27. T.G. Nieh, J. Wadsworth, Hall-Petch relation in nanocrystalline solids, *Scripta Metall. Mater.* 25 (1991) 955–958.
28. C.C. Koch, Optimization of strength and ductility in nanocrystalline and ultrafine grained metals, *Scripta Mater.* 49 (2003) 657–662.
29. L.E. Toth, *Transition Metal Carbides and Nitrides*, Academic, New York, 1971.
30. J. Marschall, F. Milstein, Theoretical strain-induced elastic anomalies in metals and alloys, *J. Phys. F* 18 (1988) 1913–1925.
31. E.A. Almond, Aspects of various processes for coating and surface hardening, *Vacuum* 34 (1984) 835–842.
32. D.W. Hoffman, Book Review: Preparation and properties of thin films, *J. Vac. Sci. Technol.* A1 (1983) 1884–1885.
33. J.A. Thornton, D.W. Hoffman, The influence of discharge current on the intrinsic stress in Mo films deposited using cylindrical and planar magnetron sputtering sources, *J. Vac. Sci. Technol.* A3 (1985) 576–579.

34. R. Abermann, R. Koch, In situ study of thin film growth by intrinsic stress measurement under ultrahigh vacuum conditions: Silver and copper under the influence of oxygen, *Thin Solid Films* 142 (1986) 65–76.

35. H. Oettel, R. Wiedemann, Residual stresses in PVD hard coatings, *Surf. Coat. Technol.* 76–77 (1995) 265–273.

36. A. Winkelmann, J.M. Cairney, M.J. Hoffman, P.J Martin, A. Bendavid, Zr-Si-N films fabricated using hybrid cathodic arc and chemical vapour deposition: Structure vs. properties, *Surf. Coat. Technol.* 200 (2005) 4213–4219.

37. M. Nose, Y. Deguchi, T. Mae, E. Honbo, T. Nagae, K. Nogi, Influence of sputtering conditions on the structure and properties of Ti-Si-N thin films prepared by r.f.-reactive sputtering, *Surf. Coat. Technol.* 174–175 (2003) 261–265.

Chapter 4

Me-Si-N Nanocomposite Films With a Low and Intermediate Si Content (<25 at.%)

An idea to improve mechanical and functional properties of the transition metal nitrides by the addition of silicon into the films has arisen from previous studies dealing namely with the addition of boron [1–4]. The refinement effect of boron through the formed boron nitride amorphous matrix, hindering the grain growth and forming a nanocomposite structure, results in reaching superhardness in Ti-B-N films [5,6]. Such effect of structure refinement and thus enhancement of the functional film properties through the nanocomposite nature of the films preventing the dislocation movement may be obtained by the addition of several alloying elements. However, it does not always lead to the improvement of the thermal stability and oxidation resistance of the films. Especially at high working temperatures close to 900°C, the loss of the enhanced mechanical properties has often been observed [7]. Another disadvantage of these hard and superhard nanocomposites is the high stress level generated during the film growth. It may cause troubles while the coated tools are subjected to high loads in cutting applications. The adhesion problems and film delamination under applied mechanical loads mainly restrict the use of coated tools in practical applications. For that reason, a great interest has been devoted to find other alternative alloying element which might overcome the observed limitations and extend the usage of the transition nitrides into other branches of industry.

Novel Nanocomposite Coatings: Advances and Industrial Applications
Rostislav Daniel and Jindřich Musil

ISBN 978-981-4411-17-2 (Hardcover), 978-981-4411-18-9 (eBook)
www.panstanford.com

In this regard, silicon seems to be a very promising candidate for tool applications. Its refinement effect is relatively strong and closely related to significant improvement of the thermal stability and oxidation resistance, even at low silicon content in the films. Moreover, its presence results in reducing the residual stresses in the films, which is very important for the good performance of coated tools in industrial applications. Si itself is, however, not responsible for the improved structure of the transition metal nitride films. Its combination with nitrogen results in the formation of an amorphous silicon nitride phase which is known for its unique characteristics. Si_3N_4-based ceramics exhibit excellent properties, such as high strength, hardness and toughness, wear-corrosion, oxidation and thermal shock resistance, low thermal expansion, self-lubricating effect, heat insulation or electricity insulation. Besides hard and tough Si_3N_4 films, transition metal nitrides represent a great interest for various applications in manufacturing industries [8]. Among the different transition metal compounds, the cubic TiN has been intensively investigated during the last years in many laboratories for their exceptional properties [9–13]. One of the drawbacks of TiN is, however, its limited oxidation resistance at high temperatures that can be reached during the process application [14,15]. Therefore, many studies have been focused on the Ti-Si-N system to further improve the mechanical and high-temperature properties of the TiN films. It has been found that the films with hardness up to 55 GPa can be synthesized [16–18]. Special attention has been also paid in improving the properties of other MeN transition metal nitrides by addition of silicon; here Me = Ti, Zr, Hf, V, Nb, Ta, Cr, Mo, W, etc., are transition metals of IVB, VB, and VIB group. Such nanocomposite hard films have shown higher hardness and toughness compared to their single-phase counterparts. Their superior characteristics rise from the coexistence of both MeN and Si_3N_4 phases growing competitively in the film and the combination of excellent properties of both phases. Right material selection of specific transition metal nitrides combined with amorphous matrix makes it possible to optimize the film properties for given applications. It has been found that the growth of MeN may be hindered by thin amorphous tissue covering the grains. Consequently, the formed nanocomposite structure strongly inhibits the crack formation and propagation under mechanical loads and yields enhanced mechanical properties.

Moreover, the amorphous silicon nitride phase provides high-temperature stability. Following sections deal in more detail with the effect of Si concentration in Me-Si-N film on its structure evolution related to the stresses, mechanical, and thermal properties as well as the resistance against oxidation. It will be shown how an improvement of individual functional properties for a specific application significantly depends on certain Si and N content. A characteristic route of mechanical properties improvement at low Si content, stabilization of the films structure at intermediate Si content, and improvement mainly of thermal stability and oxidation resistance of Me-Si-N films will be described in more detail.

Due to the unique property combination of transition metal nitrides and silicon nitride, nanocomposite Me-Si-N-based films have found commercial use in many applications. The Me-Si-N films show much higher hardness values and better stability at elevated temperatures than the undoped MeN films, and therefore, they are widely used in mechanical [19–21] and electrical [21,22] fields, as well as diffusion barrier films against Al or Cu [23–25]. However, the optimization of the films is extremely complicated due to the strong relationship between the phase composition, film structure, and film properties. The increasing requirements of advanced applications have stimulated a systematic investigation of the effect of different deposition conditions on the film behavior, either physical or mechanical, which is not sufficiently explained yet. A complete understanding of the origin of the outstanding properties of Me-Si-N films and the relationship to the developed nanocomposite structure is therefore of the highest interest. For that purpose, it is necessary to perform detailed structural analysis, namely to identify the type of phases that are developed at specific deposition conditions. The investigation of its relation with the deposition techniques is the most important issue to be clarified.

4.1 Structure, Morphology, and Phase Composition

Nanostructured films of the type nc-MeN/a-Si_3N_4 have been intensively investigated in the last decade due to the promising properties for various industrial applications given by their nanocomposite structure. For that purpose, the films have been prepared using miscellaneous deposition processes including chemical vapor deposition, plasma-assisted chemical vapor deposition, magnetron sputtering, plasma vapor deposition using a cathodic arc source, and combinations of the above such as combined cathodic arc and magnetron sputtering, as well as other techniques. The properties of systematically investigated Me-Si-N nanocomposite films with specific Si concentration slightly vary among different studies. However, the observed trends can be generalized and are thoroughly described in the following sections. Basically, Me-Si-N films exhibit variable structure, morphology, mechanical, and tribological properties as well as thermal stability and resistance against oxidation as a function of the silicon content in the film. In general, all the differences in the film properties may be explained by the different phase composition of the deposited film. The role of silicon and other alloying elements on the structure and properties of the transition nitride films will be discussed in more detail in this section.

The comparison of the structure and properties of the films with variable Si concentration allows dividing the full compositional range between pure silicon-free MeN nitride film and silicon nitride as the only phase into three basic zones "A," "B," and "C" characterized by a small, intermediate, and high silicon content. It is schematically sketched in Figs. 4.1 and 4.2. This morphological zone model proposed by Patscheider et al. for nc-TiN/a-Si_3N_4 [26] is, however, also basically valid in the case of other nc-MeN/a-Si_3N_4 nanocomposite films. Some parallelisms have been also found in nc-MeN/metal nanocomposites composed of one hard phase and a soft second phase represented by nc-ZrN/Cu [27], nc-ZrN/Ni [28] or nc-CrN/Cu films [29].

4.1.1 Zone "A"

The zone "A" is related to the Me-Si-N films with a silicon content lower than approximately 4 at.%. This referred upper limit of the silicon content can, however, strongly vary depending on the deposited material as well as on the deposition technique employed. The specific part of the zone "A" comprises pure, nearly stoichiometric MeN nitride films with the structure, which has in most cases been found to be cubic of fcc NaCl type, related to Fm3m space group [30–33]. Only in some particular cases the films exhibit hexagonal hcp [30,33,34] or tetragonal structure [35]. Depending on the deposition conditions, the nitride films can also grow to have both fcc and hcp structure [33].

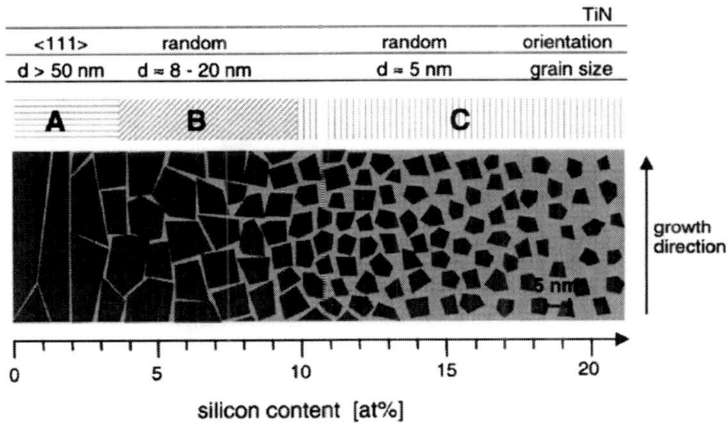

Figure 4.1 Morphological zone model for nc-TiN/a-Si_3N_4 as a function of the silicon content proposed by Patscheider et al. The structure variation from the preferentially oriented <111> structure to nearly amorphous one is attributed to the decreasing nc-TiN/a-Si_3N_4 phase ratio. The dark areas represent nanocrystals, the lighter zones stand for the amorphous phase. Reprinted from Ref. [26], Copyright (2001), with permission from Elsevier.

4.1.1.1 Film structure and morphology

The growth of the grains in silicon-free transition metal MeN nitride films is not hindered by the presence of silicon atoms or silicon nitride amorphous phase, which is known to very effectively stop the grain growth already at its very low content in the deposited film. Therefore, pure nitride films exhibit well-defined crystalline microstructure mostly preferentially oriented in certain direction, with large elongated crystallites growing commonly from the substrate to the top of the film [38] (see Fig. 4.3). The size and shape of the formed grains are, however, determined by the deposition conditions and can strongly differ depending on the energy delivered into the growing film by the substrate heating and impinging atoms and ions from the discharge. This essential parameter very rapidly influences the atom mobility and so the formed structure. It is influenced by the shadowing effects, variable mobility of atoms on the substrate surface, different surface diffusion, and grain boundary migration. Nevertheless, it fully corresponds to the structure zone models proposed by Movchan and Demchishin [39], Thornton [40–42] (considering the deposition atmosphere) or Messier [43] (also taking into account the energy of incident ions causing ion bombardment-induced mobility of the surface atoms).

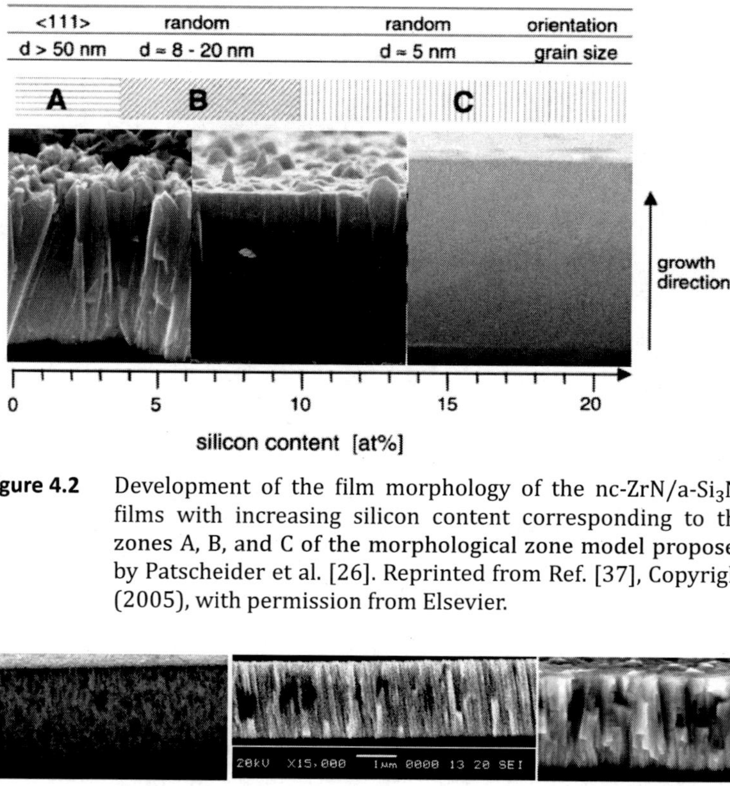

Figure 4.2 Development of the film morphology of the nc-ZrN/a-Si_3N_4 films with increasing silicon content corresponding to the zones A, B, and C of the morphological zone model proposed by Patscheider et al. [26]. Reprinted from Ref. [37], Copyright (2005), with permission from Elsevier.

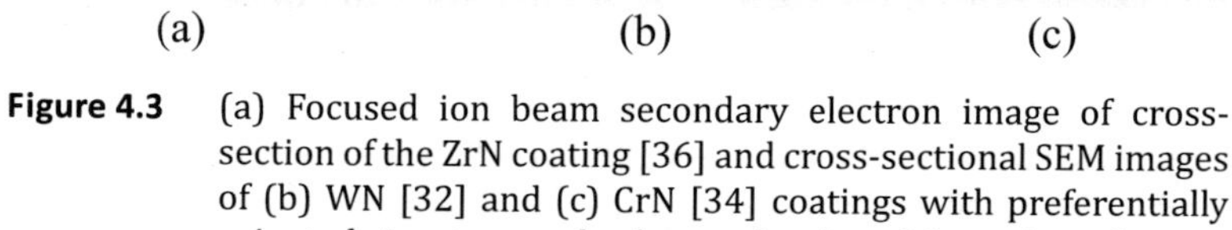

(a) (b) (c)

Figure 4.3 (a) Focused ion beam secondary electron image of cross-section of the ZrN coating [36] and cross-sectional SEM images of (b) WN [32] and (c) CrN [34] coatings with preferentially oriented structure and columns developed from the substrate to the coating surface. Reprinted from Refs. [32], [34], and [36] Copyright (2011 and 2006), with permission from Elsevier.

When a small amount of silicon is incorporated into the MeN film, a distinctive change in the structure is observed. It is in accordance with an extended Grovenor structure zone model, additionally considering the influence of the impurities and additives incorporated into films [44]. In that case, the structure of transition nitride films, predominantly preferentially oriented in the (111) direction, changes with increasing Si concentration into the (200) orientation [21,45–54,57] or the Me-Si-N films exhibit a structure composed of a mixture of randomly oriented grains [45,51,52,58–65]. These changes in the grain growth have been reported for different Me-Si-N systems prepared by various deposition processes, such as PACVD deposition technique [52] and have also been observed for the arc-magnetron hybrid processes [50] or unbalanced magnetron sputtering [49,53]. The complete structural changes, when the grain growth in the (111) direction is inhibited by the introduction of silicon atoms, is attended by a disappearing of the diffraction peaks reflected from those grains. The film structure consisting of grains with different crystallographic orientation or preferentially oriented structure in the (200) plane, after a small amount of silicon is added into the MeN film, is depicted in Figs. 4.4b and 4.4c. While the (200) film texture becomes dominated, the intensity of the (111) peak decreases and can also fully disappear. In such a case, we can talk about the complete change in the preferential orientation. This effect is also observed if some other additives are incorporated into the growing film [66–68].

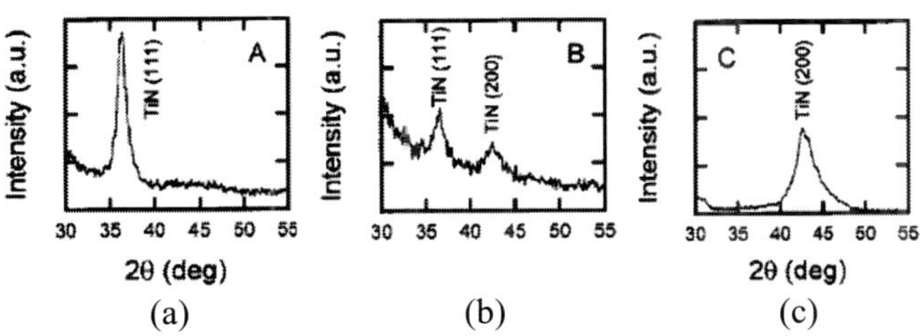

Figure 4.4 XRD patterns of (a) TiN and (b, c) Ti-Si-N films with increasing silicon content showing modification of the film structure from the (111) preferred orientation into (200) one. The growth of TiN (111) planes is inhibited by the silicon atoms incorporated into the film. Reprinted from Ref. [63], Copyright (1969), with permission from Elsevier.

Although the microstructure of silicon-contained films of the zone A is still columnar, the columnar feature becomes slightly blurred [65,69]. The observed sudden loss of the (111) orientation is bordered between 0 and approximately 3 at.% of silicon content [26,45]. Modification of the film structure associated with the incorporation of silicon atoms into the film has also been observed in the case of Me_2N nitrides with hexagonal hcp structure [70,71]. The reason for the changes in the orientation of grains is caused by obvious limitations in atomic movement [55]. It will be shown further how incorporating of Si into the MeN lattice or the presence of Si_3N_4 phase at higher Si content can effectively influence the film growth.

Although the phase diagrams of most of the transition metal nitrides show full immiscibility of the MeN phase with silicon [72], the incorporation of the silicon atoms into the fcc lattice of MeN nitrides has been described as a possible reason for the change of the crystallites growth [49,65]. Owing to relatively high kinetic energy of incoming particles, a significant atomic shot peening occurs during film deposition and the impinging atoms are forced into the lattice evoking obvious structural changes. Since no Si–N bonds are involved in the films at Si content lower than approximately 2 at.%, as it has been observed by XPS and FTIR measurements [48,57], and X-ray diffraction (XRD) patterns revealed just the presence of MeN phase in this concentration range, it can be assumed that the Si atoms are most probably incorporated in the MeN lattice by such a mechanism. This fact can also explain the structural changes in the films and the increase in the hardness and residual stress already at very low Si concentration. The generation of interstitially incorporated atoms has also been described for growth conditions, where thermodynamic diffusion processes are limited by high growth rates or relatively low temperatures. On the other hand, sufficiently high energy provided by a high substrate temperature enables the films to grow in a near equilibrium conditions. However, in that case a drop in residual stress and hardness decrease is observed [21].

Magnetron sputtering is a deposition technique characterized by a highly non-equilibrium state of the process at which materials with unique structure and properties can be formed. Compared to nearly equilibrium processes used for preparation of conventional bulk materials, sputtered thin films exhibit manifold higher hardness and exceptional physical and functional properties. Nitride films of transition metals prepared by physical vapor deposition exhibit typical nonconventional lattice distortion with obvious expansion of the lattice parameter. Except commonly assumed explanation that this effect might be due to the rhombohedral distortion of the cubic MeN lattice [73], Goldfarb et al. proposed a new alternative model, which is related to a selective trapping of interstitial atoms in various crystallographic planes of crystal lattice and is valid for both fcc and hcp structures [74]. It has been found that nitrogen or argon atoms from plasma discharge readily occupy nonequilibrium interstitial sites in the lattice of sputtered films. Moreover, the entrapment of those high energetic impinging particles is even favorable in specific planes. For instance in the case of fcc MeN cubic lattice (Fig. 4.5), a more favorable plane has been found to be (111) in comparison to other ones. This plane is the closest packed plane with the lowest energy for fcc lattice and is often observed as a dominant crystallographic plane in deposited MeN films [26,32,69]. It is important to note that the observed lattice expansion caused by entrapped atoms varies not only with different orientations of the grains but also with different incorporated atoms [74]. However, not only argon or nitrogen atoms can occupy the interstitial positions, but the same proposed model has also been used for the explanation of the lattice distortion in Me-X-N films after addition of small amount of alloying elements particularly of a small size; here X represents metal atoms such Cu, Ni, Mo, Ag, or Cr [68,75,76].

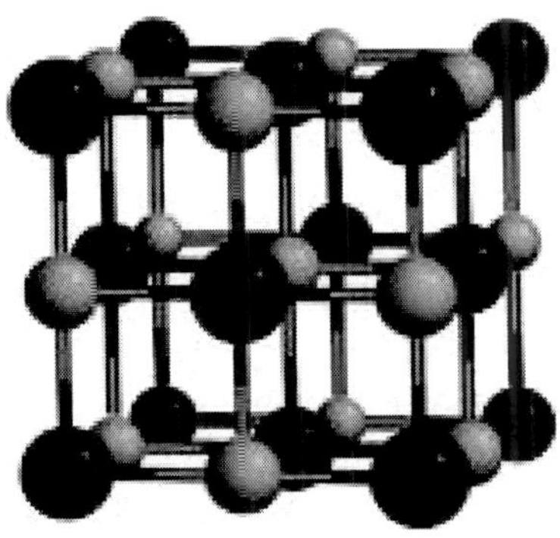

Figure 4.5 Schematic illustration of NaCl type fcc MeN lattice related to Fm3m space group consisted of transition metal atoms (dark) and N atoms (light) in lattice positions.

In the case of Me-Si-N films, many important investigations for understanding the role of silicon in film growth have been made. At present, current research findings can explain changes in the structure and properties of Me-Si-N films as a consequence of increasing silicon content corresponding to the range of zone A in two ways. In the first case, the silicon atoms might occupy the interstitial positions in the MeN lattice, if its amount is entirely low [32,55]. This finding is in accordance to the calculation results made by Goldfarb et al. taking into account the nonequilibrium state of the sputtering process [74]. On the contrary, a fairly large difference in the atomic radii of N (0.053 nm) and Zr (0.162 nm), Ti (0.140 nm), or W (0.135 nm) together with an arrangement of the atoms in the fcc MeN lattice should not allow Si atoms (0.117 nm) to be dissolved interstitially or to occupy the nitrogen sites [56]. It is known that an atom may occupy interstitial sites only in the case if its atomic size is relatively small in comparison with the size of atoms located in lattice positions. On the other hand, a substitutional solid solution is generally found between elements of similar size and it occurs as there is a reasonable degree of matching of the chemical natures and of atomic sizes of the two constituents. Hence, it is suggested that Si more probably substitutes the Me sites in the lattice. Incorporation of individual atoms into the lattice is, however, not in many cases a thermodynamically driven process set in an equilibrium state. During deposition, it is stimulated by high energy of impinging particles forming the film. This effect is more intense if the particles are additionally accelerated in the field of the biased substrate. All the facts lead into the consideration that the observed lattice distortion might be caused by both mentioned mechanisms. The amount of Si atoms incorporated into interstitial or substitutional positions is, however, fairly limited. The solubility limit of Si in MeN is referred to phase diagram studies [77] and results in a consequential formation of new phases if it is exceeded. Segregation of the forming Si_3N_4 phase into MeN grain boundaries, once an amount of incorporated Si exceeds its limit of solubility, is typical for the Me-Si-N system. A very limited range of solid solution in Me-Si-N films might be explained by a different chemical nature of Si_3N_4 from that of MeN, which leads to the separation of both forming phases [52]. The film finally consists of two coexisting phases: MeN crystalline grains embedded within the amorphous silicon nitride tissue.

Modification of the MeN lattice together with observed structural changes is, however, not induced by Si atoms only. Defects and the distortion of the lattice are also generated during the film growth by high energetic particles. Formation of vacancies, dislocations, and other lattice defects is typical for a magnetron sputtering process, especially while the substrate is biased in a dense low-pressure plasma discharge, where the main free path of the impinging particles is long enough [78]. Thermal energy introduced directly into the surface region under bombardment conditions enhances the surface mobility of adatoms and can be essential for the phase segregation and formation of the films with exceptional properties.

4.1.1.2 Phase formation

The XPS measurement of the chemical state of elements included in the Me-Si-N films shows several steps at which the phases are formed in the film with increasing Si content. First, while very low Si amount is added, the films mostly show the Me-N bonds revealing that the crystalline MeN phase is dominant in that case [54]. The structure, very similar to that of pure MeN films with preferred orientation mostly in the (111) direction changes into the (200) one or into the mixture of the grains with different orientation and the lattice, is distorted as the Si atoms are incorporated into the lattice. An effect of incorporation of Si into the MeN lattice can also be accompanied by an introduction of a higher amount of nitrogen over the film stoichiometry [21]. It has been observed in Me-Si-N films with increasing silicon content close to the upper limit of the zone A. A possible explanation might be an effort to relieve the strain induced by the lattice distortion after the silicon atoms are incorporated into the lattice [56]. Further addition of silicon into the MeN films is accompanied with the formation of new phases. Besides Me–N bonds, Si–N ones also have been observed by XPS in Me-Si-N films at the Si concentration higher than 2 at.% [30]. A detail analysis of the measured spectra shows that the chemical bonding state of Si is mainly in the form of the Si_3N_4 phase. Why there are no other phases formed in the film can be easily explained in a thermodynamical point of view. A comparison of the enthalpy of formation of individual phases which can be possibly formed of Me, Si, and N atoms, that is, the energy necessary for bond formation, clearly shows that the formation of Si_3N_4 is energetically strongly favored over that of $MeSi_x$ (see Table 4.1). Also the presence of silicon in the free state is energetically unfavorable and therefore there is no $MeSi_x$ or unreacted Si detected in most of the Me-Si-N films with a low or intermediate Si content [38,69]. The replacement of the Si–Si bonds by Si–N ones reflects that the silicon became more nitrided in the film with increasing amount of Si added [30,79]. The XPS results reveal that silicon atoms are fully nitrided in the films already at Si content higher than 3 at.% [30,48]. It indicates that Me-Si-N films with intermediate and high silicon content are mainly consisted of crystalline MeN phase coexisted with Si_3N_4. Silicon nitride has been observed in amorphous state only since it crystallizes in thin Me-Si-N films at temperatures higher than 1100°C [53]. Finally, depending on increasing amount of silicon and nitrogen incorporated into the film the amount of the amorphous Si_3N_4 phase increases further.

Which phases will be finally formed, however, also depends on the conditions at which the films are deposited—particularly on the composition and energies of particles in the sputtered flux, substrate temperature, and energy of bombarding ions at the biased substrate. It influences the elemental composition of films, atom mobility at the surface and consequently the phase composition. For instance, the appearance of free unreacted Si is ascribed to the deficit of nitrogen [38]. An amount of nitrogen incorporated into the film can, however, be increased by an enhanced ionization of nitrogen gas in the plasma discharge [37]. In that case, more Si-N bonds may be formed as free Si spontaneously reacts with nitrogen [51]. In the same way, the presence of $MeSi_x$ phase is surprising in the Me-Si-N films with a low Si content since the formation of MeN and Si_3N_4 would seem more likely. Formation of silicides occurs in the films where silicon content exceeds the amount needed for the forming Si_3N_4 phase with included nitrogen, whose content is insufficient to form bonds with all Si atoms. Free silicon is then bonded with residual unbounded Me atoms and forms $MeSi_x$ phase [37]. Despite this fact, formation of nanocrystalline and amorphous $TiSi_2$ phase has been reported in the Ti-Si-N films with the low and intermediate silicon content (5–20 at.%) prepared by chemical vapor deposition [59]. Here, the appearance of the $TiSi_2$ phase is attributed to a lower activity of nitrogen at the lower discharge current density (<1 mA/cm^2). The chemical state of silicon present in Me-Si-N composite films can also reflect the chemical state of the sputtered target. The films

Table 4.1 Enthalpy of formation of selected phases formed of Me, Si, and N atoms in [kJ/mol].

Phase	Enthalpy of formation ΔH_f	Phase	Enthalpy of formation ΔH_f	Phase	Enthalpy of formation ΔH_f	Phase	Enthalpy of formation ΔH_f	Phase	Enthalpy of formation ΔH_f
ZrN	-365.5[2]	TiN	-336.6[3]	WN	-24[4]	TaN	-252.4[3]	CrN	-123.1[3]
$ZrSi_2$	-159[1]	$TiSi_2$	-134.4[3]	WSi_2	-91.9[1]	$TaSi_2$	-119.3[3]	$CrSi_2$	-80[3]
ZrSi	-154.8[1]	TiSi	-129.8[3]	W_2N	-71.9[4]	Ta_2Si	-125.6[3]	CrSi	-53.2[3]
Si_3N_4	-745.1[3]								

[1]See Ref. [72].
[2]*Lange's Handbook of Chemistry* (15th Ed.), by J.A. Dean, McGraw-Hill, Inc., New York, 1999.
[3]*Smithells Metals Reference Book* (6th Ed.), by E.A. Brandes, Butterworths, London, 1983.
[4]*Hartstoffe*, by R. Kieffer, F. Benesovsky, Springer-Verlag, Wien, 1963.

sputtered from predominantly elemental silicon target can contain more silicon in elemental form compared to the films sputtered from already nitrided targets [54]. If the Si concentration is higher than 18 at.%, the free unreacted silicon atoms [38] or silicide phases [37] have been observed depending on the elemental composition of the film. Me-Si-N film with a low Me content and relatively low N content are mainly consisted of the Si_3N_4 phase with free Si atoms. On the other hand, Me-Si-N films are composed of $MeSi_x$, MeN, and Si_3N_4, if Me content is relatively high.

The formation of silicon nitride leads to an encapsulation of the growing MeN crystallites and restriction of their further growth. This fact can also explain observed sudden change in the crystallites orientation and the decrease of the mean crystallite size from well above 100 nm to values below 10 nm that is reflected by weaker and broader diffraction peaks corresponding to MeN grains [57]. The encapsulation of the crystallites by the Si_3N_4 phase is also associated with an observed isotropic growth of MeN grains. At approximately 2 at.% of silicon in the film, the average coverage of the MeN crystallites is about half a monolayer of Si_3N_4. Even if silicon nitride would be uniformly distributed over the crystallite surface, its amount is not high enough to cover the entire crystallite. Therefore, such insufficient encapsulation of MeN grains enables their continuous, however limited growth. Statistically, the entire crystallite surface is fully covered with a monolayer of the amorphous silicon nitride phase at approximately 4 at.% of silicon content in the film. However, the crystallite coverage is still incomplete at this silicon content because the formation of a continuous monolayer on the crystallite surface is unfavorable state owing to the finite surface energy of the Si_3N_4 phase. Therefore, the grains are just partially covered by the silicon nitride tissue and can still grow, although their growth is obviously limited. Diffraction patterns reflected from MeN grains of Me-Si-N film become weaker and narrower with increasing silicon concentration compared to those of silicon-free MeN films [49]. This is characteristic for all Me-Si-N films corresponding to the zone A.

Similar structural changes are also induced after the addition of other elements into MeN films. Change of preferential growth in certain direction and deformation of the lattice has also been observed in MeXN films; here X = Cu, Cr, Ag, Ni, or Sn [27,28,68]. The same mechanism of covering MeN grains by a thin layer has been enforced, resulting in effective hindering of their growth, although this layer is consisted of pure metals.

4.1.2 Zone "B"

The following zone labeled with "B" contains Me-Si-N films with a relatively wild range of silicon content between 4 and 10 at.%. Silicon as a selective additive plays a very important role in the structure and film properties if it is incorporated into the film at certain minimum amount. This effect is more pronounced if silicon is in a form of amorphous Si_3N_4 phase coexisting with MeN grains. In contrast to the films within zone A, where the MeN grains are not fully covered by an amorphous Si_3N_4 phase and the structure is similar to the pure MeN films, increasing Si content above the limit of the zone A causes dramatic changes in the film structure. It also subsequently influences the mechanical properties, stresses, and thermal stability of deposited films. Since above approximately 5 at.% of Si content the volume fraction of the amorphous Si_3N_4 phase is already high enough to fully cover MeN grains during their growth, films with typical nanocomposite structure are formed. Thus, the films within the zone B are characterized by the structure consisted of randomly oriented equiaxed grains embedded within the amorphous Si_3N_4 matrix (see Fig. 4.6).

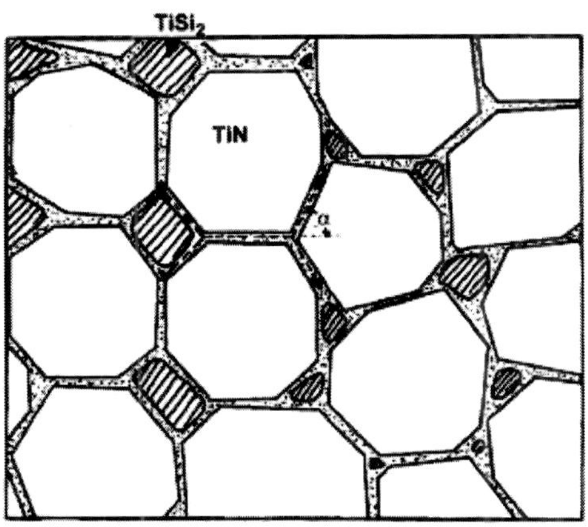

Figure 4.6 Schematic of a nanocomposite structure of the nc-TiN/a-Si3N4/a- and nc-TiSi2 films. Reprinted from Ref. [59], Copyright (2000), with permission from Elsevier.

4.1.2.1 Film structure

From a thermodynamical point of view, the formation of a continuous Si_3N_4 matrix around the crystallites is related to the reduction of the system energy. Hence, MeN grains covered by the amorphous Si_3N_4 phase have been observed already at relatively low Si content in the film. The experimental observations indicate that Si_3N_4 adheres preferentially on the grain boundaries. However, formation of a thin amorphous tissue between individual grains occurs just when the driving force caused by the energy difference $\Delta J = J_C - J_{AC}$ is large enough. Here, J_{AC} scales the interfacial energy between amorphous Si_3N_4 and crystalline MeN phase and J_C is proportional to the grain boundary energy [69].

If the MeN grain is even partially covered by an amorphous tissue, obvious decrease of its size is observed. It has been found that the reduction in the grain size strongly depends on the volume fraction of Si_3N_4. Thus, a decrease in the size of MeN grains becomes more pronounced while an amount of amorphous phase in the film increases. The average grain size in the Me-Si-N films with very low Si content is relatively high and large grains of about 30 nm are developed [50]. The addition of Si and subsequent formation of an amorphous phase give rise to a significant decrease of the grain sizes down to 5 nm at 5 vol.% of Si_3N_4 [69]. Initial rapid reduction in the grain size, however, slows down with increasing volume fraction of Si_3N_4 according to a power law [36] (see Fig. 4.7a). This finding has also been confirmed by the theoretical calculation with the help of the Monte Carlo Potts model simulation [63]. Reported values for the minimum grain size can be extremely small and vary from 2 to 15 nm at Si concentration between 5 and 10 at.% [16,26,38,49,50,80,81]. Beyond this concentration, the grain size remains almost constant even if silicon is added further. In spite of this finding, the minimum size of grains has been observed for silicon content within the range of 5–7 at.% followed by a significant increase at higher content [82]. However, it might be caused by different conditions at which the films have been produced. Lower activity of nitrogen or atom mobility and diffusion on the film surface can subsequently very strongly influence the grain growth [59].

The size reduction of MeN crystallites is reflected by significant XRD peak broadening [8,38,58,63]. Increasing volume fraction of an amorphous tissue formed around the grains also results in their random orientation [65]. At the same time, diffraction peaks corresponding to MeN crystallites disappear [58] and at high silicon concentration no peak is observed as the film becomes nearly amorphous [48,56].

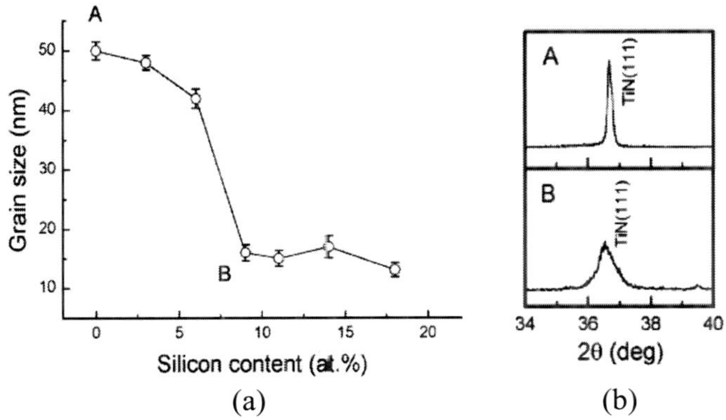

Figure 4.7 (a) Development of the size of TiN grains as a function of increasing silicon concentration in Ti-Si-N films reflected by broadening of diffraction patterns (b). Reprinted from Ref. [38], Copyright (2004), with permission from Elsevier.

4.1.2.2 Film morphology

Understanding how the MeN grains are limited in their growth, while they are covered with the Si_3N_4 tissue, is very important for finding the principle of the structure forming and origin of improved mechanical properties of Me-Si-N films. Basically, silicon can prevent the grain growth in several ways. It acts as a roadblock to the diffusing Me and N atoms, it can trap the diffusing atoms, and silicon has also been found as a possible nucleation site for lattice defects [69]. Consequently, the grains are limited in their growth because grain forming atoms are not sufficiently mobile to migrate over the surface to the preferred sites for perfect crystallization. In principle, all changes in the surface diffusion caused by present silicon nitride phase affects distinctively the film growth. It results in a rapid modification of film morphology according to the structure zone models developed by Grovenor et al. [44] and Barna and Adamik [83].

It has been observed in all Me-Si-N films that the film morphology changes significantly with increasing silicon content. Columnar growth of the grains [53] or fiber structure [57] is typical for the Me-Si-N films containing low amount of silicon in concentration range up to 5 at.% corresponding to the zone A. The films are primarily consisted of long columnar grains oriented in the direction of film growth. Since newly formed grains are reduced in their size at higher Si concentration, a fine-grained film structure is subsequently established. Moreover, this structure refinement is more pronounced at higher volume fraction of an amorphous phase resulting in the lowering of grain size up to approximately 5 nm at about 10 vol.% of silicon nitride. Besides the hindering of grain growth, the additives at certain amount also stimulate renucleation of the grains. This fact results in the formation of globular and amorphous-like structure at medium and high silicon content, respectively. The tendency to form a more dense structure with reduced grain size with increasing Si content can be seen from Fig. 4.8.

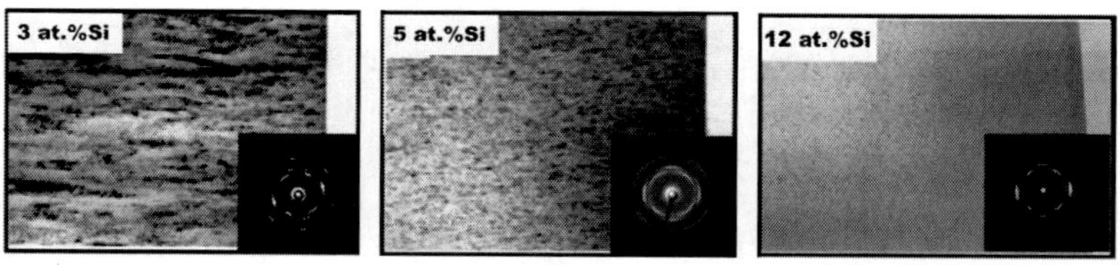

(a) (b) (c)

Figure 4.8 Cross-sectional TEM micrographs of the Zr-Si-N films with silicon content of (a) 3 at.%, (b) 5 at.% and 12 at.% and corresponding electron diffraction patterns illustrating the influence of silicon incorporation on the film morphology. Addition of increasing amount of Si changes the columnar structure of pure ZrN film into (a) a dense finely grained, (b) globular, and (c) nearly amorphous structure. Reprinted with permission from Ref. [55]. Copyright [1969], American Vacuum Society.

The film microstructure is, in general, given by several competing growing mechanisms. The geometrical shadowing effect is ascribed to preferred growth in vertical direction, while enhanced surface diffusion of adatoms tends to smooth structures. Both effects can be pronounced or limited by the impact energy through strong bombardment of sputtering ions at biased substrates [31] or by shadowing [38]. It can dramatically enhance or reduce atomic mobility, which consequently effectively influences the film growth. The competition of those factors strongly drives morphology evolution of growing surfaces. For instance, the columnar microstructure consisting of large columnar grains with average column width of 60–80 nm is developed in the MeN films sputtered at low pressures and at high ion bombardment [63]. In the silicon-contained film, the columnar nature is less manifested and the grains quickly become smaller and equiaxed [26]. Those films still exhibit the columnar-like morphology, however, with much smaller column width [38]. An increase in the grain size with increasing Si content up to 3 at.% has also been observed by XRD revealing a different behavior from that stated by others [55]. This appearance is, however, caused by the growth of grains in a columnar fashion in the direction of growth. In fact, the lateral grain size decreases spontaneously with increasing Si content also in that case, as observed by transition electron microscopy (TEM). It reveals some limitations of X-ray diffraction analysis at Bragg–Brentano configuration used for determination of the film structure. It has been found that a columnar structure of Me-Si-N films with fine grains exhibits very dense grain boundaries [31]. Each column is here formed by an agglomeration of MeN nanocrystallites embedded within the amorphous Si_3N_4 phase [30]. Once sufficiently high amount of Si_3N_4 is formed, the width of columnar MeN grains is distinctively reduced due to its impeding effect. As the silicon concentration increases further, the limited atom mobility and surface diffusion, together with inhibited grain growth induced by the silicon nitride phase formed, results in vanishing of the columnar structure and fine-grained structure with equiaxial grains of several nm in diameter is developed instead [38,55].

The morphology of Me-Si-N films has also been systematically investigated by the means of TEM and high-resolution transition electron microscopy (HRTEM) to determine the size and distribution of the crystallites in amorphous matrix with its increasing volume fraction (Fig. 4.9).

Cross-sectional HRTEM examinations consistently reveal the presence of an amorphous region between differently oriented MeN crystallites inside each column. Thus, the microstructure changes to a nanocomposite one, having much finer MeN crystallites uniformly embedded within an amorphous tissue, as the Si content increases in the film. The crystallites with partly aligned microstructure have been found to be penetrated by an amorphous phase, however, not distributed homogeneously in this amorphous matrix [26,32,63]. By using TEM micrographs in Me-Si-N films with intermediate Si concentration, it can be clearly seen that the crystalline columns with 5–10 nm in width and in some cases more than 10 nm in height are separated by amorphous regions with just several nanometers in thickness. Moreover, the films seem to exhibit no dislocations at boundaries between individual MeN nanograins [63] and most grain boundaries appear to be atomically sharp [81]. It is characteristic for all Me-Si-N films, where the microstructure consists of nanocrystalline MeN grains situated in the amorphous Si_3N_4 phase. Amorphous regions between two columns have been found to be richer in Si than the columns themselves. It confirms the XPS observations, which have identified the amorphous tissue as Si_3N_4 [30]. The effect of incorporated silicon through increasing volume fraction of the amorphous silicon nitride tissue is obviously reflected in structural changes of the films. At Si content higher than 15 at.%, the electron diffraction patterns become diffused and at even higher content MeN crystallites are no longer observed in HRTEM images (see Fig. 4.9c). This is consistent with the results of XRD analysis, where no peaks corresponding to MeN are detected in the same silicon concentration range. Thus, diffraction spectra correspond mainly to amorphous state of the films at high silicon content [37,58].

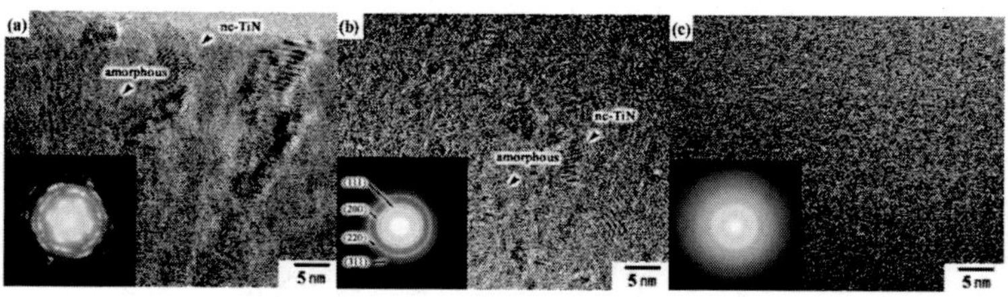

Figure 4.9 Cross-sectional HRTEM images and corresponding electron diffraction patterns of the Ti-Si-N films with Si content of (a) 8 at.%, (b) 11 at.%, and (c) 28 at.%. Crystalline TiN grains distributed over the film are separated by an amorphous phase forming nanocomposite nc-MeN/a-Si_3N_4 structure. Its volume fraction increases significantly with increasing Si concentration in the film. Reprinted from Ref. [58], Copyright (2002), with permission from Elsevier.

In general, it is difficult to determine by XRD analysis whether the Si_3N_4 phase or any elemental Si is present in the film. Although no XRD peak corresponding to the crystalline Si or Si_3N_4 has been found, XPS results clearly indicate that Me-Si-N films contain the amorphous Si_3N_4 phase already at a low Si concentration [8,36]. In the same way, elemental silicon has been determined by XPS in the films with a high Si and low N content suggesting its low concentration or amorphous state. Additionally, it is well known that ceramics, such as silicon nitride, need much higher temperature for crystallization due to their characteristic high viscosity [51]. For instance, the silicon nitride grains have been found to grow at temperatures higher than 1100°C [53,84]. It suggests that a combination of several analytic methods for determination of phase composition of Me-Si-N films is needed.

4.1.2.3 Surface roughness

A film growth influenced by increasing content of silicon incorporated into the growing film also results in obvious changes of surface roughness. Since surface diffusion-dominated film growth becomes ascendant at certain amount of formed Si_3N_4 volume fraction, the columnar growth vanishes and fine-grained structure is formed. In Fig. 4.10, typical evolution of film roughness reflecting structural changes in Me-Si-N film with increasing Si content is depicted.

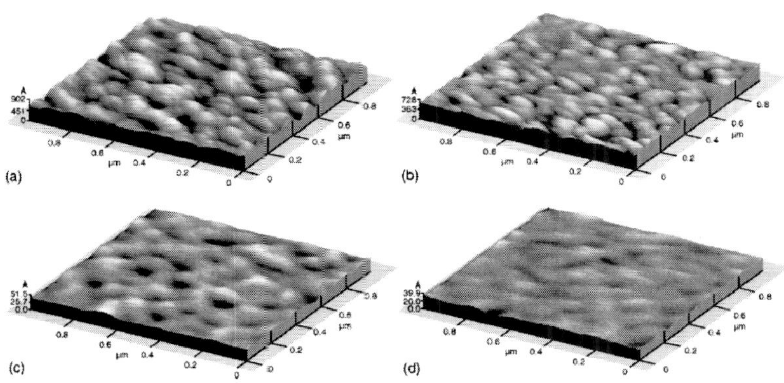

Figure 4.10 AFM surface images of the Ti-Si-N films with four different Si content (a) 0 at.%, (b) 9 at.%, (c) 14 at.%, and (d) 18 at.%. Reprinted from Ref. [38], Copyright (2004), with permission from Elsevier.

Surface roughness is strongly related to the film structure and directly depends on the size and orientation of the crystallites. Therefore, crystalline films developed with large columnar grains exhibit very rough surfaces with a continuous distribution of mounds [51]. At higher Si contents, surface smoothing is obviously observed in all Me-Si-N films (see Fig. 4.10). This effect is usually explained as a consequence of a decreased average grain size since more nuclei are formed as silicon is incorporated into the film [38]. Also enhanced surface diffusion tends to smooth out roughness in a growing surface [63]. The nanocrystalline and amorphous films have especially low roughness, which is related to the strong impeding effect of the amorphous Si_3N_4 phase on grain growth resulting in very fine-grained structure [31].

4.1.2.4 Structure development with atom mobility

Magnetron sputtering is well known as a unique technique for the film sputtering with variability in the process conditions and high versatility, which is mainly given by an easy control over the sputtering process [85]. Its flexibility has been proved by deposition of films with a wide range of composition and properties [86–93]. Well-structured films and the films with the phase composition, which is commonly hardly formed by any other deposition technique, have been produced by magnetron sputtering [85]. It is attributed to the strong influence of process parameters on the structure, composition, and properties of deposited films. In terms of coating formation, the density, ionization rate, composition, and energy of depositing fluxes are decisive factors, which need to be controlled. It is well known that which structure will be formed strongly depends on the energy of impinging particles. It directly influences the surface diffusion and consequent arrangement of arriving atoms on the substrate surface. If the atom mobility at the surface is not sufficiently high atypical phases may be formed, compared to those originating at higher atom energies. Typically formed nanocomposite structure of Me-Si-N films, composed of the crystalline MeN phase and the amorphous silicon nitride tissue covering the crystals, completely changes once the atom mobility is significantly reduced. In the Ti-Si-N system deposited by reactive magnetron sputtering at a low bias voltage and low substrate temperatures, a metastable TiSiN phase has been identified [49]. It is typically formed at the deposition conditions leading to the low surface atom diffusion, while a nanocomposite structure no longer develops. Although increasing amount of Si incorporated into the MeN lattice generally results in increasing lattice distortion [21,50,55–57,65], the lattice parameter associated with this new phase is almost constant in the investigated concentration range. In Fig. 4.11, an empirical model for phase formation in Ti-Si-N system proposed by Vaz et al. is given [61]. It is assumed that low deposition temperatures and an absence of the ion bombardment do not provide necessary atom mobility at the surface or the bulk atomic diffusion for the complete phase segregation and the consequent formation of polycrystalline grains and the amorphous matrix [80]. Formation of the nanocomposite nc-MeN/a-Si_3N_4 structure therefore requires additional energy delivered to the film through the impinging particles with high energies or through the substrate heating in order to ensure sufficient superficial mobility of the species at the growing film [5,53]. The energetic particle bombardment provides an important contribution to phase segregation and also to surface smoothing via the surface diffusion of adatoms during growth process [38].

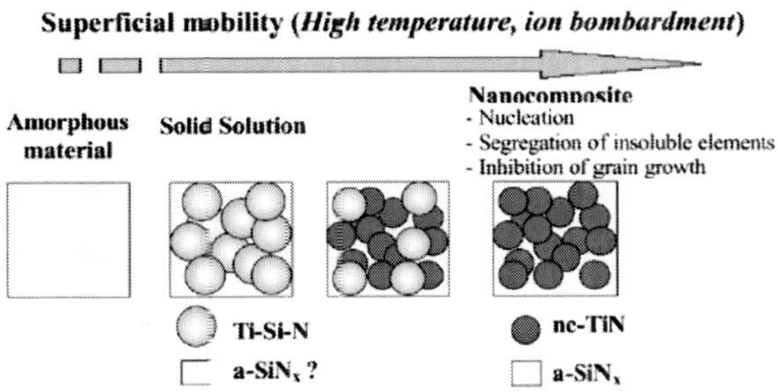

Figure 4.11 Empirical model for phase formation in Ti-Si-N films. Reprinted from Ref. [61], Copyright (2001), with permission from Elsevier.

Besides the essential role of the energy of impinging particles, an effect of deposition temperature on the growth of the Me-Si-N films has been intensively studied as well. A general effect of enhanced grain growth as a consequence of increasing deposition temperature has also been observed in the case of Me-Si-N system, where additionally the growth rate significantly differs with various Si concentrations. The grains grow much faster in pure MeN films than that in Me-Si-N ones [21]. It suggests that the silicon present in MeN films markedly stabilizes the film structure. Here the stabilization of the film structure proceeds by two different mechanisms. At the low Si content, silicon atoms incorporated into the lattice tend to lock diffusion paths and thus stabilize the structure, while at higher Si concentration the amorphous Si_3N_4 phase substitutes, even more effectively, the role of atomic silicon. Finally, an effect of target power has been investigated. A different power delivered to the Si_3N_4 and Ti targets in dual magnetron system has been used for the deposition of Ti-Si-N films in order to control the film structure. In this way, films with specific properties and with completely different structures have been synthesized [8]. With the increase in Si_3N_4 target power, the silicon ion's mobility and partial pressure increases significantly, reducing the titanium ion's partial pressure. It gives rise to higher probability of Si_3N_4 phase formation, compared to TiN, that results in hindering of TiN crystal growth and consequently in a decrease in the crystallite size. If the Ti target power significantly decreases, the Ti ions do not have enough mobility and energy to form crystalline TiN and thus the films deposited are basically totally amorphous [8].

All those findings show a great importance of additional energy delivered to the film during its growth in forming individual phases and consequently the film structure. It has been experimentally proved that, in principle, several possible ways exist in order to improve the adatom mobility and to enhance a phase formation through the high surface energy of impinging particles: by increasing the target power, by reduction the amount of atomic collisions in the plasma discharge, by the substrate biasing, and by the substrate heating. All the effects listed above significantly influence the film growth, the phase composition, and thus the film structure. It basically arises from the enhanced chemical processes taking place among incoming particles on the substrate surface as well as from the enhanced plasma chemistry in the discharge by means of the high-energetic ionized particles of working gas and the sputtered atoms. Under such conditions, the formation of Me-Si-N films with a nanocomposite structure of the type nc-MeN/a-Si_3N_4 is significantly promoted. Since the amorphous Si_3N_4 phase is formed, the growth of the MeN grains is markedly hindered through their encapsulation, resulting in unique film properties. It was already mentioned that just one monolayer of the amorphous silicon nitride tissue surrounding the individual grains is sufficient to slow down the grain growth and completely change the structure of the deposited film. This is the reason why the structure and properties of the MeN films are so sensitive to change in very narrow compositional range. The growth inhibiting phenomenon caused by the increasing content of the Si_3N_4 phase in the films is responsible for altering almost all functional film properties. Besides the hardness and elastic modulus, electrical and optical properties also change significantly with the structure variation and changes in the phase composition [37,94]. An essential role of the silicon nitride phase on the film properties arising exactly from the nanocomposite nature of the Me-Si-N films has been investigated by many authors and will be discussed in more detail in following sections.

4.1.3 Zone "C"

4.1.3.1 Film structure and morphology

The third zone labeled with "C" is related to the insignificant decrease of the grain size and only minimum change of the grain orientation. An increasing amount of the amorphous silicon nitride phase in Me-Si-N films to the detriment of the volume fraction of the MeN crystallites results in a complete amorphous-like structure. It is represented in a high silicon content range of the zone model, schematically illustrated in Fig. 4.1. The MeN grains with the size below 5 nm randomly distributed over the film are no longer detectable by XRD analysis that suggests an overall amorphous state of the films. However, a presence of crystallite grains has been proved by XPS and FTIR measurements [30,45]. Such as amorphous-like nature of the Me-Si-N films within the zone "C" with extremely fine grains dispersed in an amorphous matrix results additionally in a smooth and glassy appearance in cross-section (Fig. 4.9c). Ultra-fine grain film structure also involves a surface roughness reaching very low level typically under several nm (Fig. 4.10d). It may vary depending on the deposition technique used, however, in all cases the surface roughness exhibit a minimum value at the highest amount of the amorphous Si_3N_4 phase.

It has been experimentally verified that according to the zone model proposed by Patscheider et al., the film morphology varies with increasing amount of the amorphous silicon nitride phase as follows [26]: large elongated grains at very low silicon contents corresponding to the zone A lose their preferred orientation once a higher Si content is incorporated into the MeN lattice; as the amorphous Si_3N_4 phase is formed in the film, the full interpenetration of both phases is observed within the zone B resulting in development of structure consisting of isolated MeN nanocrystals embedded within an amorphous tissue. It finally develops in nearly amorphous structure of the zone C at the highest Si content.

An understanding of the role of Si and N atoms in Me-Si-N films with their various contents is entirely important for understanding of specific film properties appearing at certain elemental and thus phase compositions. It has been verified in many nanocomposite systems composed of nanocrystalline and amorphous phases that unique film properties can be obtained just if the optimum volume fraction ratio of both phases is reached [95,96]. Significant role of the amount of the Si_3N_4 phase on the enhanced mechanical properties of Me-Si-N films has been systematically studied and is demonstrated in next sections.

4.2 Development of Hardness and Macrostress

The proposed structure zone model classifying Me-Si-N films into three groups A, B, and C is useful to describe the evolution of film hardness and residual macrostress as a function of increasing Si concentration as well. In fact, two different trends of hardness variation have been found: a continuous hardness decrease [48,81] or a hill-type with a hardness maximum at intermediate silicon content [16,36,55–57] (see Fig. 4.12). The elastic modulus of Me-Si-N films showed a similar development. It is important to note that the nonlinear hill-type trend has been found to be typical for Me-Si-N films, since it has been observed in almost all films.

Mechanical properties of Me-Si-N films obviously reflect a development of film structure and its phase composition. Therefore, a change in hardness in both cases significantly varies with the size of the grains, their orientation, and the MeN/Si_3N_4 fraction volume ratio. Increasing silicon content in the film gives rise to different hardening mechanisms which influences response of the material to the mechanical load. So far, no exact optimal amount of silicon concentration has been found for obtaining maximum hardness in deposited Me-Si-N films. Maximum hardness typically between 30 and 50 GPa have been achieved ranging at silicon contents of 5–12 at.%, while discrepancies in the results reported in the literature arise mainly from the different deposition techniques and deposition parameters used in various studies. In general, the greatly enhanced hardness of Me-Si-N films in stated silicon range is considered to be due to its improved microstructure.

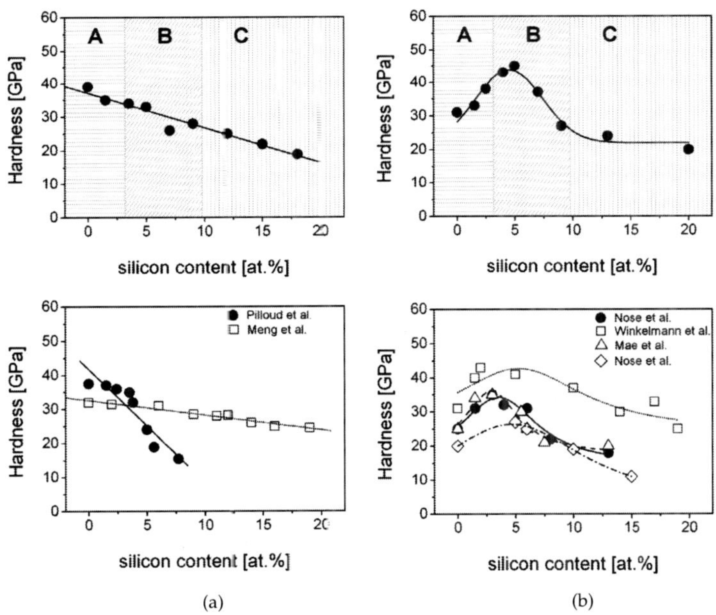

Figure 4.12 Schematic illustration of two trends in hardness development with increasing silicon concentration in Me-Si-N films: (a) a continuous hardness decrease and (b) a typical nonlinear hill-type trend. Results reported by various authors on different Me-Si-N films and three characteristic zones A, B, and C are also depicted.

Beside the film structure, the internal residual macrostress is also known as a possible contributing factor, which might determine the film hardness. Its value is mainly affected by deposition conditions. Bombardment of the growing film by ionized particles with high energy causes peening effect of sputtered atoms in the film inducing the lattice distortion. Changes in atomic arrangement are thought to be another possible reason for the changes in lattice constant and macrostress and consequently of hardness and elastic modulus. It is even pronounced if other additives are additionally incorporated into the crystallographic lattice. Difference in the atomic radii among the film constituents together with all arising defects in the lattice contributes to an increase in total macrostress. Similarly, macrostress of the film is influenced by the T_S/T_m ratio and by a difference in thermal expansion coefficient of the substrate and deposited film, which determines thermal component of macrostress σ_{th}; here T_S is deposition and T_m melting temperature, respectively. A general trend in the stress development observed in Me-Si-N films as a function of increasing silicon content is displayed in Fig. 4.13a.

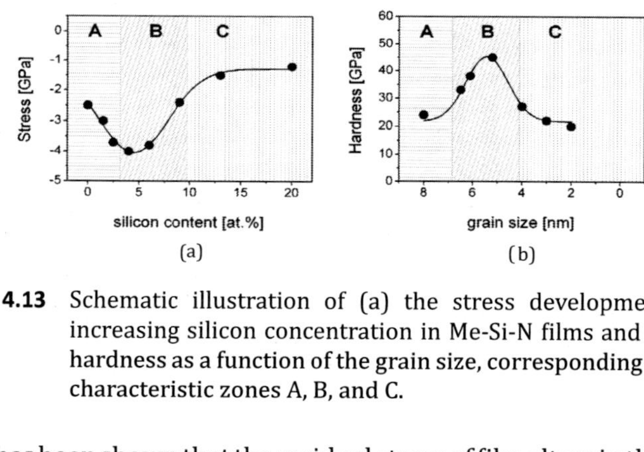

(a) (b)

Figure 4.13 Schematic illustration of (a) the stress development with increasing silicon concentration in Me-Si-N films and (b) film hardness as a function of the grain size, corresponding to three characteristic zones A, B, and C.

It has been shown that the residual stress of film alters in the same manner as the hardness. Namely, a compressive stress achieved its maximum value at maximum film hardness commonly at the same silicon content; compare Fig. 4.12b and 4.13a. Moreover, in most Me-Si-N films the relationship between film hardness and residual stress can be expressed by a straight line regardless on the silicon content [56]. It indicates close relationship between macrostress and mechanical properties of Me-Si-N films especially at low and intermediate silicon content.

Plotting film hardness as a function of MeN crystallite size demonstrates strong correlation as well. Significant effect of Si content on the grain size reflected by changes in film hardness is represented in Fig. 4.13b. Here, hardness increases dramatically as crystallite size decreases below approximately 15 nm and reaches its maximum at certain crystallite size between 5 and 10 nm. This development can be explained by the Hall–Petch relationship [97] (see Chapter 1). On the other hand, further decrease in crystallite size is followed by a strong decrease in hardness, which is in accordance with the so-called "anti-Hall–Petch" effect [98], and the hardness tails off to the one of an amorphous phase as crystallite size diminishes [8]. This matches well with Schiotz's theoretical prediction [99]. Since there is no optimum silicon concentration found to provide the highest hardness of Me-Si-N films within the structure zones A, B, or C, an origin of enhanced hardness seems to be a complex problem of many factors. Only a complex relation between the size of the grains and their orientation, lattice distortion, residual stress, and phase composition might explain observed improvement of the mechanical properties.

4.2.1 Zone A

The present studies showed that significant hardening of the Me-Si-N films related to the zone A takes place if just a small amount of silicon is added into the film. It tends to an increase in residual stress and lattice constant together with a concomitant decrease of the mean crystallite size followed by an increase in hardness and elastic modulus. Increasing tendency of film hardness is characteristic for Me-Si-N films within zone A (see Fig. 4.12b).

It was already mentioned that as a consequence of increasing amount of silicon atoms incorporated into the MeN lattice during formation of Me-Si-N film, an obvious lattice distortion is observed. This effect, however, cannot fully explain increase in the hardness in Me-Si-N films since it has been shown that for higher Si content (>5 at.%), even if lattice distortion increases, the film hardness decreases [17]. It gives a rise into the consideration that increasing interstitial concentration results in a type of solid solution hardening [21,30,55]. It is replaced by completely different hardening mechanism once new phase is formed in the film at higher Si concentration. The effect of two coexisting phases in nanocomposite Me-Si-N films on film hardness is discussed in next paragraphs. Incorporation of Si atoms into the interstitial positions contributes to strengthen the structure by making more difficult further lattice distortion when an external stress is applied. The distortion of the lattice also probably hinders the dislocation motion through the crystals and act as its energy barrier [100].

Owing to the fact that nitride films of transition metals can admit just a limited number of atoms in interstitial positions, there must exist a concentration limit for incorporation of silicon into the MeN lattice leading into formation of new phases if it is exceeded. Such a limit is specific for various MeN materials, that is also reflected in the zone model revealing a boundary shift between zones A and B for individual Me-Si-N films. Since typically no ternary compound is reported in Me-Si-N systems [72], an excess of silicon above given limit leads to the changes in phase composition of the films at certain silicon concentration. Beside the crystalline MeN phase, the amorphous silicon nitride phase appears located at the grain boundaries, partially covering the growing grains. There exist two possible explanations how the Si_3N_4 phase is formed resulting in distinctive changes of film properties. The first one is based on the assumptions that silicon segregates to grain boundaries if its maximum quantity in the lattice is surpassed. It makes the grain boundaries silicon-rich and hardens the structure slightly [34]. Consequently, the silicon nitride phase is formed, as silicon starts to react with nitrogen delivered from plasma discharge. The silicon nitride phase then continuously grows and starts to cover MeN grains as an amorphous tissue. In the second case, the silicon nitride phase is considered to be formed immediately as silicon and nitrogen arrive to the surface due to the high enthalpy of formation of Si_3N_4. It might explain why the silicon nitride phase is also detected at the silicon content, which is lower than the one corresponding to the maximum possible amount of silicon admitted in the MeN lattice. The silicon nitride tissue located at the grain boundaries has been found to influence the grain growth and moderately contribute to the improvement of film hardness already at its relatively low amount. It dramatically influences film morphology and texture, which prove to be essential in determination of film hardness. However, the more pronounced effect is observed if the grains are fully covered by the amorphous tissue.

It should be considered that the orientation and size of MeN grains significantly influence the mechanical properties and can also contribute to enlighten the observed hardness variation. It has been shown that films with NaCl-type structure oriented in the (111) direction show higher hardness than films oriented in the (200) one [101]. It can be explained by means of plastic deformation of materials. It is known that film hardness is determined by the resistance to plastic deformation. The higher is the resistance to plastic deformation, the harder the material. If normal load is applied to the film surface, as during indentation hardness measurement, the plastic deformation begins while the applied stress σ_a reaches its critical value. It strongly depends on the orientation of the grains, which varies with the silicon concentration. The magnitude of σ_a to reach a critical shear stress value is given by the so-called Schmid factor $m = \cos\phi \cos\lambda$ and can be obtained from $\tau = \sigma_a.m$, where σ_a is the applied stress, τ the shear stress in a slip plane, λ the angle between the directions of the applied stress and the normal to slip plane, and λ the angle between the directions of the applied stress and the slip [17]. The slipping system of fcc or hcp materials is well defined [102,103], and it has been shown that for selected orientations parallel to the surface a higher stress must be applied to reach the critical shear stress and, consequently, a higher hardness value is obtained. This might explain higher hardness of randomly oriented Me-Si-N films compared to the ones with preferentially orientated MeN grains in the (200) direction [26,48].

In general, a rise of hardness is mainly related to the morphological film evolution due to the Si addition. Thus, the maximum hardness is observed at vanishing of preferred orientation and development of equiaxed grains [26] or at the most prominent columnar growth [34] while small amount of Si is incorporated. On the other hand, a change in phase composition, where a formation of harder hexagonal phase is prevented by addition of a small Si content and cubic phase is developed instead, leads to a slight decrease of the film hardness [30]. So far as the Si_3N_4 phase is formed in the Me-Si-N films at a low Si content, its amount is commonly not high enough for hindering the deformation through dislocation movement between large grains. The hardness is therefore just slightly enhanced over the bulk value of the main MeN phase and the film properties are just moderately changed. However, completely different situation occurs if higher amount of the Si_3N_4 phase is developed at higher Si content (see Sections 5.8 and 6.7). In that case, the change in the film properties is related to the development of the nanocomposite structure. Another factor influencing the film hardness is the amount of defects and voids in the boundaries of the columnar structure. If the boundaries are weak owing to high concentration of defects, cracks can more easily propagate along these weak boundaries under the applied load [32].

Together with the changes of structural parameters, the stress also changes considerably within the zone A. In fact, the residual compressive stress follows a similar trend to hardness. Its increasing tendency with increasing silicon concentration seems to be induced by increasing lattice distortion and peening effects as higher amount of incorporated silicon atoms are dissolved in the MeN lattice. It is known that compressive stress is strongly related to this lattice distortion and can contribute to the hardening effect. Therefore, an increase in hardness of Me-Si-N films is frequently attributed to a high compressive stress level [36]. The residual stress can also vary with the changes of the structure and film morphology. It has been shown that the crystalline films with dense structure exhibit a higher compressive stress compared to the films with columnar one [32]. Although the stress maximum corresponds to the maximum of hardness and it is considered as a main factor determining the film hardness, it has been shown that the relationship between hardness and stress is not always proportional [36]. Relatively slight change in hardness compared to a high stress variation, which can be in some cases more than 3 GPa, indicates that the enhanced hardness of Me-Si-N films observed with the addition of Si cannot be attributed just to the level of compressive stress [48]. It is believed that compressive stress participates to hardness just partially and belongs to the other factors determining the mechanical properties of Me-Si-N films. Also a decrease in the grain size cannot fully explain the appearance of the maximum in hardness in the range of silicon concentration within the zone A because the effect of the grain refinement is too small for the magnitude of the hardening observed. Therefore, it is assumed that the main reason for the film hardening within the zone A is a synergetic effect of several factors discussed above and their combination.

4.2.2 Zone B

Characteristic for the zone B is a further increase in hardness with increasing silicon concentration achieving its maximum mostly in "superhard" range ($H \geq 40$ GPa). Exceptional properties of Me-Si-N films with intermediate silicon content are given by a nanocomposite structure, which is developed if the MeN grains are covered by certain amount of an amorphous tissue, effectively hindering their growth. The microstructure consisting of nanocrystals embedded within an amorphous phase is responsible for unique mechanical properties resulting in very high hardness and improved wear resistance [30,80]. It has been reported that there is a close relationship between the hardness, grain size, and volume fraction of the amorphous Si_3N_4 phase. In general, enhanced properties of the Me-Si-N films strongly depend on both the nanograin size and composition of the amorphous matrix [90,91]. For instance, the very high hardness of Me-Si-N films up to 50 GPa with nc-MeN/a-Si_3N_4 structure requires crystallites smaller than 5 nm separated with less than 1 nm thick amorphous matrix [84]. At these conditions, both dislocation formation and incoherence stress relaxation are suppressed, providing superhardness. Superhard films can, however, also be formed consisting of the grains with size between 10 and 20 nm [38].

4.2.2.1 Hardening effect in nanocomposites

While the maximum of hardness within the zone A corresponds to the high quantity of silicon atoms admitted inside the MeN lattice [57], a coexistence of the amorphous silicon nitride phase with MeN crystallites can be interpreted as the reason for enhanced hardness of Me-Si-N films with an intermediate silicon content. In the first case, the film hardening is related to the distortion of the lattice acting as an energy barrier for dislocation movement, while in the second case the hardening effect is caused by a two-phase nanocomposite structure. It assists to the stress relaxation and through as formed amorphous phase significantly reduces grain boundary sliding under applied load. The amorphous tissue, moreover, restricts the generation of dislocations and defects in the film and also plays an important role as a barrier to dislocation propagation [63].

From the hardness–silicon content dependence can be clearly seen that the hardness maximum is located in relatively small range of silicon concentration (see Fig. 4.12b). Exceeding the required amount of silicon consequently results in an obvious decrease in hardness. It is thought that a maximum of hardness is closely related to certain optimum silicon concentration in the film corresponding to a calculated percolation threshold, that is, the point where a full interpenetration of the two phases is achieved. At this situation, the whole system exhibits a minimum free energy as the surface area of the interface between the MeN nanocrystals and amorphous tissue reaches its maximum [104]. Then the surfaces of the grains are just covered by 1 to 4 monolayers of Si_3N_4 [5,26]. Such amount of amorphous phase is apparently sufficient to further promote the hardness of the nanocomposite.

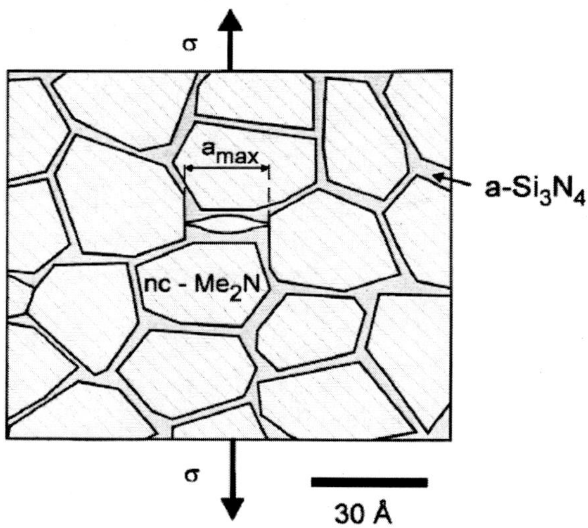

Figure 4.14 Suggested, idealized structure of a nanocrystalline/amorphous composite in which the growth of microcracks is limited due to the small thickness of the amorphous tissue. Reprinted with permission from Ref. [104]. Copyright [1969], American Vacuum Society.

A hardening effect in nanocomposite structure is also attributed to the fact that nanocrystals are free of dislocations. Although dislocations could be eventually formed under applied load, they cannot move through the amorphous phase located at the grain boundaries even if it is very thin. Thus, enhanced hardness of nanocomposite coatings is given mainly by the difficulty of crack propagation, as the crack size is limited to the grain size [59] (Fig. 4.14). The investigations of deformed TiN-based nanocomposite films have shown that the films deform in a quasi-plastic manner via boundary cracking [105]. The main prerequisite for precluding the dislocation movement and crack propagation has been found to be the volume fraction of the second phase. If all grains are fully percolated, deformation is obviously limited [8]. Response of the nanocomposite structure to the applied load, however, also strongly depends on the state of the amorphous phase. The existence of unbound silicon within the amorphous phase is accompanied with reduced film hardness. It is caused by poor properties of silicon-rich tissue situated at the grain boundaries. Incomplete nitridation of silicon owing to insufficient amount of nitrogen in the film leads to the formation of amorphous phase, which enables to propagate cracks more easily, in comparison to fully nitrided one [38].

4.2.2.2 Hardness dependence on the grain size

Reported results reveal that the enhanced hardness of Me-Si-N films coincides obviously with reduced grain size. The addition of silicon into MeN films has been shown to refine the grains in consequence of the restraint role of amorphous phase on the grain growth. It implies a high density of grain boundaries hindering the dislocation multiplication and movement. Also the grain boundary sliding and the incoherence stress relaxation are suppressed at the decreased grain size.

Significant hardness dependence of Me-Si-N films on the size of the MeN grains is characteristic for this type of nanocomposite material. Typically, hardness distinctively increases with increasing silicon content and reaches maximum at the point where the grain size becomes values around 5–10 nm [56,58,100]. This behavior corresponds well with the Hall–Petch relation where hardness is inversely proportional to the size of the grains [97]. Although reported maximum hardness varies between 35 and 105 GPa [5,106], it is commonly around 40–50 GPa [26]. In fact, it has also been found almost no dependence of the hardness on the grain size [56]. It might be due to the fact that this grain size obtained from XRD analysis represents the size of the grains just in a direction perpendicular to the film surface. The hardness is, however, affected also by the in-plane size of the grains, which can strongly differ [56]. It should be noted that elastic modulus follows the same trend as the hardness within the zone B revealing strong correlation.

Besides the grain size effect, the correlation of residual stress and hardness has also been observed. The compressive stress exhibits a similar dependence on the silicon content achieving its maximum in the range of silicon concentration corresponding well to that related to maximum of hardness, compare Figs. 4.12b and 4.13a. The highest compressive stress of up to 10 GPa [50] could be interpreted in terms of lattice distortion [57], however, it cannot explain the variation in hardness completely. It has been shown that the hardness of Me-Si-N films reveals increasing tendency with increasing silicon concentration, although the compressive stress is already decreased [32]. Thus, even if residual stress does not exhibit its maximum, the highest hardness can be provided by the development in the structure and morphology [57]. It also suggests that compressive internal stress in the Me-Si-N films within the zone 'B' contributes to the film hardness just partially and belongs to the other essential factors determining the film properties.

It has been shown that the region of highest values of the hardness of Me-Si-N films is closely related to the formation of a nanocomposite structure with optimum volume of fully nitrided amorphous phase situated at the grain boundaries. Although the nanocomposite structure is still preserved in Me-Si-N films at higher silicon content, this fact cannot justify the high hardness anymore. Just relatively narrow range of silicon content, at which certain optimum volume fraction of amorphous tissue is formed, provides very high hardness attended by relatively high compressive stress level and very small size of MeN grains. Once this optimum volume of amorphous phase is exceeded with a further increase in silicon content in the film, both values of hardness and elastic modulus decrease drastically [36,51,56,63,80,107]. This steep decrease in hardness can only be interpreted by the synergetic effect of several factors: observed further decrease in the grain size and changes of its shape, the increase in the silicon nitride amorphous phase content, and the decrease in the compressive stress level. All these factors seem to play an essential role in determination of properties of Me-Si-N films at the higher end of the silicon concentration range within the zone B [48]. Similar decreasing tendency of the film hardness with addition of higher amount of alloying elements has been observed for sputtered films of other systems, such as W-C-Co [108], Cr-Ni-N, Ti-Mo-N [76], Zr-Cu-N [75], Ti-Cu-N [66], Zr Ni-N [28], Ti-Ni-N [109], Zr-Y-N [110], or W-Ni-N [111]. In that case it also suggests that enhanced mechanical properties of such nanocomposite films composed of two different phases are provided just at their specific volume ratio.

In Me-Si-N films with intermediate and high silicon content, the nanocrystalline structure reveals dispersion of grain size between 2 and 4 nm. As the grain size decreases with increasing silicon concentration, the shape of crystal grain also significantly changes from long columns to spherical grains [55]. This fact results from significant increase in the volume fraction of amorphous phase, which no longer allows the grains to form long crystallites as it is at lower silicon content. The effect of grain size below 5 nm on the hardness of Me-Si-N films has been already reported [8]. The so-called anti-Hall–Petch effect, where the resistance to plastic deformation is proportional to the size of grains, results in low hardness if the grains reach that size [98]. In materials with relatively large grains, a dislocation multiplication process can dominate and lead to material hardening following the normal Hall–Petch relation. The very fine-grained structure, however, results in dominant dislocation annihilation process at grain boundaries, which gives rise to softening of the material and breakdown of the Hall–Petch relation. Moreover, the highest hardness has been shown to be attributed to the maximum volume crystallite fraction in nanocomposite Me-Si-N films [8]. The films with very small grains and relatively high volume fraction of amorphous tissue exhibit weak grain boundaries and therefore deform relatively easily by grain boundary sliding under applied load. Thus, the grains effectively hindered in their growth by high amount of atoms situated at grain boundaries no longer provide enhanced hardness. The effect of a hardness decrease with increasing the volume fraction of amorphous phase to the detriment of the crystalline one is typically reported in Me-Si-N films [16,59]. However, the hardness reduction is not related just to the thickening of the Si_3N_4 phase. Independently of the amount of amorphous phase, it also strongly depends on its state. As it was already discussed above, the weakening of amorphous matrix occurs if free silicon exists simultaneously with Si_3N_4 in the nanocomposite. If this phenomenon is more pronounced, the higher amount of the Si_3N_4 phase and free Si atoms is involved in the film. Therefore, insufficient nitrification of incorporated silicon also has to be considered responsible for a decrease of film hardness. Although the thickening of amorphous phase affects the hardness, in the means of its reduction, much more than the effect of free Si, it is of high importance to ensure sufficiently high amount of dissociated nitrogen for full nitrification of both phases. The amorphous matrix becomes harder and stable just if it is completely nitrided which avoids deformation through weak grain boundaries and together with coexisted MeN grains provides enhanced properties. It will be shown in the next section how this effect can influence the thermal properties of films and their resistance against oxidation at elevated temperatures.

A segregation of silicon atoms to grain boundaries induces significant decrease of the compressive stress in the film while certain high volume of an amorphous matrix is formed [48]. It is quite linear with the silicon concentration as can be seen from Fig. 4.13a. This stress reduction corresponds well with the hardness decrease, and therefore, it is believed that it also partially contributes to the determination of film hardness.

4.2.3 Zone C

The zone C follows a decreasing trend in hardness of the zone B from its higher end of the silicon concentration and is characterized by a relatively low film hardness appearing once the silicon content is raised above 10 at.% (see Fig. 4.12). A possible reason of the loss of enhanced mechanical properties of the nanocomposite Me-Si-N films might be explained by reaching the percolation threshold of the amorphous Si_3N_4 phase in the films as it is mentioned above. Once the volume fraction of the amorphous phase separating the MeN grains exceeds a certain value, the hardness of the nanocomposite is no longer promoted by an interaction of both phases and the unique properties observed in zone B are suddenly lost. All these findings show that the enhanced hardness in films with nanocomposite structure of the c-MeN/a-Si_3N_4 type is strongly related to the optimum value of the crystallite to amorphous phase volume ratio [63,80]. Thus, so far as the size of the grains is reduced further and an amorphous phase increases in its volume, finally containing main portion in the structure, the hardness markedly decreases. In the films with a high Si content, a deformation mechanism of the dominating amorphous phase alone determines the final mechanical properties of the film since the nanocomposite structure no longer governs the enhanced film hardness within the compositional range of the zone C [30,54]. Therefore, the hardness of the films with an amorphous-like structure composed of a high amount of the amorphous Si_3N_4 phase exhibits hardness of approximately 18 GPa, which is a value close to the commonly reported hardness of silicon nitride [112]. Besides hardness, other properties are also related to the properties of the dominating amorphous phase. The formation of an amorphous-like structure leads to a reduction of the film macrostress. The compressive stress of nearly amorphous films is very low and almost constant (1 GPa) [48]. The film structure of Me-Si-N films consists of a mixture of two phases, where the silicon nitride phase markedly dominates, and additionally provides extremely high thermal stability and resistance against oxidation [113]. More details are given in the following section and in Chapters 5 and 6.

4.2.3.1 Summary

In many papers, the greatly improved mechanical properties of Me-Si-N films have been simply explained by the formation of a nanocomposite structure or simple solid solution hardening mechanism [55]. From recent investigations it can, however, be assumed that there does not exist one main parameter, which determines enhanced properties of Me-Si-N films. The experimental results suggest that the film hardness and its elasticity are attributed to the synergetic effect of several factors. Thus, just a specific combination of the level of the compressive stress, the fine-grained microstructure with randomly oriented nanocrystalline grains embedded within the amorphous Si_3N_4 phase at certain crystalline-to-amorphous phase volume fraction ratio together with lattice distortion tends to enhanced mechanical properties of Me-Si-N films.

4.3 Oxidation Resistance

Presently, the main driving force in finding the most promising candidates for industrial applications is based on investigation of thin films with enhanced mechanical properties, particularly with enhanced hardness and elasticity. However, besides the right selection of the H/E ratio [114], thermal stability of films and their resistance against high-temperature oxidation in air is also an important criterion for technical application on machining and shaping tools. The higher the temperature which the film can sustain, the longer is the lifetime of the coated tools at high machining speeds. Moreover, coated systems are in many applications exposed to stress and corrosive media or other hostile environments at elevated temperature. Under such conditions significant changes may occur at the coating surface, which then consequently affect the performance of the system as a whole. The widespread application of transition metal nitrides in wear-protective coatings, diffusion barriers, or electrical contacts causes growing interest in their improved diffusion and thermal stability and resistance against oxidation by addition of various alloying elements. Better performance of coated working tools at extreme conditions can be, however, attained just at right material combination.

4.3.1 Material Selection

It is known that many of transition MeN films decompose or oxidize already at relatively low temperatures [115]. However, the usage of films, for instance in cutting applications where coated tools are exposed at high speeds to high mechanical loads, requires their very stable structure and mechanical properties even at elevated temperatures. Stabilization of the film structure by addition of alloying elements has been already reported [116–120]. The most promising way how to enhance thermal stability of hard films proved to be a formation of a nanocomposite structure. Although nanocomposite structure is also formed if an amorphous matrix is composed of soft metals [28,29,75,76,109,110], high thermal stability and resistance against oxidation is given mainly by a stability of all the film components. It is partially limited in this type of nanocomposites. By an incorporation of Si atoms into the film, much more stable amorphous Si_3N_4 phase is easily formed, as Si reacts with atomic nitrogen from a plasma discharge. Such an amorphous tissue covering MeN grains acts as an effective barrier for thermal activated grain growth and diffusion of oxygen through the film. In this way, thermal stability even of easily decomposing and oxidizing WN or W_2N films has been significantly enhanced [121]. Considerably improved resistance against oxidation of transition metal nitrides by a factor of 6–10 has been already reported at Si content slightly exceeding 2 at % [36,54,122].

4.3.2 Oxidation Mechanisms

Due to higher oxidation resistance of pure TiN compared to ZrN, CrN or WN [123,124], most investigations have been focused mainly on Ti-Si-N nanocomposite films [36,54,125–127]. Significantly enhanced thermal stability together with specific mechanical properties resulting from incorporation of silicon also makes, however, other Me-Si-N films very perspective in many special applications. Although just limited information have been reported in literature on thermal properties of Me-Si-N films especially with their intermediate or high silicon content, the detailed experiments clearly revealed several different oxidation mechanisms involved depending on the oxidation temperature and phase composition of the film, that is, particularly on the amount of silicon and nitrogen incorporated into the film.

4.3.2.1 Low Si content

Although transition metal nitride films are thermodynamically stable, they relatively readily oxidize already at moderately high temperatures. This fact causes mechanical troubles in tool applications since the film properties are completely modified when the films are exposed to ambient air at elevated temperatures. Investigations of the oxidation properties and the film stability of various nitride coatings show that oxidation, in principle, proceeds in the same way, however, the differences result from the oxidation products which are finally formed. In general, during the oxidation process, oxygen substitutes nitrogen which leaves the solids in gaseous state. Subsequently formed oxide phase commonly exhibits various properties depending on its elemental composition, structure, and location in the film which reveals the main difference already before the film completely oxidizes.

Typical oxidation mechanism presents the formation of an oxidation reaction product on the film surface [119,128,129]. It grows as higher oxygen amount is incorporated into the film via its diffusion through the oxide layer. Although the resistance of the MeN film against oxidation should be relatively high since it might be protected by the product of oxidation located on the film surface, it limits the oxygen diffusion just moderately. It is due to the polycrystalline nature of MeO_x oxides providing rapid transport of oxygen to the film through the voids and defects in the structure. On the other hand, the resistance to oxidation of MoN or WN films is very low due to their tendency to decompose during oxidation into less stable components which oxidizes very intensively [130,131]. Moreover, the formed oxides exhibit a nonprotective effect against diffusion and strong inclination to volatilize. The volatilization reaction is subsequently followed by a decrease of the oxide layer thickness and a formation of fissures and cracks on the surface [131].

In order to improve the oxidation resistance of transition nitride films, besides alloying elements, such as V, Al, Y, or Hf, Si has also been added to the investigated MeN film systems. It has been shown that the silicon addition into MeN-based films reduces the oxidation rate and serves as significant means for increased oxidation resistance already at its relatively low content [122]. Crystallographic nature of the MeO_x oxide layer located on the top of the air-annealed sample is obviously influenced by the silicon mainly through the Si–N bonds formed in the MeN lattice. The silicon atoms dissolved in the film stabilize the formed oxide phase and, in this way, partially inhibit further oxidation. Thus, the onset of oxidation is distinctively enhanced. In principle, the stabilization of MeO_x phases has been related to the nanometric crystallites size in the oxidized layer. It is caused by the presence of Si in the films inhibiting the coalescence of grains during the oxidation process [122]. To that effect might, in addition, also contribute a presence of amorphous silicon nitride. It may be formed at the grain boundaries already at very low Si concentration in the film [30]. Its amount is, however, not sufficiently high to cover MeN grains completely and the film oxidation is just slowed down [57]. An enhanced oxidation resistance has been observed only when the compact Si_3N_4 phase completely surrounding the grains is formed, acting as a barrier against oxygen diffusion. Therefore, the major effect on the improvement of the oxidation resistance of MeN films by a small amount of silicon has been found to be provided through the stabilized product of oxidation together with homogeneous and fine structure composed of the mixture of randomly oriented grains. Such a structure is indeed developed in the MeN films at a low Si content and is considered to suppress the oxidation reaction due to the inhibited transport of oxygen within the film. Although the oxidation resistance of MeN films has been enhanced by addition of small Si amount, the onset of oxidation is improved just with a several tens of degrees in that low concentration range. By exceeding the critical temperature limit, the grains even in stabilized oxide layer started to grow and form a massive product of oxidation to the detriment of the remaining film. A polycrystalline nature of the oxide phase, moreover, provides enhanced oxygen diffusion along the voids formed during the growth of the oxide layer which subsequently results in an oxidation mechanism typical for pure MeN films. An onset temperature of oxidation, however, shifts obviously towards higher temperatures with increasing silicon concentration, as it will be shown in next sections.

4.3.2.2 Intermediate Si content

Me-Si-N films with intermediate Si content behave in an entirely different manner due to the complete changes in their morphology and phase composition as increasing amount of silicon is incorporated into the film. The amorphous silicon nitride phase formed in the film covers and separates crystalline MeN grains completely and acts as an efficient barrier against oxygen diffusion. In this way, it protects MeN grains from oxidation [16]. Silicon nitride is, for the same reason, used as a stable passive layer against oxidation in microelectronic devices [132].

The experimental investigations of three sets of Me-Si-N films with the low, intermediate, and high content of the silicon nitride phase revealed two different oxidation mechanisms proving an essential role of the amount of the amorphous Si_3N_4 phase on the film resistance against oxidation. If the MeN grains are covered by the amorphous silicon nitride tissue of just a few monolayers, the protection of the film proved to be partially limited. Although a small amount of amorphous tissue ensures enhanced mechanical properties due to hindered dislocation movement, it allows oxygen to diffuse from the oxidation atmosphere into the film through the grain boundaries where it reacts with silicon nitride. The amorphous SiO_2 oxidized from Si_3N_4 is known for its antidiffusion properties and plays a major role as the oxidation barrier [80,133]. Through so formed passivating silicon oxide layer oxygen migration is prevented, which results in reduced oxidation of the MeN phase and a subsequent growth of the MeO_x phase [57]. Although in this way, MeN grains are protected against oxidation up to certain critical temperature limit, the formed thin silicon oxide layer can no longer provide enhanced mechanical properties of the oxidized films. Under the applied load dislocations may move through the weakened grain boundaries and, thus, present silicon oxide layer results in a significant degradation of the mechanical properties.

An addition of higher amount of silicon into MeN-based films induces formation of higher volume of the amorphous Si_3N_4 phase which in consequence obviously influences the oxidation behavior of the film. Investigations of the kinetics of the oxide layer growth in Me-Si-N films with a silicon concentration above 8 at.% reveal lower oxidation rates by approximately an order of magnitude compared to pure MeN films and the onset of oxidation at higher temperatures due to the highly protective effect of the silicon nitride tissue since it is relatively thick [16,122]. Its oxidation results in an increase in SiO_x amount which is still located at the grain boundaries; however, through its higher volume the MeN grains are protected more effectively. The main reason why certain alloying elements are added into MeN films in order to enhance their oxidation resistance is due to their ability to form their own oxides on the surface in preference to those produced typically in pure transition metal nitrides. In the case of Me-Si-N films, relatively high amount of silicon has to be incorporated into the film to form the silicon nitride tissue providing a protective effect. This is because only at sufficiently high amount of the amorphous Si_3N_4 phase oxygen cannot diffuse through the grain boundaries deeply into the film. Moreover, a top oxide layer is formed on the film surface, acting as a barrier against further oxidation. Thus, from the surface toward the substrate, three zones can be identified with increasing depth: the oxide upper layer separated by a sharp interface from the Me-Si-N film and the substrate. An increasing thickness of the silicon oxide layer on the film surface despite remaining film has been proved by XRD analysis where the peak intensity of crystalline MeN phase significantly decreases with elongated oxidation time and/or oxidation temperature [125].

The oxidation of films, however, strongly depends on thermal stability of all individual phases and for this reason the films are in many cases protected just at specific temperature range. If the MeN phase decomposes at elevated temperature, free transition metal atoms are much more susceptible to oxidation due to their high affinity to oxygen. In that case, oxygen ions diffusing from the oxidation atmosphere induce the oxidation of both the crystalline MeN and the amorphous Si_3N_4 phases leading to rapid loss of mechanical properties. Thus, the oxidation occurs by means of decomposed Me atoms diffusing to the interface between the two phases where MeO_x oxides are consequently formed [16]. This process proceeds more likely compared to the oxygen diffusion through the passivating silicon nitride or silicon oxide layer followed by a reaction with Me atoms. It also suggests lower diffusion energies for most of Me atoms [54]. The diffusion-controlled mechanism seems to be dominant not only in the films where the silicon nitride tissue is very thin but mainly where it is fully missing, as in the case of pure MeN films [16]. Besides above-mentioned oxidation mechanism, a formation of thin oxide layer of ternary Me-Si-O has been observed at the interface between nitride and oxide layers in the films with a high crystalline to amorphous phase fraction volume ratio at higher oxidation temperatures. This ternary phase subsequently segregates into immiscible two phases of MeO_x and SiO_x, once the oxidation temperature increases further [125].

Besides the stability of the individual phases of which the film is composed, the oxidation resistance of the film also depends on the stability of the formed oxides. Especially at very high temperatures, by exceeding the critical limit corresponding to the specific amount of the silicon nitride phase in the film, the passivating silicon oxide layer can no longer retain its protective barrier effect. A subsequent decomposition of the oxides on the film surface above a threshold temperature is commonly followed by a dramatic acceleration of the oxidation process and the film becomes readily fully oxidized. In that case, an oxidation of the MeN phase or decomposed Me atoms cannot any longer be hindered further by the SiO_x layer, whose structure is usually completely disturbed. It has been shown that a diffusion of Me atoms through amorphous silicon nitride or formed silicon oxide layer is very fast once the diffusion barrier is no longer effective [16]. The loss of enhanced film properties and additionally the oxidation of the substrate alone indicate the limits of the resistance of films against air oxidation above the critical threshold temperature. Therefore, a special attention is paid to the stated activation energies and oxidation rates apparently characterizing which oxidation mechanism occurs in Me-Si-N films in the investigated temperature range and also how effectively the Si_3N_4 phase acts as a diffusion barrier reducing the oxidation of individual phases.

4.3.2.3 Oxidation kinetics

The oxidation experiments performed on the silicon nitride-rich Me-Si-N films indicate that the oxidation mechanism involves two steps depending on the annealing temperature. Below specific temperature, which is characteristic for given individual material and has been reported, for instance, as approximately 850°C for Ti-Si-N films with 14 at.% silicon content [16], the oxidation proceeds very slowly. Exceeding the critical temperature limit characterizing the stability of individual phases formed in the film, the oxidation significantly accelerates achieving relatively high rates. An initial stage of oxidation can be expressed by the linear law where the growth of oxide layer is proportional to time and the oxidation rate is given by the slope of the time–layer thickness curve. The linear law of oxidation describing the rate growth of the oxide layer is an empirical relationship that is typically associated to the formation and build-up of a nonprotective oxide layers or to the oxidation of the surfaces before the oxide layer is thick enough to operate as a protective diffusion barrier. Thus, the diffusion of oxygen is not limited and the surface oxidizes relatively easily. Once the oxide layer is formed on the surface, a film oxidation is distinctively slowed down since oxygen diffusion, from the oxidation atmosphere to the interface with the film surface, is significantly reduced. The formed oxide layer thickens with time according to a parabolic law where its growth on a sample is proportional to the square root of time. Lower oxidation rate is commonly observed in the thin films characteristic by a formation of a stable surface oxide layer acting as a passive barrier against oxygen diffusion (Fig. 4.15).

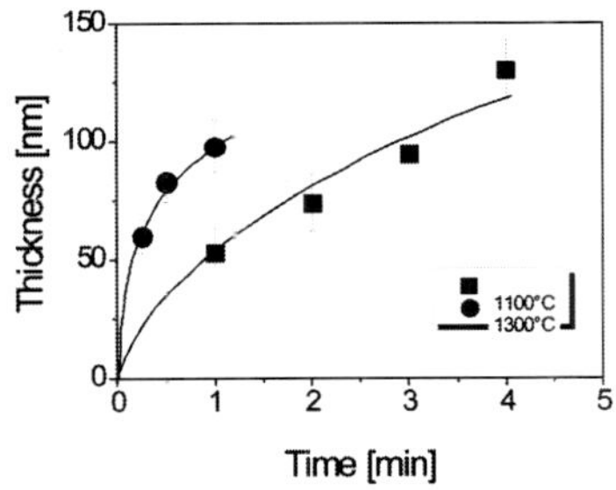

Figure 4.15 Oxidation rates plotted as a function of the thickness of oxidized layer and oxidation time corresponding to various oxidation mechanisms expressed by linear, parabolic, and logarithmic laws.

The parabolic growth of the passive oxide layer is also widely obeyed at higher oxidation temperatures if the layer still proves the protective antidiffusion effect. It, however, depends on the phase composition of annealed Me-Si-N thin films, oxidation temperature, and time at which the film is exposed to the oxidation atmosphere. It has been observed that highly essential for the resistance of the films against oxidation is the amount of the stable silicon nitride tissue, which is consequently transformed to the oxide phase. All the effects determine whether both oxygen and metal ion diffusion are involved in the oxide growth and thus the final oxidation rate. As long as the MeN phase decomposes at higher temperatures, the diffusion of Me atoms increases that subsequently accounts for an increase in the oxidation rate. Since the growth of an oxide layer is thermally activated process, higher temperatures enhance the oxygen diffusion and increase the rate at which the oxide layer grow. Oxidation at constant temperature gives into a rise a similar effect if the film is exposed to the annealing temperature at prolonged time. Thus, the oxidation rate is given by the annealing temperature while an oxide layer thickens following the parabolic law.

As the metal atoms of the film are consumed during oxidation by reaction with oxygen, the interface between oxide layer and nonoxidized film surface shifts towards the substrate. Once the whole remaining film completely oxidizes, a weak interface between the oxide layer and substrate may cause subsequent film delamination especially if the system is exposed to the mechanical loads. The delamination is even promoted by the large residual stress induced in the oxide layer during its growth [134]. Moreover, by exceeding the critical oxidation temperature a subsequent decomposition of the passivating oxide layer results in a formation of the porous structure with cracks. Since the film-substrate system can be protected against oxidation just if the formed passive oxide layer is sufficiently stable and dense, the rapid diffusion of the oxygen throughout the damaged oxide layer causes in both cases an acceleration of the oxidation of the remaining film or the substrate. Enhanced thermal movement of oxygen ions through extended diffusion paths in disturbed oxidized layer leads to the breakdown of the parabolic law. The oxidation rate follows further the linear law as a consequence of failed diffusion barrier effect of the passivating oxide layer. An XPS depth profile of the oxide layer formed on the surface of Me-Si-N coatings reveals mainly the presence of SiO_x and MeO_x phases. Compared to enhanced resistance against oxidation of Ti-Al-N films associated with a strong migration of aluminum atoms to the surface while the passive layer is formed, no silicon enrichment in the oxide layer has been observed in Me-Si-N films. This marked difference in the oxidation behavior reveals an essential role of elemental and phase composition on the thermal and oxidation stability of the films.

In general, a protection of the substrate against corrosive environment requires the film to exhibit relatively good mechanical properties even if it is oxidized. Just in that case the coated tool can effectively operate in the corrosive atmosphere at elevated temperatures. A strong dependence of the oxidation resistance and the critical limit for film decomposition on the elemental and phase composition of the film is evident from all the reported experimental investigations. It principally determines the efficiency of the substrate protection.

4.3.2.4 High Si content

A significant improvement of the oxidation resistance and Si concentration-dependent oxidation rate results particularly from the nanocomposite nature of the Me-Si-N films. Films composed of a massive silicon nitride layer, which separates the crystalline MeN grains, showed the slowest oxidation due to the so formed efficient barrier against oxygen diffusion. Therefore, the most promising materials for high-temperature applications have been found within Me-Si-N films with a Si content higher than 25 at.%. A detailed investigation of enhanced thermal stability and oxidation resistance performed at the Zr-Si-N and W-Si-N films up to 1000°C, however, revealed an importance of very stable phases of which the film is composed [94,113]. Moreover, it has been shown that at the high Si content in the film, an amount of nitrogen dissolved in the films plays an essential role on the resistance against oxidation since just fully nitrified Si_3N_4 phase can sustain the oxygen attack at very high temperatures due to its extreme thermal stability. The oxidation behavior can be, in this case, characterized by the logarithmic rate law associated mainly to highly protective films. The effect of a high amount of the stable silicon nitride phase and un-nitrified Si on the thermal stability and oxidation behavior of the Me-Si-N films will be discussed in more detail in Sections 5.11, 5.12, and 6.9.

4.3.3 Roughness

Mechanical properties and particularly the wear resistance of Me-Si-N films are attributed to the optimum value of the hardness to the Young's modulus ratio [135]. The potential of the films in tribological applications is, however, also conditioned by their good performance at elevated temperatures. For these reasons, ternary Me-Si-N film systems have been actively explored to overcome the problems related to a degradation of mechanical properties during annealing. In practical applications, the tools are, moreover, often exposed to high temperatures in various environments where the film surface readily oxidizes. A formation of oxide phases results typically in a strong roughening of the film surface which may greatly affect the performance of the coated tools. A significant increase in surface roughness is related to the formation of facetted well-crystallized grains of an oxide phase namely at higher temperatures [16] (see Fig. 4.16).

Finely structured surfaces of Me-Si-N films provide better tribological properties compared to rough ones [64]. Therefore, there is a huge interest to increase the thermal stability and oxidation resistance of the films to ensure that the film roughness remains low even if a thin passive oxide layer is formed on the film surface. It improves tribological performance of the films in wide temperature range that is very important for machine tool applications. The most promising candidates for high-temperature applications have been found in the Me-Si-N films with a high silicon concentration which are thermally stable and resistant against oxidation up to 1000°C [37,113].

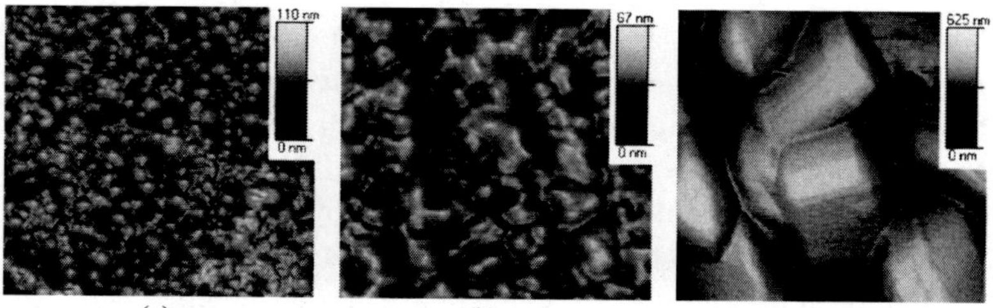

(a) (b) (c)

Figure 4.16 AFM images of the surface of the Ti-Si-N film with 14 at.% of Si oxidized in air at different temperatures. The mean roughness R_a develops from (a) 7 nm at RT through (b) 5 nm at 770°C to (c) 100 nm at 940°C as a consequence of a formation of the TiO_x phase. Reprinted from Ref. [16], Copyright (1999), with permission from Elsevier.

4.3.4 Dependence on the Structure, Morphology, and Amount of Defects

Successful strategy for finding perspective material candidates for working tools also has to take into account a strong dependency of a high temperature stability and resistance against oxidation on the microstructure and morphology of the films. MeN films exhibit mainly a well-developed columnar microstructure characterized by voids between columns, in specific cases connecting directly the surface of the film with the substrate. Through so formed diffusion paths, an oxygen migration proceeds at high rates resulting in a rapid film oxidation. In general, all the defects among the columns and grains have been reported as a reason of promoted diffusion of oxygen through the film and subsequent fast film oxidation [57]. The essential role of the film structure is represented by activation energy for the oxidation process. It varies in relatively wide range just due to the differences in film morphology, defect density, or stoichiometry [123,133]. In principle, a significant improvement of the oxidation resistance has been observed if a dense, homogeneous, and fine-grained microstructure is formed. The reason has been found in a pronounced suppression of the oxidation reactions since the oxygen diffusion is inhibited through the strengthened grain voids-free boundaries. A presence of massive grain boundaries, resulting from the nanometer-scaled microstructure, thus ensures not only the enhanced hardness but also the enhanced oxidation stability of films [16]. As a highly effective means for the structure refinement and densification proved to be an addition of silicon as an alloying element. Silicon atoms added into MeN-based films form with nitrogen stable amorphous silicon nitride tissue located around the MeN grains and thereby effectively inhibits their growth and simultaneously the atomic movement within the film. However, not only the volume and density of the silicon nitride tissue but also the grain size and compactness of MeN phase significantly affects the oxidation resistance [136]. An effect of the film structure on the oxidation behavior has been demonstrated for Ti-Si-N films with various Si content revealing the lowest oxidation resistance in the film with the columnar microstructure [137]. Besides lower oxidation rates, the films with compact defects-free structure exhibit the onset of oxidation at higher temperatures compared to the films with a lower content of the alloying element [122].

The main reason for the improvement of the oxidation resistance of Me-Si-N films has been found in a formation of a stable oxide layer on the film surface. However, also the products of a high temperature oxidation have to be defects-free. Polycrystalline oxides cannot act as a passive barrier against oxygen diffusion if they contain a high amount of point and line defects and voids at the grain boundaries which provide paths for easy oxygen ion movement. Enhanced oxygen diffusion is directly responsible for an increase in the oxidation rate, especially if the structure is additionally disturbed. Thus, the high-temperature oxidation resistance can be obtained just if a compact, stable, passive oxide layer is formed on the film surface through a massive silicon nitride phase. Moreover, it has been observed that the onset temperature of the oxidation is proportional to the amount of the silicon nitride phase. If the oxide layer is thin, the higher volume of the amorphous Si_3N_4 phase is formed in the film [94,113].

Experimental results clearly show that one of the possible ways to improve the resistance against oxidation seems to be preparation of dense fine-grained films with defects and voids-free microstructure consisting of thermally stable phases. An addition of a sufficient amount of certain alloying elements, such as silicon or aluminum, has been found to be extremely effective for this purpose. The alloying elements stabilize the film microstructure and contribute to the formation of thin amorphous oxide layer on the film surface, which serves as an efficient barrier for the penetration of oxygen into the film. Thereby, the high-temperature oxidation resistance of the films is finally strongly improved [113].

4.4 The Problems of the Film Reproducibility

The experimental results regarding the structure and properties of the Me-Si-N films, published in literature, demonstrate an obvious lack of film reproducibility (see Fig. 4.12). The differences in results, which can be found in published papers [45–65], give rise to questions on possible reasons resulting in various properties of films with the same elemental composition. It seems that the problem is very complex and a finding of all possible reasons is not an easy task at all. At present, it is known that even in a pure binary system the decisive factor influencing the final structure formed is a sufficiently high chemical activity of the system to provide a thermodynamic driving force for phase formation. Thus, relatively high deposition temperature as well as the ion bombardment has to be used to enhance surface or bulk diffusion of condensing adatoms forming the film. Just under such conditions the formation of nanostructured Me-Si-N films proceeds resulting in a stable structure and enhanced film hardness. That can be realized if the films are prepared in relatively intense glow discharge with an optimal pressure of nitrogen, high deposition temperature, and under intensive ion bombardment when high energy delivered into the growing film is ensured. While the mobility of atoms on the substrate surface is low, especially at low and intermediate Si content, the film morphology remains columnar and the maximum hardness appears at a different total content of silicon. It is thought that it is closely related to the different crystallite sizes and shapes and different crystallite-to-amorphous phase volume fraction ratio. A strong dependence of the structure and mechanical properties on the phase composition of the Me-Si-N films related to the deposition conditions has been reported [61,106,138]. For instance, the hardness maximum within the Ti-Si-N system has been typically found at higher total Si content if $TiSi_2$ phase is present in the film, in addition to Si_3N_4.

Besides the various deposition conditions, the poor reproducibility found in many papers might, in addition, also correspond to the presence of impurities. Recently, a huge discussion has been started on this topic. Although the role of the deposition conditions on the properties of the films is quite well known, the effect of the impurity content, mainly of oxygen, is not yet explained in a satisfactory manner. It seems that indeed the amount of oxygen in the film might likely be the reason for the low hardness and differences in the development of the forming phases. Mainly the hardness might be highly affected by Me–O or Si–O bonds formed in the film. Taking into account that the maximum hardness of the films is achieved at approximately one monolayer of Si_3N_4 acting as a gluing tissue between the MeN nanocrystals, it is understandable that the presence of oxygen with larger size compared to nitrogen atoms within this interfacial layer can have such a strong effect [139]. Moreover, the oxygen content is found to be relatively high in films prepared at low deposition rates, where it may achieve several at.%. It suggests an importance of a low residual gas pressure in the deposition system through extensive and prolonged outgassing with the substrate at the deposition temperature which enables to decrease the oxygen content below 1 at.% and minimize its effect on the structure and properties of the films. However, also in order to understand fully a role of other impurities, more detailed study is absolutely necessary. Without a complex knowledge of the effect of minor impurities on the growth and composition of the films, stable Me-Si-N films with reproducible properties cannot be prepared.

References

1. M. Ürgen, A.F. Cakir, O.L. Eryilmaz, C. Mitterer, Corrosion of zirconium boride and zirconium boron nitride coated steels, *Surf. Coat. Technol.* 71 (1995) 60–66.
2. P. Hammer, A. Steiner, R. Villa, M. Baker, P.N. Gibson, J. Haupt, W. Gissler, Titanium boron nitride coatings of very high hardness, *Surf. Coat. Technol.* 68–69 (1994) 194–198.
3. W. Gissler, Structure and properties of Ti-B-N coatings, *Surf. Coat. Technol.* 68–69 (1994) 556–563.

4. H. Holzschuh, Deposition of Ti-B-N (single and multilayer) and Zr-B-N coatings by chemical vapor deposition techniques on cutting tools, *Thin Solid Films* 469–470 (2004) 92–98.
5. S. Vepřek, P. Nesládek, A. Niederhofer, F. Glatz, M. Jílek, M. Šíma, Recent progress in the superhard nanocrystalline composites: Towards their industrialization and understanding of the origin of the superhardness, *Surf. Coat. Technol.* 108–109 (1998) 138–147.
6. C. Mitterer, P.H. Mayrhofer, M. Beschliesser, P. Losbichler, P. Warbichler, F. Hofer, P.N. Gibson, W. Gissler, H. Hrubý, J. Musil, J. Vlček, Microstructure and properties of nanocomposite Ti-B-N and Ti-B-C coatings, *Surf. Coat. Technol.* 121 (1999) 405–411.
7. D.B. Lee, M.H. Kim, Y.C. Lee, S.C. Kwon, High temperature oxidation of TiCrN coatings deposited on a steel substrate by ion plating, *Surf. Coat. Technol.* 141 (2001) 232–239.
8. S. Zhang, D. Sun, Y. Fu, H. Du, Effect of sputtering target power on microstructure and mechanical properties of nanocomposite nc-TiN/a-SiN_x thin films, *Thin Solid Films* 447–448 (2004) 462–467.
9. V. Valvoda, R. Kužel, R. Černý, J. Musil, Structure of TiN coatings deposited at relatively high rates and low temperatures by magnetron sputtering, *Thin Solid Film* 156 (1988) 53–64.
10. J. Musil, S. Kadlec, Reactive sputtering of TiN films at large substrate to target distances, *Vacuum* 5 (1990) 435–444.
11. J.-E. Sundgren, Structure and properties of TiN coatings, *Thin Solid Films* 128 (1985) 21–44.
12. I. Petrov, L. Hultman, J.-E. Sundgren, J.E. Greene, Polycrystalline TiN films deposited by reactive bias magnetron sputtering: effects of ion bombardment on resputtering rates, film composition and microstructure, *J. Vac. Sci. Technol.* A10 (1992) 265–272.
13. D.S. Rickerby, P.J. Burnett, Correlation of process and system parameters with structure and properties of physically vapour deposited hard coatings, *Thin Solid Films* 157 (1988) 195–222.
14. J.-W. He, Ch.-D. Bai, K.-W. Xu, N.-S. Hu, Improving the anticorrosion and mechanical behaviour of PACVD TiN, *Surf. Coat. Technol.* 74–75 (1995) 387–393.
15. I. Milošev, H.H. Strehblow, B. Navinšek, Comparison of TiN, ZrN and CrN hard nitride coatings: Electrochemical and thermal oxidation, *Thin Solid Films* 303 (1997) 246–254.
16. M. Diserens, J. Patscheider, F. Lévy, Mechanical properties and oxidation resistance of nanocomposite TiN-SiNx physical-vapor-deposited thin films, *Surf. Coat. Technol.* 120–121 (1999) 158–165.

17. A. Cavaleiro, C. Louro, Nanocrystalline structure and hardness of thin films, *Vacuum* 64 (2002) 211–218.

18. H.-D. Männling, D.S. Patil, K. Moto, M. Jílek, S. Veprek, Thermal stability of superhard nanocomposite coatings consisting of immiscible nitrides, *Surf. Coat. Technol.* 146–147 (2001) 263–267.

19. J. Musil, J. Vlček, F. Regent, F. Kunc, H. Zeman, Hard nanocomposite coatings prepared by magnetron sputtering, *Key Eng. Mater.* 230–232 (2002) 613–622.

20. M.-A. Nicolet, Reactively sputtered ternary films of the type TM-Si-N and their properties (TM = early transition metal), *Vacuum* 59 (2000) 716–720.

21. M. Zhou, M. Nose, Y. Deguchi, T. Mae, K. Nogi, Influence of sputtering conditions on microstructure and mechanical properties of Zr-Si-N films prepared by radio-frequency-reactive sputtering, *J. Vac. Sci. Technol.* A21 (2003) 1791–1795.

22. J.S. Reid, E. Kolawa, C.M. Garland, M.-A. Nicolet, F. Cardone, D. Gupta, R.P. Ruiz, Amorphous (Mo, Ta, or W)-Si-N diffusion barriers for Al metallizations, *J. Appl. Phys.* 79 (1996) 1109–1115.

23. O.H. Gokce, S. Amin, N.M. Ravindra, D.J. Szostak, R.J. Paff, J.G. Fleming, C.J. Galewski, J. Shallenberger, R. Eby, Effects of annealing on X-ray-amorphous CVD W-Si-N barrier layer materials, *Thin Solid Films* 353 (1999) 149–156.

24. Y. Wang, Ch. Zhu, Z. Song, Y. Li, High temperature stability of Zr-Si-N diffusion barrier in Cu/Si contact system, *Microelectron. Eng.* 71 (2004) 69–75.

25. Ch. Lee, Y.-H. Shin, Ta-Si-N as a diffusion barrier between Cu and Si, Mat. Chem. Phys. 57 (1998) 17–22.

26. J. Patscheider, T. Zehnder, M. Diserens, Structure-performance relations in nanocomposite coatings, *Surf. Coat. Technol.* 146–147 (2001) 201–208.

27. J. Musil, P. Zeman, H. Hrubý, P.H. Mayrhofer, ZrN/Cu nanocomposite film—A novel superhard material, *Surf. Coat. Technol.* 120–121 (1999) 179–183.

28. J. Musil, P. Karvánková, J. Kasl, Hard and superhard Zr-Ni-N nanocomposite films, *Surf. Coat. Technol.* 139 (2001) 101–109.

29. J. Musil, I. Leipner, M. Kolega, Nanocrystalline and nanocomposite CrCu and CrCu-N films prepared by magnetron sputtering, *Surf. Coat. Technol.* 115(1) (1999) 32–37.

30. M. Benkahoul, C.S. Sandu, N. Tabet, M. Parlinska-Wojtan, A. Karimi, F. Lévy, Effect of Si incorporation on the properties of niobium nitride films deposited by DC reactive magnetron sputtering, *Surf. Coat. Technol* 188–189 (2004) 435–439.

31. T. Fu, Z.F. Zhou, K.Y. Li, Y.G. Shen, Characterization of sputter deposited W-Si-N coatings based on α-W structure, *Mater. Lett.* 59 (2005) 618–623.

32. T. Fu, Z.F. Zhou, K.Y. Li, Y.G. Shen, Structure, stress and hardness of sputter deposited nanocomposite W-Si-N coatings, *Surf. Coat. Technol.* 200 (2005) 2525–2530.

33. P. Hones, N. Martin, M. Regula, F. Lévy, Structural and mechanical properties of chromium nitride, molybdenum nitride and tungsten nitride thin films, *J. Phys. D: Appl. Phys.* 36 (2003) 1–7.

34. E. Martinez, R. Sanjinés, O. Banakh, F. Lévy, Electrical, optical and mechanical properties of sputtered CrNy and $Cr_{1-x}Si_xN1.02$ thin films, *Thin Solid Films* 447–448 (2004) 332–336.

35. V.P. Anitha, S. Major, D. Chandrashekharam, M. Bhatnagar, Deposition of molybdenum nitride thin films by r.f. reactive magnetron sputtering, *Surf. Coat. Technol.* 79 (1996) 50–54.

36. A. Winkelmann, J.M. Cairney, M.J. Hoffman, P.J Martin, A. Bendavid, Zr-Si-N films fabricated using hybrid cathodic arc and chemical vapour deposition: Structure vs. properties, *Surf. Coat. Technol.* 200 (2005) 4213–4219.

37. J. Musil, R. Daniel, O. Takai, Structure and properties of magnetron sputtered Zr-Si-N films with a high (≥25 at.%) Si content, *Thin Solid Films* 478 (2005) 238–247.

38. N. Jiang, Y.G. Shen, Y.-W. Mai, T. Chan, S.C. Tung, Nanocomposite Ti-Si-N films deposited by reactive unbalanced magnetron sputtering at room temperature, *Mater. Sci. Eng.* B106 (2004) 163–171.

39. B.A. Movchan, A.V. Demchishin, Study of the structure and properties of vacuum condensates of Ni, Ti, W, Al_2O_3 and ZrO_2, *Phys. Metals Metallogr.* 28 (1969) 83–85.

40. J.A. Thornton, Influence of apparatus geometry and deposition conditions on the structure and topography of thick sputtered coatings, *J. Vac. Sci. Technol.* A11 (1974) 666–670.

41. J.A. Thornton, High rate thick films, *An. Rev. Mater. Sci.* 7 (1977) 239–260.

42. J.A. Thornton, The influence of bias sputter parameters on thick copper coatings deposited using a hollow cathode, *Thin Solid Films* 40 (1977) 335–344.

43. R. Messier, A.P. Giri, R.A. Roy, Revised structure zone model for thin film physical structure, *J. Vac. Sci. Technol.* A2 (1984) 500–503.

44. C.R.M. Grovenor, H.T.G. Hentzell, D.A. Smith, The development of grain structure during growth of metallic films, *Acta Metall.* 32 (1984) 773–781.

45. P.J. Martin, A. Bendavid, J.M. Cairney, M. Hoffman, Nanocomposite Ti-Si-N, Zr-Si-N, Ti-Al-Si-N, Ti-Al-V-Si-N thin film coatings deposited by vacuum arc deposition, *Surf. Coat. Technol.* 200 (2005) 2228–2235.

46. S. Carvalho, E. Ribeiro, L. Rebouta, C. Tavares, J.P. Mendonça, A.C. Monteiro, N.J.M. Carvalho, J.Th.M. De Hosson, A. Cavaleiro, Microstructure, mechanical properties and cutting performance of superhard (Ti,Si,Al)N nanocomposite films grown by d.c. reactive magnetron sputtering, *Surf. Coat. Technol.* 177–178 (2004) 459–468.

47. B-S. Yau, J.-L. Huang, D.-F. Lii, P. Sajgalik, Investigation of nanocrystal-(Ti,Al)N_x/amorphous-SiN_y composite films by co-deposition process, *Surf. Coat. Technol.* 177–178 (2004) 209–214.

48. D. Pilloud, J.F. Pierson, A.P. Marques, A. Cavaleiro, Structural changes in Zr-Si-N films vs. their silicon content, *Surf. Coat. Technol.* 180–181 (2004) 352–356.

49. F. Vaz, L. Rebouta, B. Almeida, P. Goudeau, J. Pacaud, J.P. Rivière, J. Bessa e Sousa, Structural analysis of $Ti_{1-x}Si_xN_y$ nanocomposite films prepared by reactive magnetron sputtering, *Surf. Coat. Technol.* 120–121 (1999) 166–172.

50. P.J. Martin, A. Bendavid, Properties of $Ti_{1-x}Si_xN_y$ films deposited by concurrent cathodic arc evaporation and magnetron sputtering, *Surf. Coat. Technol.* 163–164 (2003) 245–250.

51. K.H. Kim, S.R. Choi, S.-Y. Yoon, Superhard Ti-Si-N coatings by a hybrid system of arc ion plating and sputtering techniques, *Surf. Coat. Technol.* 161 (2002) 243–248.

52. B.H. Park, Y.-I. Kim, K.H. Kim, Effect of silicon addition on microstructure and mechanical property of titanium nitride film prepared by plasma-assisted chemical vapor deposition, *Thin Solid Films* 348 (1999) 210–214.

53. M. Diserens, J. Patscheider, F. Lévy, Improving the properties of titanium nitride by incorporation of silicon, *Surf. Coat. Technol.* 108–109 (1998) 241–246.

54. F.-J. Haug, P. Schwaller, J. Wloka, J. Patscheider, A. Karimi, M. Tobler, Stoichiometry dependence of hardness, elastic properties and oxidation resistance in TiN/SiN_x nanocomposites deposited by a hybrid process, *J. Vac. Sci. Technol.* A22 (2004) 1229–1234.

55. M. Nose, W.A. Chiou, M. Zhou, T. Mae, M. Meshii, Microstructure and mechanical properties of Zr-Si-N films prepared by rf-reactive sputtering, *J. Vac. Sci. Technol.* A20 (2002) 823–828.

56. T. Mae, M. Nose, M. Zhou, T. Nagae, K. Shimamura, The effects of Si addition on the structure and mechanical properties of ZrN thin films deposited by an r.f. reactive sputtering method, *Surf. Coat. Technol.* 142–144 (2001) 954–958.

57. M. Nose, M. Zhou, T. Nagae, T. Mae, M. Yokota, S. Saji, Properties of Zr-Si-N coatings prepared by RF reactive sputtering, *Surf. Coat. Technol.* 132 (2000) 163–168.

58. S.H. Kim, J.K. Kim, K.H. Kim, Influence of deposition conditions on the microstructure and mechanical properties of Ti-Si-N films by DC reactive magnetron sputtering, *Thin Solid Films* 420–421 (2002) 360–365.

59. S. Veprek, A. Niederhofer, K. Moto, T. Bolom, H.-D. Männling, P. Nesládek, G. Dollinger, A. Bergmaier, Composition, nanostructure and origin of the ultrahardness in nc-TiN/a-Si_3N_4/a- and nc-$TiSi_2$ nanocomposites with HV=80 to ≥105 GPa, *Surf. Coat. Technol.* 133–134 (2000) 152–159.

60. J.L. He, C.K. Chen, M.H. Hon, Wear of Ti-Si-N coated ceramic cutting inserts, *Wear* 181–183 (1995) 189–193.

61. F. Vaz, L. Rebouta, P. Goudeau, T. Girardeau, J. Pacaud, J.P. Riviere, A. Traverse, Structural transitions in hard Si-based TiN coatings: The effect of bias voltage and temperature, *Surf. Coat. Technol.* 146–147 (2001) 274–279.

62. C.H. Zhang, X.C. Lu, H. Wang, J.B. Luo, Y.G. Shen, K.Y. Li, Microstructure, mechanical properties, and oxidation resistance of nanocomposite Ti-Si-N coatings, *Appl. Surf. Sci.* 252 (2006) 6141–6153.

63. C.H. Zhang, Z.-J. Liu, K.Y. Li, Y.G. Shen, J.B. Luo, Microstructure, surface morphology and mechanical properties of nanocrystalline TiN/amorphous Si_3N_4 composite films synthesized by ion beam assisted deposition, *J. Appl. Phys.* 95 (2004) 1460–1467.

64. D. Pilloud, J.F. Pierson, J. Takadoum, Structure and tribological properties of reactively sputtered Zr-Si-N films, *Thin Solid Films* 496 (2006) 445–449.

65. J.W. Nah, S.K. Hwang, C.M. Lee, Development of a complex heat resistant hard coating based on (Ta, Si)N by reactive sputtering, *Mat. Chem. Phys.* 62 (2000) 115–121.

66. J.L. He, Y. Setsuhara, I. Shimizu, S. Miyake, Structure refinement and hardness enhancement of titanium nitride films by addition of copper, *Surf. Coat. Technol.* 137 (2001) 38–42.

67. J. Musil, R. Daniel, Structure and mechanical properties of magnetron sputtered Zr-Ti-Cu-N films, *Surf. Coat. Technol.* 166 (2003) 243–253.

68. J.G. Han, H.S. Myung, H.M. Lee, L.R. Shaginyan, Microstructure and mechanical properties of Ti-Ag-N and Ti-Cr-N superhard nanostructured coatings, *Surf. Coat. Technol.* 174–175 (2003) 738–743.

69. Y.G. Shen, Z.-J. Liu, N. Jiang, H.S. Zhang, K.H. Chan, Z.K. Xu, Effect of Si addition on surface morphology and structural properties of TiN films, *J. Mater. Res.* 19 (2004) 523–534.

70. S.H. Ahn, Y.S. Choi, J.G. Kim, J.G. Han, A study on corrosion resistance characteristics of PVD Cr-N coated steels by electrochemical method, *Surf. Coat. Technol.* 150 (2002) 319–326.

71. R. Watanabe, I. Yazaki, Y. Igarashi, N. Toyota, K. Noto, Preparation of superconducting molybdenum nitride fine powder by the plasma spraying method, *J. Mater. Sci. Lett.* 5 (1986) 255–256.

72. P. Rogl, J.C. Schuster, *Phase Diagrams of Ternary Boron Nitride and Silicon Nitride Systems*, ASM International, Materials Park, Ohio, 1992.

73. A. Laor, L. Zevin, J. Pelleg, Anisotropy in residual strains and the lattice parameter of reactive sputter-deposited ZrN films, *Thin Solid Films* 232 (1993) 143–148.

74. I. Goldfarb, J. Pelleg, L. Zevin, N. Croitoru, Lattice distortion in thin films of IVB metal (Ti, Zr, Hr) nitrides, *Thin Solid Films* 200 (1991) 117–127.

75. P. Zeman, R. Čerstvý, P.H. Mayrhofer, C. Mitterer, J. Musil, Structure and properties of hard and superhard Zr-Cu-N nanocomposite coatings, *Mater. Sci. Eng.* A289 (2000) 189–197.

76. F. Regent, J. Musil, Magnetron sputtered Cr-Ni-N and Ti-Mo-N films: Comparison of mechanical properties, *Surf. Coat. Technol.* 142–144 (2001) 146–151.

77. S. Sambasivan, W.T. Petuskey, Phase chemistry in the Ti-Si-N system: Thermochemical review with phase stability diagrams, *J. Mater. Res.* 9 (1994) 2362–2369.

78. D.M. Mattox, Particle bombardment effects on thin-film deposition: A review, *J. Vac. Sci. Technol.* A7 (1989) 1105–1114.

79. R. Kärcher, L. Ley, R.L. Johnson, Electronic structure of hydrogenated and unhydrogenated amorphous SiN_x (0≤x≥1.6): A photoemission study, *Phys. Rev.* B30 (1984) 1896–1910.

80. F. Vaz, L. Rebouta, P. Goudeau, J. Pacaud, H. Garem, J.P. Riviere, A. Cavaleiro, E. Alves, Characterisation of $Ti_{1-x}Si_xN_y$ nanocomposite films, *Surf. Coat. Technol.* 133–134 (2000) 307–313.

81. W.J. Meng, X.D. Zhang, B. Shi, J.C. Jiang, L.E. Rehn, P.M. Baldo, R.C. Tittsworth, Structure and mechanical properties of Ti-Si-N ceramic nanocomposite coatings, *Surf. Coat. Technol.* 163–164 (2003) 251–259.

82. S. Veprek, S. Reiprich, A concept for the design of novel superhard coatings, *Thin Solid Films* 268 (1995) 64–71.

83. P.B. Barna, M. Adamik, Formation and characterization of the structure of surface coatings, in Y. Paleau, P.B. Barna (Eds.), *Protective Coatings and Thin Films*, Kluwer Academic Publisher, 1997.

84. S. Vepřek, The search for novel superhard materials, *J. Vac. Sci. Technol.* A17 (1999) 2401–2420.

85. B. Window, Issues in magnetron sputtering of hard coatings, *Surf. Coat. Technol.* 81 (1996) 92–98.

86. R. Manaila, D. Biro, A. Devenyi, D. Fratiloiu, R. Popescu, J.E. Totolici, Structure of nitride film hard coatings prepared by reactive magnetron sputtering, *Appl. Surf. Sci.* 134 (1998) 1–10.

87. H. Gleiter, Nanocrystalline materials, *Progr. Mater. Sci.* 33(1989) 223–315.

88. J. Musil, J. Vlček, Magnetron sputtering of films with controlled texture and grain size, *Mater. Chem. Phys.* 54 (1998) 116–122.

89. J. Musil, Hard and superhard nanocomposite coatings, *Surf. Coat. Technol.* 25 (2000) 322–330.

90. H. Gleiter, Nanostructured materials: Basic concepts and microstructure, *Acta Mater.* 48 (2000) 1–29.

91. J. Musil, J. Vlček, Magnetron sputtering of hard nanocomposite coatings and their properties, *Surf. Coat. Technol.* 142–144 (2001) 557–566.

92. A.A. Voevodin, J.S. Zabinski, Supertough wear-resistant coatings with "chameleon" surface adaptation, *Thin Solid Films* 370 (2000) 223–231.

93. C. Mitterer, J. Komenda-Stallmaier, P. Losbichler, P. Schmölz, H. Störi, Decorative boride coatings based on LaB_6, *Surf. Coat. Technol.* 74–75 (1995) 1020–1027.

94. R. Daniel, J. Musil, P. Zeman, C. Mitterer, Thermal stability of magnetron sputtered Zr-Si-N films, *Surf. Coat. Technol.* 201 (2006) 3368–3376.

95. T. Zehnder, J. Patscheider, Nanocomposite TiC/a-C:H hard coatings deposited by reactive PVD, *Surf. Coat. Technol.* 133–134 (2000) 138–144.

96. A.A. Voevodin, J.M. Schneider, C. Rebholz, A. Matthews, Multilayer composite ceramic-metal-DLC coatings for sliding wear applications, *Tribol. Int.* 29(7) (1996) 559–570.

97. E.O. Hall, The deformation and ageing of mild steel: III Discussion of results, *Proc. Phys. Soc.* B64 (1951) 747–753.

98. G.A. Malygin, Break-down of the Hall-Petch law in micro- and nanocrystalline materials, *Phys. Solid. State* 37(8) (1995) 1248–1253.

99. J. Schiotz, in *Proceeding of the 22nd Riso International Symposium on Materials Science*, Roskilde, Denmark, 2001.

100. M. Nose, Y. Deguchi, T. Mae, E. Honbo, T. Nagae, K. Nogi, Influence of sputtering conditions on the structure and properties of Ti-Si-N thin films prepared by r.f.-reactive sputtering, *Surf. Coat. Technol.* 174–175 (2003) 261–265.

101. H. Ljungcrantz, M. Oden, L. Hultman, J.E. Greene, J.-E Sundgren, Nanoindentation studies of single-crystal (001)-, (011)-, and (111)-oriented TiN layers on MgO, *J. Appl. Phys.* 80 (1996) 6725–6733.

102. J.V. Fernandes, J.H. Schmitt, Dislocation microstructures in steel during deep drawing, *Philos. Mag.* A48 (1983) 841–870.

103. E.V. Kozlov, N.A. Koneva, L.A. Teplyakova, D.V. Lychagin, L.I. Trishkina, Contact and barrier dislocation resistance and their effect on characteristics of slip and work hardening, *Mat. Sci. Eng.* A319–321 (2001) 261–265.

104. S. Christiansen, M. Albrecht, H.P. Strunk, S. Veprek, Microstructure of novel superhard nanocrystalline-amorphous composites as analyzed by high resolution transmission electron microscopy, *J. Vac. Sci. Technol.* B16 (1998) 19–22.

105. J.M. Cairney, M.J. Hoffman, P.R. Munroe, P.J Martin, A. Bendavid, Deformation and fracture of Ti-Si-N nanocomposite films, *Thin Solid Films* 479 (2005) 193–200.

106. W.J. Meng, X.D. Zhang, B. Shi, R.C. Tittsworth, L.E. Rehn, P.M. Baldo, Microstructural and mechanical properties of Ti-Si-N coatings, *J. Mater. Res.* 17 (2002) 2628–2632.

107. X. Hu, Z. Han, G. Li, M. Gu, Microstructure and properties of Ti-Si-N nanocomposite films, *J. Vac. Sci. Technol.* A20 (2002) 1921–1926.

108. A. Cavaleiro, M.T. Vieira, Evaluation of hardness of sputtered W-C-Co thin films, *Surf. Eng.* 10 (1994) 147–151.

109. M. Mišina, J. Musil, S. Kadlec, Composite TiN-Ni thin films deposited by reactive magnetron sputter ion-plating, *Surf. Coat. Technol.* 110 (1998) 168–172.

110. J. Musil, H. Poláková, Hard nanocomposite Zr-Y-N coatings: Correlation between hardness and structure, *Surf. Coat. Technol.* 127 (2000) 99–106.

111. A. Cavaleiro, B. Trindade, M.T. Vieira, Deposition and characterization of fine-grained W-Ni-C/N ternary films, *Surf. Coat. Technol.* 116–119 (1999) 944–948.

112. H. Holleck, Material selection for hard coatings, *J. Vac. Sci. Technol.* A4 (1986) 2661–2669.

113. P. Zeman, J. Musil, Difference in high-temperature oxidation resistance of amorphous Zr-Si-N and W-Si-N films with a high Si content, *Appl. Surf. Sci.* 252 (2006) 8319–8325.

114. A. Leyland, A. Matthews, On the significance of the H/E ratio in wear control: A nanocomposite coating approach to optimised tribological behaviour, *Wear* 246 (2000) 1–11.

115. L. Hultman, Thermal stability of nitride thin films, *Vacuum* 57 (2000) 1–30.

116. D. McIntyre, J.E. Greene, G. Håkansson, J.-E. Sundgren, W.D. Münz, Oxidation of metastable single-phase polycrystalline $Ti_{0.5}Al_{0.5}N$ films: Kinetics and mechanisms, *J. Appl. Phys.* 67 (1990) 1542–1553.

117. C.W. Kim, K.H. Kim, Anti-oxidation properties of Ti-Al-N film prepared by plasma-assisted chemical vapor deposition and roles of Al, *Thin Solid Films* 307 (1997) 113–119.

118. F. Vaz, L. Rebouta, M. Andritschky, M.F. da Silva, J.C. Soares, Thermal oxidation of $Ti_{1-x}Al_xN$ coatings in air, *J. Eur. Ceram. Soc.* 17 (1997) 1971–1977.

119. A. Mitsuo, S. Uchida, N. Nihira, M. Iwaki, Improvement of high-temperature oxidation resistance of titanium nitride and titanium carbide films by aluminum ion implantation, *Surf. Coat. Technol.* 103–104 (1998) 98–103.

120. C. Héau, R.Y. Fillit, F. Vaux, F. Pascaretti, Study of thermal stability of some hard nitride coatings deposited by reactive magnetron sputtering, *Surf. Coat. Technol.* 120–121 (1999) 200–205.

121. C. Louro, A. Cavaleiro, The oxidation behaviour of mixed tungsten silicon sputtered coatings, *Thin Solid Films* 343–344 (1999) 51–56.

122. D. Pilloud, J.F. Pierson, M.C. Marco de Lucas, M. Alnot, Stabilisation of tetragonal zirconia in oxidized Zr-Si-N nanocomposite coatings, *Appl. Surf. Sci.* 229 (2004) 132–139.

123. M. Wittmer, J. Noser, H. Melchior, Oxidation kinetics of TiN thin films, *J. Appl. Phys.* 52 (1981) 6659–6664.

124. J.-N. Tu, J.-G. Duh, S.-Y. Tsai, Morphology, mechanical properties, and oxidation behavior of reactively sputtered Cr-N films, *Surf. Coat. Technol.* 133–134 (2000) 181–185.

125. J.B. Choi, K. Cho, M.-H. Lee, K H. Kim, Effects of Si content and free Si on oxidation behavior of Ti-Si-N coating layers, *Thin Solid Films* 447–448 (2004) 365–370.

126. X.D. Zhang, W.J. Meng, W. Wang, L.E. Rehn, P.M. Baldo, R.D. Evans, Temperature dependence of structure and mechanical properties of Ti-Si-N coatings, *Surf. Coat. Technol.* 177–178 (2004) 325–333.

127. S. Veprek, M. Haussmann, S. Reiprich, L. Shizhi, J. Dian, Novel thermodynamically stable and oxidation resistant superhard coating materials, *Surf. Coat. Technol.* 86–87 (1996) 394–401.

128. J. Almer, M. Oden, L. Hultman, G. Hakanson, Microstructural evolution during tempering of arc-evaporated Cr-N coatings, *J. Vac. Sci. Technol.* A18 (2000) 121–130.

129. D.B. Lee, Y.C. Lee, S.C. Kwon, High temperature oxidation of a CrN coating deposited on a steel substrate by ion plating, *Surf. Coat. Technol.* 141 (2001) 227–231.

130. Y.G. Shen, Y.W. Mai, W.E. McBride, Q.C. Zhang, D.R. McKenzie, Structural properties and nitrogen-loss characteristics in sputtered tungsten nitride films, *Thin Solid Films* 372 (2000) 257–264.

131. N. Solak, F. Ustel, M. Urgen, S. Aydin, A.F. Cakir, Oxidation behavior of molybdenum nitride coatings, *Surf. Coat. Technol.* 174–175 (2003) 713–719.

132. F.H.P.M. Habraken, A.E.T. Kuiper, Silicon nitride and oxy-nitride films, *Mater. Sci. Eng. R* 12 (3) (1994) 123–175.

133. G. Llauro, F. Gourbilleau, F. Sibieude, R. Hillel, Oxidation behavior of CVD TiN-Ti_5Si_3 composite coatings, *Thin Solid Films* 315 (1998) 336–344.

134. F.-H. Lu, W.-Z. Lo, Degradation of ZrN films at high temperature under controlled atmosphere, *J. Vac. Sci. Technol.* A22 (2004) 2071–2076.

135. C. Rebholz, A. Leyland, J.M. Schneider, A.A. Voevodin, A. Matthews, Structure, hardness and mechanical properties of magnetron-sputtered

titanium-aluminium boride films, *Surf. Coat. Technol.* 120–121 (1999) 412–417.

136. T.P. Thorpe, A.A. Morish, S.B. Qadri, Effect of grain size on the oxidation kinetics of sputtered titanium nitride films, *J. Vac. Sci. Technol.* A6 (1988) 1727–1729.

137. J.B. Choi, K. Cho, Y. Kim, K.H. Kim, P.K. Song, Microstructure effect on the high-temperature oxidation resistance of Ti-Si-N coating layers, *J. Appl. Phys.* 42 (2003) 6556–6559.

138. E.-A. Lee, K.H. Kim, Deposition and mechanical properties of Ti-Si-N coated layer on WC-Co by RF inductively coupled plasma-enhanced chemical vapor deposition, *Thin Solid Films* 420–421 (2002) 371–376.

139. J. Procházka, P. Karvánková, M.G.J. Vepřek-Heijman, S. Vepřek, Conditions required for achieving superhardness of ≥45 GPa in nc-TiN/a-Si_3N_4 nanocomposites, *Mater. Sci. Eng.* A384 (2004) 102–116.

Chapter 5

Novel Nanocomposite Zr-Si-N Films with a High Si Content (≥25 at.%)

In recent years, there is a great interest to synthesize thin films with enhanced properties for various industrial applications, such as cutting and machining operations, forming dies, diffusion barriers, protective coatings against oxidation, and chemical corrosion. Improved performance in next-generation hard coatings is being sought by enhancing the physical properties of the coatings, especially hardness. Such requirements have been found in hard transition metal nitride films, which are widely used in many applications as high-performance protective coatings [1–3]. However, although MeN films exhibit excellent mechanical properties, they cannot be used at high temperature in air due to their poor chemical stability [4,5]. The limitations of their usage at elevated temperatures and in hostile and aggressive environments attract intensive effort to improve performance of the MeN films by addition of various alloying elements. Among many different multi-component materials, which have been actively explored, mainly additives, such as aluminum (Al), chromium (Cr), and silicon (Si), have been frequently reported to improve the mechanical properties and overcome the oxidation problems of transition metal nitride films [6–8].

Due to an extremely high hardness reported by some researchers [9], Ti-Si-N films are attracting much interest; also other two-phase

Novel Nanocomposite Coatings: Advances and Industrial Applications
Rostislav Daniel and Jindřich Musil

ISBN 978-981-4411-17-2 (Hardcover), 978-981-4411-18-9 (eBook)
www.panstanford.com

systems of the Me-Si-N type form a superhard material with better stability at elevated temperature than the undoped MeN films [10–12]. One of the representative transition metal nitride films, which have been shown to exhibit similar properties to that of TiN, is ZrN. Compared to TiN, ZrN has a higher negative free energy of formation [13]; it is formed much easily and is even more stable than TiN. Although the chemical stability of ZrN at high temperatures is not sufficient for tool applications, after the addition of Si, the Zr-Si-N nanocomposite films composed of a nanocrystalline material embedded within an amorphous matrix, exhibit improved mechanical properties and thermal stability already at relatively low Si content [14]. Many papers report on the effect of a small Si content on the structure, phase composition, or film hardness of the Zr-Si-N films [15–17]; however, there is almost no reference on the properties of those films with a high (>25 at.%) Si content. Increasing interest in this kind of material is related to unique properties of Si_3N_4, which is formed in the film with higher Si content in the form of an amorphous matrix covering the ZrN grains. Si_3N_4-based ceramics have been shown to have excellent properties of high strength, hardness and toughness, wear-corrosion, oxidation and thermal shock resistance, low thermal expansion, self-lubricating or heat, and electrical insulation [18]. It makes them interesting for many applications in tools.

The origin of the hardness enhancement of the Zr-Si-N films over the hardness corresponding to the bulk ZrN has been found in the relationship between the film nanostructure, strong chemical bonding between elements of the nanocomposite, macrostress σ generated in the film during its growth, or their combined action. However, many questions remain unanswered regarding the relation between the properties of the films and the size of their grains and parameters which control the thermal stability and oxidation resistance of nanocomposite films. In this chapter, detailed study of Zr-Si-N films with a high (≥25 at.%) Si content composed of two stable ZrN_x and Si_3N_4 phases is present with the aim to answer some of these questions. Further discussion on why Me-Si-N films with X-ray amorphous structure in the as-deposited state and a high content of the Si_3N_4 phase can exhibit high hardness up to or even exceeding 30 GPa and excellent thermal stability and oxidation resistance above 1000°C is discussed.

5.1 Deposition Rate

The dependences of the deposition rate a_D of the Zr-Si-N films deposited at (i) I_d = 1 A, i_s = 0.5 mA/cm^2; (ii) I_d = 1 A, i_s = 1 mA/cm^2; and (iii) I_d = 2 A, i_s = 1 mA/cm^2 on the partial pressure of nitrogen p_{N_2} are given in Fig. 5.1. It can be seen that the deposition rate a_D of the Zr-Si-N films considerably depends on the combination of the process parameters under which the films were prepared, particularly on the I_d, i_s, and p_{N_2}. The development of the deposition rate a_D, as depicted in Fig. 5.1, is specific for the films reactively sputter deposited at increasing p_{N_2}. Typically, a_D is approximately constant in the metallic mode of sputtering and dramatically decreases in the transition mode. This radical drop in a deposition rate can be explained by the formation of nitrided compounds on the target surface. This process is usually called a "target poisoning". Since the sputtering yield of Si or ZrN is much lower compared to that of original metal, a decrease in deposition rate is obvious at p_{N_2}> 0.1 Pa, when the nitrided compounds are formed. However, another reason for the change of the deposition rate of the films sputtered at constant total pressure, $p_T = p_{Ar} + p_{N_2}$ can be found. While the partial pressure of N increases to the detriment of the Ar (the p_{N_2}:p_{Ar} ratio increases), the overall sputtering yield of the target might decrease. In fact, N_2 and N are preferably ionized in the plasma discharge having an ionization potential lower than that of Ar (Ar) (15.58, 14.53, and 15.76 eV). Thus, a target is sputtered mainly by N ions at higher p_{N_2} since a preferred ionization results in an increase in the number of N_2^+ ions. While the ions hit the target surface, they are subjected to the reaction where $N_2^+ \rightarrow N^+ + N$. The effective sputtering of the target is then lower because the sputtering yield of much lighter N^+ ion is considerably lower compared to heavier Ar^+. It subsequently leads to a decrease in deposition rate [19,20]. The results revealed that the deposition rate decreases only very slowly in the nitride mode of sputtering at p_{N_2}> 0.3 Pa, what is related both to the limits for the N ionization and the formation of the compounds on the target surface.

As expected, the higher the deposition rate a_D is, the higher is the I_d used for film deposition and it also slightly increases with increase in the ion current density i_s, as the target sputtering enhances and the plasma discharge becomes denser. In both cases, higher N content is incorporated into the film forming with Si and zirconium (Zr) higher amount of nitrided compounds. Together with increase in sputtering yield of target material, the deposition rate a_D increases as well. Besides more intensive ion bombardment of the growing film at i_s = 1 mA/cm^2, a_D is higher compared to a_D of films sputtered at i_s = 0.5 mA/cm^2. However, the enhanced resputtering of the growing film by more intensive ion bombardment leads into a decrease in the difference between the deposition rate a_D of films deposited in the metallic and nitride mode of sputtering. Therefore, the deposition rate a_D is about four times higher in the metallic mode than in the nitride mode when films are sputtered at i_s = 0.5 mA/cm^2 and I_d = 1 A but is only about three times higher in the case of film sputtering at i_s = 1 mA/cm^2 and I_d = 1 and 2 A, that is, in more intensified discharge.

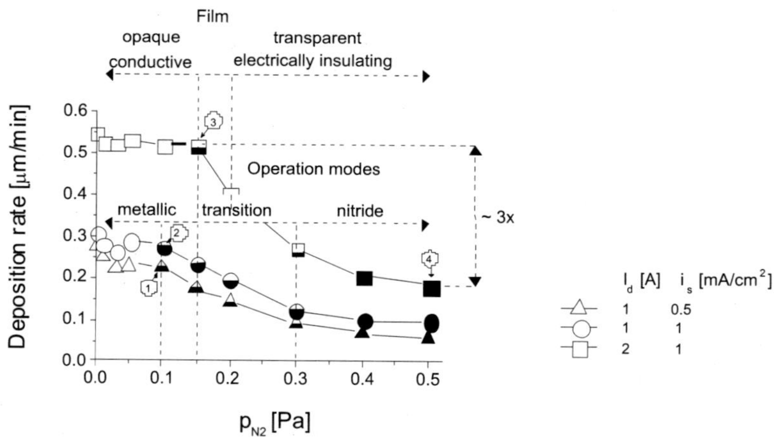

Figure 5.1 Deposition rate of the Zr-Si-N films, prepared by direct current (d.c.) reactive magnetron sputtering at two values: (i) I_d = 1 and 2 A and (ii) substrate ion current density, i_s = 0.5 and 1 mA/cm^2, as a function of partial pressure of nitrogen p_{N_2}. Constant deposition conditions: U_s = −100 V, T_s = 500°C, $d_{s\text{-}t}$ = 50 mm, and p_T = 0.7 Pa. The labels 1, 2, 3, and 4 correspond to films referred in the text. Reprinted from Ref. [28], Copyright (2005), with permission from Elsevier.

It should be noted that an increase in deposition rate a_D with increase in the discharge current I_d and the substrate ion current density i_s occurs only when the films are sputtered using the conventional magnetron sputtering. In the case of magnetron discharges with enhanced ionization the discharge voltage U_d decreases due to a lowering of plasma impedance, which results in the decrease of the target sputtering yield and so in the decrease of a_D.

5.2 Elemental Composition

The development of the elemental composition of the Zr-Si-N films reactively sputtered from the $ZrSi_2$ target at three combinations of magnetron discharge current I_d and ion current density i_s, with increase in partial pressure of nitrogen p_{N_2}, is shown in Fig. 5.2. Here, a strong dependence of the elemental composition of the Zr-Si-N films on the deposition conditions (p_{N_2}, I_d, and i_s), under which the films were prepared, can be seen.

Regardless on the magnitude of I_d and i_s, the amount of N incorporated into the growing film strongly increases with an increase in p_{N_2} and achieves its maximum value, which is in some cases higher than 55 at.%, already at relatively low p_{N_2}. It suggests a high affinity of sputtered atoms to N resulting with a decrease in their content in the Zr-Si-N film while they are bonded with an increase in amount of N. The effect of high affinity of Si to N was demonstrated in the case of films sputtered using Zr target with different number of Si chips stuck on the target surface. Here, the amount of N incorporated into the film deposited at constant value of p_{N_2} increases just owing to an increase in the Si content in the film [14]. It suggests that the increment of Si itself induces the incorporation of N to the films [21]. Figure 5.2 shows that an increase in p_{N_2} from 0.1 Pa up to 0.5 Pa does not result in a further increase in the N content in the film. However, although its amount is constant in the films, in the given p_{N_2} range, the physical properties of those films obviously change. It can be explained by a different phase composition of the film sputtered at increasing p_{N_2}. The role of a high affinity of Si to N on the phase composition of the Zr-Si-N film is discussed in detail in Section 5.3 below.

A more intensive sputtering of the target at higher target power results in a shift of the transition mode of sputtering into higher values of p_{N_2}. It can be seen in Fig. 5.2 as a more gradual increase in the N content with an increase in p_{N_2} in the Zr-Si-N films sputtered at I_d = 2 A (compare Fig. 5.2b and c). Therefore, higher p_{N_2} is necessary to obtain the same elemental composition and the amount of nitrided compounds formed in the film sputtered at I_d = 2 A compared to that sputtered at lower I_d = 1 A. Also the maximum value of the N content in the film is reached firstly at p_{N_2} = 0.2 Pa, while the films are prepared at I_d = 2 A. Furthermore, the amount of Si in the Zr-Si-N films sputtered in the transition and nitride mode at I_d = 1 A is lower due to the more intensive poisoning effect, compared to the films sputtered at I_d = 2 A. This fact results with a decrease in the sputtering yield of the target material, particularly of Si atoms.

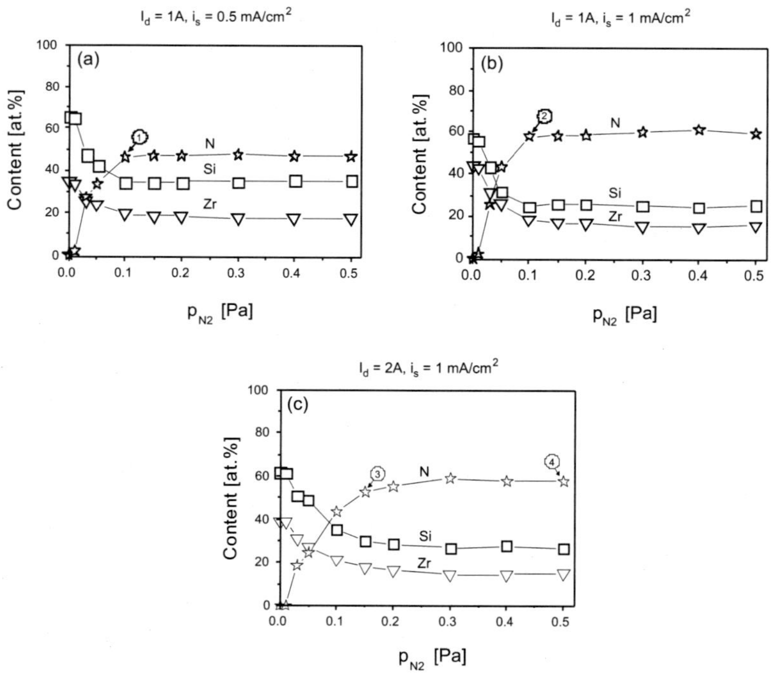

Figure 5.2 Elemental composition of the Zr-Si-N films, sputtered at U_s = −100 V and (a) I_d = 1 A, i_s = 0.5 mA/cm^2, (b) I_d = 1 A, i_s = 1 mA/cm^2, and (c) I_d = 2 A, i_s = 1 mA/cm^2, as a function of partial pressure of nitrogen p_{N_2}. Reprinted from Ref. [28], Copyright (2005), with permission from Elsevier.

The Zr-Si-N films sputtered at higher values of the substrate ion current density i_s exhibit obviously higher N content (compare Fig. 5.2a and b). In the principal, the formation of nitride phases in the films is, given by two potential reaction paths, both depending on the deposition conditions. It appears that the nitridation of Si and Zr is achieved through direct sputtering of the compounds formed on the target surface. However, more effective way seems to be sputtering of elemental atoms and reaction with N on the sample surface. Since the nitrogen, N_2 molecules cannot be directly incorporated into the forming film during its deposition, atomic nitrogen, N and N_2^+ ions have to be produced in the discharge. They are the only N species, which can be chemisorbed on the substrate surface, or on the Si and Zr atoms and so incorporated into the film [22]. Their amount is higher in more intensified discharge where the molecular N is intensively and more effectively dissociated and ionized. The acceleration of positive ions in the field of the biased sample serves for enhanced impact dissociation and the production of reactive atomic N [23]. Hence, the N content increases in the Zr-Si-N film sputtered with an increase in the ion current density i_s (see Fig. 5.3). Beside the two reaction paths, the possibility of reactions in the plasma should also be considered. However, the reaction efficiency is closely related to the conditions of the plasma discharge and the configuration of deposition system. The target-to-substrate distance, discharge power, or the mean free path for elastic collisions between Ar and N, given by the total pressure of both working gases, play an essential role. The ionization or dissociation of N_2 molecules by collisions or the electron impact requires threshold energies of at least 9.7 eV corresponding to the dissociation energy of N. This value can, however, be above the mean electron energies which are typically around 5 eV or less in magnetron discharges [24]. Also, the cross-sections of ionizing and dissociative collisions are about one order of magnitude lower than those of elastic collisions [23]. Thus, dissociation of N in regions other than the cathode sheath of the target or the vicinity of the substrate seems unlikely for the deposition conditions. Here, it can be seen how the configuration of the deposition chamber and plasma discharge itself is very important for the efficiency of reactive processes during film deposition.

The saturation of the N content in the Zr-Si-N film sputtered at increasing p_{N_2} also bears on the amount of atomic N or N_2^+ ions which can be produced in the discharge and react with Zr and Si. It has been shown that the ion density is considerably lower in the discharge where the Ar content is relatively low compared to the pure Ar discharge. A possible explanation is, a difference in the electron energy distribution function, in particular, in the high-energy tail as indicated by the Langmuir probe measurements [25]. Lower secondary electron yields at the target in the N_2 discharge and thus, a lower number of high-energy electrons probably cause a saturated amount of ionized N which can be incorporated into the growing film.

Figure 5.3 presents the development of the Si and Zr content and the Si:Zr atomic ratio of the Zr-Si-N films as a function of their N content. The trend for the Si and Zr content is identical in all Zr-Si-N films deposited with increase in p_{N_2} regardless on the magnitude of the ion current density, i_s. Here, the amount of Si and Zr in the film linearly decreases with increase in the N content; however, the value of the Si:Zr ratio remains constant.

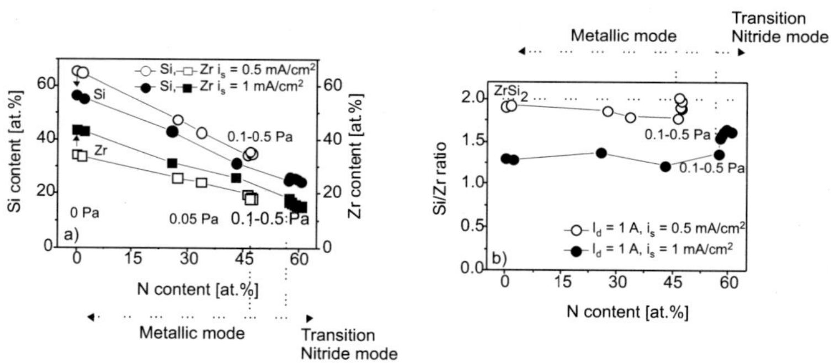

Figure 5.3 Development of (a) the Si and Zr content and (b) the Si:Zr ratio of the Zr-Si-N films, sputter deposited at I_d = 1 A, U_s = −100 V, i_s = 0.5, and 1 mA/cm^2 and increasing p_{N_2}, as a function of the N content in the film. The Si:Zr = 2 ratio of the alloyed target is also depicted.

In general, the films with the elemental composition of the original alloyed target can be obtained by non-reactive sputtering. However, it is possible only if the composition of the sputtered target is homogenous and does not change in time for instance due to diffusion processes. All such effects can influence the composition of the elements in the sputtered flux during film deposition and so the elemental composition of sputtered films as well. In Fig. 5.3b, the Si:Zr atomic ratio of the Zr-Si-N films sputtered at two different values of the ion current density i_s is displayed. Compared to the elemental composition of the alloyed $ZrSi_2$ target with the atomic ratio Si:Zr equal to 2, the non-reactively sputter deposited films prepared at p_{N_2} = 0 Pa and i_s = 0.5 and 1 mA/cm^2 exhibit obviously lower values of the Si:Zr ratio. It is attributed to the preferential resputtering of Si during the film growth that consequently results in its lower concentration. The deposition of the Zr-Si-N films was performed with the negative substrate bias of −100 V, which allows the ionized particles to have relatively high impacting energy. Mainly at a low N concentration in the film much of the incorporated atoms are in the elemental state. Then, the lighter element of the multicomponent film is preferentially resputtered away from the film, as here in the case of Si atoms (m_{Si} = 28 amu, m_{Zr} = 91 amu). Therefore, while the film deposited under the less-intensive ion bombardment at i_s = 0.5 mA/cm^2 exhibits the elemental composition (Si:Zr = 1.9) close to the original composition of the target (Si:Zr = 2), the enhanced bombardment at i_s = 1 mA/cm^2 results in a decrease of the Si:Zr ratio of the film down to 1.3. The decrease of the Si content from 65 at.% to 56 at.% in the Zr-Si-(N) film with increasing i_s is obviously seen in Fig. 5.3a. It is important to note that the substrate bias, and so the ion energy, was kept constant in this experiment.

For the N-containing Zr-Si-N films, the resputtering of metallic Si and Zr atoms from the film induced by an ion bombardment is attenuated through the nitrided Si_3N_4 and eventually ZrN compounds formed during the film growth in a N_2 atmosphere. It makes resputtering more difficult since the sputtering yield of nitrides is several times lower than that of pure metals [26]. Thus, the difference between resputtering level of metallic atoms in Zr-Si-N films containing various amount of N is expressive. Unreacted Si, the dominant form of Si at low nitridation level, is more intensively resputtered away by the high energetic particles than in the nitrided form. Consequently, stands for lower amount of Si incorporated into the films at low N partial pressure. Typically, the lower the N partial pressure is, the more important role plays the chosen bias voltage. The deposition at bias voltages of magnitude of −100 V or even higher and N partial pressure lower than 0.1 Pa results in the differences in the Si concentration up to 10 at.% or more [27]. On the contrary, at higher nitrogen partial pressure, no significant influence of the bias voltage on the elemental composition is observed due to the strong bonds formed between Si, Zr, and N [28]. In general, the amount of resputtered atoms strongly depends on the bond strength between individual atoms, the intensity of ion bombardment and the energy of the impacting particles (momentum, kinetic, energy, etc.).

Sputtering of the target is affected by the same phenomenon. Also in this case, the compounds of metal atoms formed with N on the target surface are sputtered with markedly lower deposition rates what results in a lowering in the amount of particles in the sputtered flux. A decrease in deposition rates is referred to higher binding energies, while the films are sputter deposited in a reactive atmosphere. In addition, lower sputtering yields per impinging ion of the compounds formed on the target surface and higher secondary electron yields, thus result in consuming much of the energy for electron emission instead of the preferred sputtering of material [29]. The effect of target poisoning is also closely related to the unstable operation and arcing during deposition mainly when the insulating compounds are formed on the target surface. In order to avoid commonly observed poisoning of the target and resulting instabilities of the deposition process the pulsed magnetron sputtering has recently become a popular method of thin film deposition. The main reasons for magnetron sputtering process instabilities are target surface charges appearing on sputtered materials. The sputtering process becomes unstable owing to the charged agglomerates resulting in a strong electron thermoemission at a high target power density. That effect can lead to micro-arcs on the target surface and in consequence to droplets and microdefects in deposited films [30,31].

Almost constant value of the deposition rate a_D of the Zr-Si-N films in the metallic mode of sputtering suggests a negligible target poisoning effect at $p_{N_2} < 0.1$ Pa. The sputtering yields of the target atoms are constant in that case as well as the amount of the Si and Zr atoms in the sputtered flux. In spite of a rapid increase in the N content in the film with increase in p_{N_2}, the films deposited in the metallic mode contain an insufficient amount of N necessary for forming fully nitrided compounds. Since not all Si and Zr can be bonded to N at lower p_{N_2}, free metal atoms, or even the components from the $ZrSi_x$ compound can be still resputtered away (see Section 5.3). As it was already mentioned, Si is preferentially resputtered in relation to Zr, with respect to their weight. Therefore, the Si:Zr atomic ratio of the films is permanently lower than that of the alloyed target up to 45 at.% or 60 at.% of the N content in the film (i_s = 0.5 or 1 mA/cm^2) until all the metal atoms are completely bonded to N forming stable nitrided compounds. Hence, an increase in the Si:Zr ratio was observed in the films deposited at the transition and nitride mode of sputtering at $p_{N_2} > 0.1$ Pa when the target poisoning effect became dominant. It results from the lowering of the sputtering yields of both components in the target together with a negligible resputtering of Si and Zr atoms, which are all bonded to N. However, it is important to note that regardless on the deposition conditions, all the Zr-Si-N films contain approximately 25 at.% of Si or more [28].

The experimental results show that there are generally two effects enforced that influence the Si and Zr content in the Zr-Si-N films during their formation. On the one hand, the amount of individual atoms sputtered from the target, and on the other hand the number of atoms resputtered from the forming film induced by ion bombardment. The relationship between both effects is, however, closely attributed to the efficiency of the ionization and dissociation of molecular N in the discharge at given p_{N_2}. It influences the amount of incorporated atomic N into the film and the degree of the target poisoning. In any case, the presence of the nitrided compounds on the target surface or in the film attenuates the (re-)sputtering yields and subsequently determines the final Si:Zr ratio of the film. If one of the two mentioned effects becomes dominant, Si:Zr may be rapidly changed. This is a case of the films sputtered at different ion current densities where the dependence of the Si and N content in the film on these two synergetic effects is relatively strong, compare Fig. 5.2a and b. Although the resputtering of Zr from the Zr-Si-N film seems to be negligible compared to Si even if the film is deposited under intensive bombardment at $U_s = -100$ V and i_s = 1 mA/cm^2, strong resputtering may be observed at different compound combination in the target and the film [32,33]. All these results reveal that the substrate bias applied during the film formation might serve as a possible control of the Zr-Si-N film composition.

5.3 Chemical Bonding and Phase Composition

It is known that the heat of formation of Si_3N_4 $\Delta H_{fSi_3N_4} = -745.1$ kJ/mol is more negative than that of ZrN $\Delta H_{fZrN} = -365.5$ kJ/mol [34,35] or ZrSi ($\Delta H_{fZrSi} = -154.8$ kJ/mol, $\Delta H_{fZrSi_2} = -159$ kJ/mol) [36]. This fact allows assuming that a preferential bonding of N to Si atoms will be established during the Zr-Si-N film formation in a reactive N_2 atmosphere. Therefore, it can be expected firstly that the Si_3N_4 phase is formed and only remaining unbounded N atoms (over Si_3N_4) will be joined to Zr forming ZrN phase. On the contrary, when Si and Zr atoms are not fully nitrided owing to the insufficient amount of N in the film to form Si_3N_4 and ZrN with all Si and Zr atoms, a formation of $ZrSi_x$ phase also occurs. Based on these assumptions, the phase composition of the Zr-Si-N films sputtered at increasing p_{N_2} can be assigned as depicted in Fig. 5.4. Here, the N:Si atomic ratio, excess of N over that amount bounded in the Si_3N_4 phase (N:Si ≥ 1.33) and the Zr content in the Zr-Si-N film as a function of I_d, i_S, and p_{N_2} are also displayed.

The development of the atomic N:Si ratio presented in this figure reveals a strong dependence of the amount of Si and N incorporated into the film during its growth on the partial pressure of nitrogen p_{N_2}, magnetron discharge current I_d, and flux of ions i_s bombarding the growing film as discussed above. The main part of the sputtered flux from the target consists mainly of the Zr and Si atoms which are enriched with N from working gas mainly in a form of N_2^+, N^+, or N [23]. The elemental composition of the films is, however, governed by reactions occurring on the target surface, directly in the plasma discharge or on the sample surface. Thus, the compounds formed on the substrate strongly depend on the deposition conditions. With increase in i_s more effective incorporation of atomic N into the film and more intensive resputtering of Si results in higher the N:Si ratio making possible formation of higher amount of the Si_3N_4 phase already at lower p_{N_2}. On the contrary, the higher target power leads to a more gradual increase in N:Si and the value N:Si = 1.33, corresponding to a formation of the Si_3N_4 phase, is then reached at higher values of the partial pressure of N (see Fig. 5.4c). Thus, the films sputtered at $I_d = 2$ A exhibit the same phase composition and an amount of nitrided compounds as the films sputtered at $I_d = 1$ A, firstly at higher p_{N_2}. Additionally, the experimental results show that it is possible to prepare films with required phase composition at higher deposition rates just using higher target power and p_{N_2} (see Figs. 5.1 and 5.4).

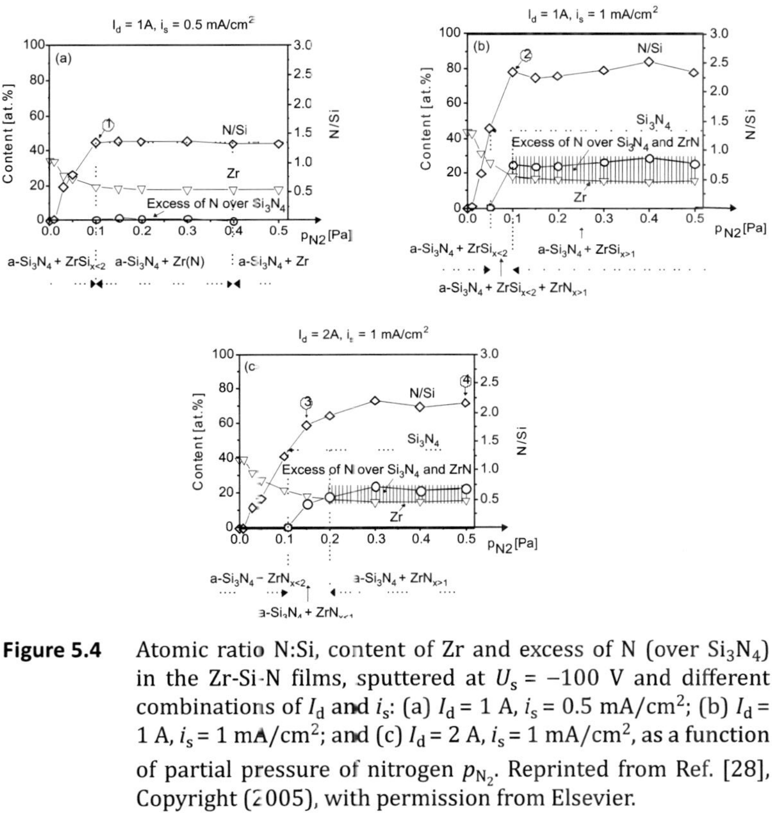

Figure 5.4 Atomic ratio N:Si, content of Zr and excess of N (over Si_3N_4) in the Zr-Si-N films, sputtered at $U_s = -100$ V and different combinations of I_d and i_s: (a) $I_d = 1$ A, $i_s = 0.5$ mA/cm^2; (b) $I_d = 1$ A, $i_s = 1$ mA/cm^2; and (c) $I_d = 2$ A, $i_s = 1$ mA/cm^2, as a function of partial pressure of nitrogen p_{N_2}. Reprinted from Ref. [28], Copyright (2005), with permission from Elsevier.

In general, changes in the deposition conditions, under which the Zr-Si-N films were prepared, expressively influence the elemental composition of the films and consequently their properties. It can be explained by the fact that the formation of phases, from which the films are finally composed, is strongly affected by the elemental composition of the Zr-Si-N films prepared at various deposition conditions. With respect to the assumptions that affinity of N to Si is higher than to Zr and is also higher than affinity of Si to Zr, that is,

Si–N bond will be establish preferentially, the phase composition of the Zr-Si-N film varies with increasing p_{N_2} as follows:

From c-$ZrSi_{x<2}$ phase at p_{N_2}=0 Pa through a-Si_3N_4 + $ZrSi_{x<2}$ to

(a) a-Si_3N_4 + $ZrN_{x<1}$ and a-Si_3N_4 + Zr at I_d = 1 A and i_s = 0.5 mA/cm^2 (Fig. 5.4a) or to

(b) a-Si_3N_4 + $ZrSi_{x<2}$ + $ZrN_{x>1}$ and a-Si_3N_4 + $ZrN_{x>1}$ at I_d = 1 A and i_s = 1 mA/cm^2 (Fig. 5.4b) or to

(c) a-Si_3N_4 + $ZrN_{x<1}$ and a-Si_3N_4 + $ZrN_{x>1}$ at I_d = 2 A and i_s = 1 mA/cm^2 (Fig. 5.4c) [28].

Here, c- and a- denotes the crystalline and X-ray amorphous structure, respectively.

The validity of the assumptions explaining the trend of formation of individual phases in the Zr-Si-N films were confirmed by experimental results obtained in the case of films with a Si content lower than 20 at.%. The development of the type of established bonds during the film formation has been investigated as a function of increasing the Si content mainly in the Zr-Si-N films sputter deposited from composite targets consisted of Zr plate and different number of Si chips [15,32]. The results obtained by X-ray photoelectron spectroscopy (XPS) clearly show that the Si–N bonds are readily formed in the films due to the high reactivity of Si with N. Thus, the Si atoms are fully nitrided in the film already at relatively low Si content forming the Si_3N_4 phase [15]. To neglect the effect of the ion bombardment realized in order to remove the surface contamination before own XPS measurement, Pilloud et al. has performed also Fourier transform infrared spectroscopy (FTIR) measurement to obtain additional information on the chemical bonds involved in the film. In that case, the results are not influenced by a possible ion-bombardment-induced preferential resputtering of one of the element contained in the film. It has been confirmed by FTIR that in the Zr-Si-N films with a Si content higher than 2.2 at.%, Si–N bonds are involved and their amount increases with increase in Si concentration. The chemical state, clarifying the phase composition of the Zr-Si-N films, has been estimated by XPS also in Nose et al. [17]. The results obtained by Nose et al. show huge differences between the bonds involved in the film in the as-deposited state and after etching for 5 min (*ca.* 20 nm). In the first case, the dominant Si_3N_4 phase has been found to be formed in the Zr-Si-N film with 10 at.%

Si. On the other hand, mainly Zr–N as well as the Zr–Si bonds has been observed after etching. This fact, however, might be affected by the ion bombardment with high energetic ions, which can strongly change the bonds between atoms when some of the element is resputtered away.

The results obtained for Zr-Si-N films with a low Si content (<20 at.%), showing the validity of the above assumptions, were further confirmed by Raman spectroscopy measurement performed on representative Zr-Si-N films with a high (≥25 at.%) Si content sputtered from an alloyed $ZrSi_2$ target. Raman spectroscopy is well-known non-invasive convenient technique commonly used for characterization of weakly crystallized and amorphous structures to estimate the chemical bonds established in the material. In Raman spectroscopy, the light from a laser is typically focused on the sample surface through a microscope and the scattered light is collected and directed into a spectrometer and detected by a charge-coupled device (CCD) detector. The coupling of a Raman spectrometer with an optical microscope enables the user to obtain high quality spectra from micro- to nanometer-sized particles. Raman scattering arises practically from an inelastic interaction between photons (light of the incident laser) and phonons (vibrations of the crystal lattice) and used to be presented as a spectrum where peak intensity (counts) versus Raman shift (cm^{-1}) is plotted. Raman spectra of high Si-containing Zr-Si-N films, sputtered at I_d = 1 A, U_s = −100 V, i_s = 1 mA/cm^2, and increase in p_{N_2} with different elemental composition, are displayed in Fig. 5.5.

Non-reactively sputtered $Zr_{43}Si_{57}$ film is characterized by several peaks located between 200 and 400 cm^{-1}. They correspond to the bonds established between Zr and Si atoms. A group of bands of the acoustic modes in the range from 200 to 400 cm^{-1} indicates that the film structure is composed of the $ZrSi_x$ grains with different stoichiometry x close to 2. Their intensities significantly decrease in the $Zr_{31}Si_{43}N_{26}$ film (p_{N_2} = 0.03 Pa) while the metallic Zr and Si atoms become nitrided, see Fig. 5.5b. The formation of nitrided phases is observed from the Raman spectra of the Zr-Si-N films already at the N:Si ratio around 0.6. The Raman spectrum corresponding to the $Zr_{18}Si_{25}N_{57}$ film with higher N content sputtered at p_{N_2} = 0.1 Pa exhibits a very weak scattering from the $ZrSi_x$ phase. One broad peak with a low intensity located at 230 cm^{-1} reflects just relatively low amount of the Zr–Si bonds which, however, are most probably responsible for the electrical conductivity of this film. The peaks related to the $ZrSi_x$ compound fully disappear in the $Zr_{15}Si_{24}N_{61}$ film what suggests that Zr and Si atoms are already fully nitrided in the Zr-Si-N films sputtered at $p_{N_2} > 0.1$ Pa. Another two peaks at 500 and 950 cm^{-1} of measured spectrum are superimposed peaks corresponding to both ZrN and Si_3N_4 phases. Indeed, the Raman phonon spectra of ZrN have been reported at 450 and 1100 cm^{-1} [37], those of Si_3N_4 are referred to 400 and 900 cm^{-1} [38]. Intensity of both peaks increased in the $Zr_{15}Si_{24}N_{61}$ film which give arise to the consideration that an amount of the Si–N and Zr–N bonds increases with increase in the N content in the film that is with increase in p_{N_2}.

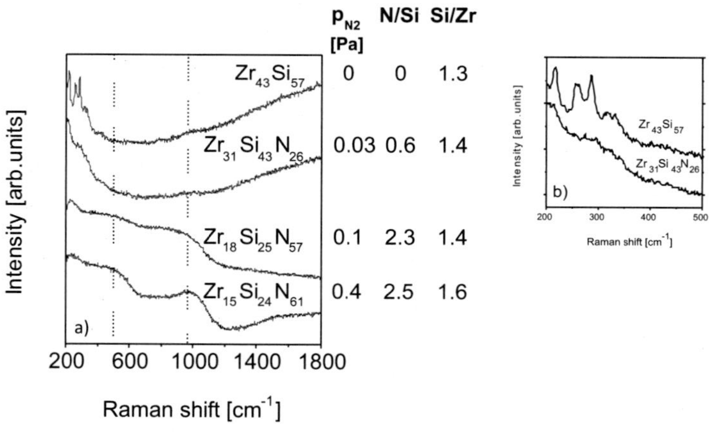

Figure 5.5 Raman spectra of the Zr-Si-N films sputtered at $I_d = 1$ A, $U_s = -100$ V, $i_s = 1$ mA/cm^2, and various partial pressure of nitrogen $p_{N_2} = 0$, 0.03, 0.1, and 0.4 Pa. The excitation wavelength used was 514 nm of an Ar^+ ion laser.

Based on the results obtained for a low (<20 at.%) and a high (≥25 at.%) Si content in the Zr-Si-N films, the phase formation can be generalized as follows: the Si–N bonds are preferentially established in the film reactively sputtered in a N_2 atmosphere forming the Si_3N_4 phase. If there exists an excess of Si over N:Si = 1.33, the Zr–Si bonds are readily formed in the film. On the other hand, while the N content is sufficiently high and exceeds over Si_3N_4, Zr atoms are bonded with incorporated N forming the ZrN_x phase. Furthermore, XPS measurements performed on the Zr-Si-N films reveal that at higher Si contents above 25 at.% also free unbounded Si can be present in the film if the N concentration is insufficiently low [39]. Thus, it may be concluded that the Si bonding status in Zr-Si-N films is in the form of Si_3N_4 with low Si and high N contents, while it can be in the form of free Si, $ZrSi_2$, or Si_3N_4 at high Si and low N contents. The validity of these assumptions and trends of the phase formation in the Zr-Si-N films described above fully corresponds to other reported studies [7,16,17] and was verified by thermal annealing experiments [40], described in Section 5.11. Hypothesis of possible formation of other type of bonds with a high enthalpy of formation between individual Zr, Si, and N atoms was not confirmed [33].

A high importance of the chemical reactivity of individual elements in the plasma discharge, or on the sample surface, has been reported as well [20]. Insufficient reactivity may result in a development of commonly unfavored bonds. From that reason, the $TiSi_2$ phase tends to be formed in the growing film at a lower activity of N although its amount should be sufficient for full nitrification of both Si_3N_4 and TiN phases [41]. This effect is simultaneously reflected by an amount of the Si_3N_4 phase formed in the films. Since the Si_3N_4 phase is typically formed as amorphous stoichiometric Si_3N_4 [42], its volume content in the film can be calculated. The development of the volume fraction of Si_3N_4 with increase in the N content in the Zr-Si-N films is illustrated in Fig. 5.6.

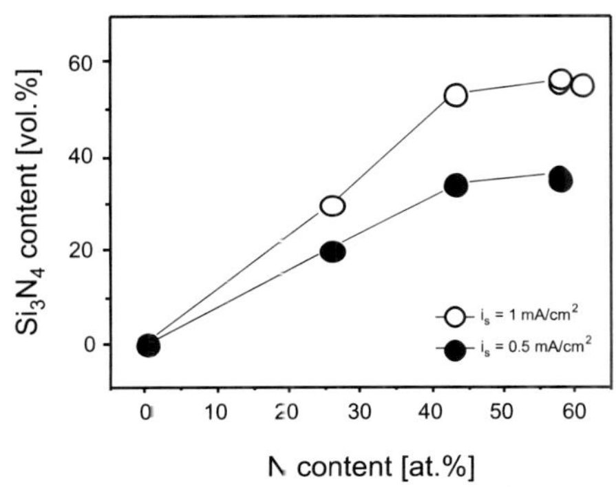

Figure 5.6 Development of the volume content of the amorphous Si_3N_4 phase with increase in amount of N incorporated into the Zr-Si-N films prepared at various ion current density i_s = 0.5 and 1 mA/cm^2.

By increasing the partial pressure of nitrogen p_{N_2}, a higher amount of the Si_3N_4 phase is formed to the detriment of the crystalline $ZrSi_x$. Considerably, that alters the volume fraction ratio between the crystalline and amorphous phase where the Si_3N_4 becomes dominant. Here, the saturation of the volume content of the Si_3N_4 phase directly results from a limited incorporation of a higher N amount into the film as mentioned above. Even if the film is sputter deposited at $p_{N_2} > 0.3$ Pa, the N content remains almost constant. However, the experimental results reveal that with increase in ion discharge, current density i_s gives rise to the higher N activity resulting in a higher amount of Si–N bonds and consequently, higher volume content of the Si_3N_4 phase in the film. The resputtering effect of condensed N at more intensive bombardment is effectively compensated by a high reactivity of the impinging N^+ ions while stable compounds are formed [20].

It is worthy to note that even small change in the phase composition of the Zr-Si-N films causes considerable changes in physical properties of those films. The variation of the process parameters under which the film is produced, was found to be an easy way how to prepare Zr-Si-N films with various structure, mechanical, optical, electrical, and thermal properties as well as resistance against oxidation. Dependence of the structure, physical properties, and thermal stability of the Zr-Si-N films on their phase composition will be discussed in detail in next sections.

5.4 Electrical and Optical Properties

The experimental results clearly show that two different groups of Zr-Si-N films may be obtained by reactive sputtering from an alloyed $ZrSi_2$ target at increasing p_{N_2}. The films with lower N content deposited in the metallic mode of sputtering exhibit completely different physical, mechanical, and thermal properties and resistance against oxidation compared to the films with the N content higher than 45 at.%, sputtered in the transition and nitride modes. This phenomenon closely connected to the phase composition of the films will be discussed in more detail.

The development of the optical properties and electric resistivity ρ of the Zr-Si-N films is schematically displayed in the Fig. 5.7. It can be seen that the crystalline Zr-Si-N films characterized by the $ZrSi_x$ phase or by the mixture of the crystalline $ZrSi_x$ and the amorphous Si_3N_4 phases sputter deposited in the metallic mode of sputtering at low values of p_{N_2} are optically opaque and electrically conductive. On the contrary, the X-ray amorphous Zr-Si-N films produced in the transition and nitride mode of sputtering at $p_{N_2} > 0.1$ Pa are optically transparent and electrically insulating.

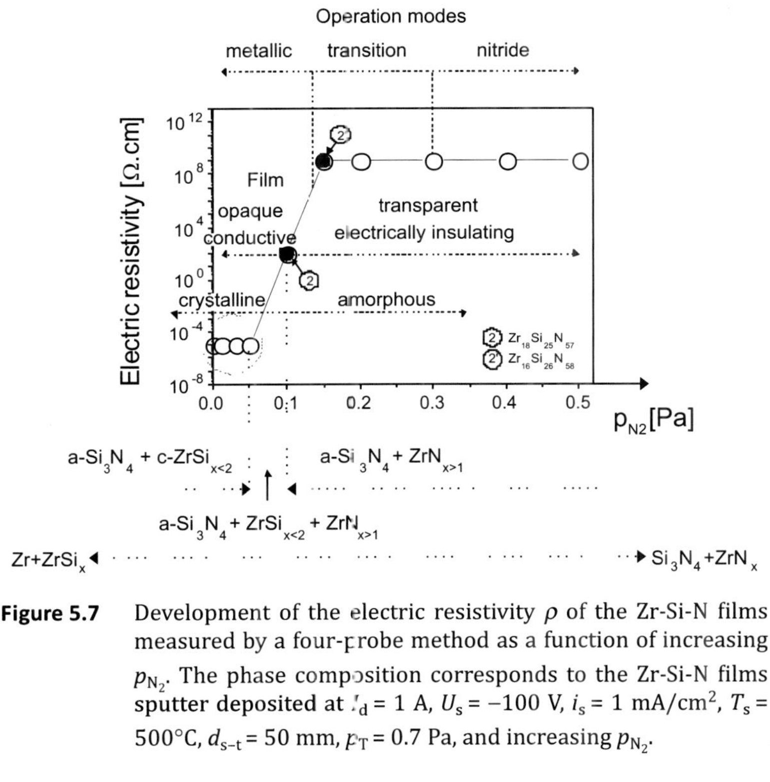

Figure 5.7 Development of the electric resistivity ρ of the Zr-Si-N films measured by a four-probe method as a function of increasing p_{N_2}. The phase composition corresponds to the Zr-Si-N films sputter deposited at $I_d = 1$ A, $U_s = -100$ V, $i_s = 1$ mA/cm^2, $T_s = 500°C$, $d_{s-t} = 50$ mm, $p_T = 0.7$ Pa, and increasing p_{N_2}.

The electric resistivity ρ of the Zr-Si-N films sputtered at low values of p_{N_2} exhibits a very low value below a measurable limit of the four probe measuring device ($\rho \approx 10^{-5}$ Ωcm). At the beginning of the transition mode of sputtering at $p_{N_2} = 0.1$ Pa, the film resistivity starts to increase and the films sputtered already at $p_{N_2} \geq 0.15$ Pa exhibit very high almost constant values showing fully dielectric behavior of the films.

That dramatic change in electrical properties correlates well with the change of the phase composition of the Zr-Si-N films. Since it depends strongly on the elemental composition of the film, only a small variation in deposition parameters may cause significant changes in the phase development which consequently affect other film properties. The dominant effect of the elemental and phase composition of the Zr-Si-N films on their physical properties is demonstrated in the case of two X-ray amorphous $Zr_{18}Si_{25}N_{57}$ and $Zr_{16}Si_{26}N_{58}$ films deposited at p_{N_2} = 0.1 and 0.15 Pa, respectively. The films, depicted in Fig. 5.7 as points 2 and 2′, strongly differ in electrical conductivity and optical transparency although their elemental composition is almost the same. This difference is due to the fact that even a very small increase of p_{N_2} results in a serious change of the film phase composition. The Zr-Si-N films sputtered at $p_{N_2} < 0.15$ Pa are composed mainly of the $ZrSi_x$ phase. Together with free Zr or Si atoms, metallic $ZrSi_x$ phase is also responsible for electrical conductivity and non-transparency of these films. While their amount decreases in the film, the electrical resistivity rapidly increases. Similar behavior has been observed also in the case of Ta-Si-N films with a high (≥40 at.%) Si content [43]. Since sufficient amount of N is available in the films sputtered at $p_{N_2} \geq$ 0.15 Pa, all Si atoms are converted into Si_3N_4 and, similarly, free Zr atoms form the ZrN_x compound. Once the nitrification level of both Zr and Si increases with a higher amount of N incorporated into the film, the over-stoichiometric $ZrN_{x>1}$ phase is formed. Compared to the stoichiometric ZrN, which exhibits a metallic behavior with a low electrical resistivity and the gold-like optical reflectance, the nitrogen-rich $ZrN_{x>1}$ phase is semi-transparent and dielectric [44]. Hence, the Zr-Si-N films sputtered at $p_{N_2} > 0.1$ Pa are no longer electrically conductive since the electrical and optical properties of the films are given by both dominating electrically insulating and in addition optically transparent $ZrN_{x>1}$ and Si_3N_4 phases. In general, the electrical resistivity of the metal nitride film is a contribution of several independent electron scattering processes due to phonons, impurity atoms, and film defects. An increase in the resistivity of the ZrN_x phase is thus, ascribed to a higher electron scattering caused by higher amount of N incorporated into the lattice. Since N plays the role of impurities in the lattice, its effect on the electrical properties is the higher the more over-stoichiometric phase is formed. Furthermore, the stretching out of the cell that increases the distance between Zr atoms leads to the weakness of the metallic bond and thus, to the increase of the resistivity [45]. It is worthy to note that the films sputtered at higher target power are electrically insulating and optically transparent only when they are sputtered at higher p_{N_2}, due to the shift of the transition and nitride mode of sputtering. The differences in the elemental and phase composition between films sputtered at $I_d = 1$ and 2 A are displayed in Figs. 5.3 and 5.4. In summary, the variations in deposition conditions prove to be an easy way how to control the properties of the Zr-Si-N films by changing their phase composition. The dependence of the thermal behavior of the Zr-Si-N film on the deposition conditions is analyzed in detail further in Section 5.11.

5.5 Structure of Zr-Si-N Films

The microstructure analysis of the Zr-Si-N films is very important, since it enables to understand the complicated relationship between the phase composition and other physical properties of the films. Taking into account previous studies on the elemental and phase composition of the Zr-Si-N films, it can be expected that the zirconium silicide will be the only phase detected by X-ray diffraction analysis at low N content in the sputtered films. This fact is closely associated to the preferential type of bonding between incorporated atoms. Since the tendency to form bonds between Zr and Si atoms during the film deposition is relatively high, it is not surprising that the structure of the non-reactively sputter deposited Zr-Si film reveals the $ZrSi_2$ phase as the only established. It is fully in an agreement with results obtained by Raman spectroscopy measurement (Fig. 5.5). The enthalpy of formation $\Delta H_{f\,ZrSi2}$ of the $ZrSi_2$ phase is reported as -159 kJ/mol [36].

In Fig. 5.8, the structure of two non-reactively sputtered Zr-Si films is shown. Both films exhibit polycrystalline structure composed of the mixture of $ZrSi_2$ grains with different crystallographic orientations. The structure, however, significantly changes corresponding to the elemental composition of the sputtered films. An amount of elements present in the film is closely related to the deposition conditions and may vary in relatively wide range. More intensive ion bombardment in the case of the $Zr_{43}Si_{57}$ film sputtered at $i_s = 1$ mA/cm^2 results in lower Si concentration due to its preferential resputtering from the growing film compared to the $Zr_{35}Si_{65}$ film ($i_s = 0.5$ mA/cm^2). Besides the change in the elemental composition, also a difference in the orientation of the $ZrSi_2$ grains has been observed in both films. Furthermore, more dense structure is formed under higher ion bombardment.

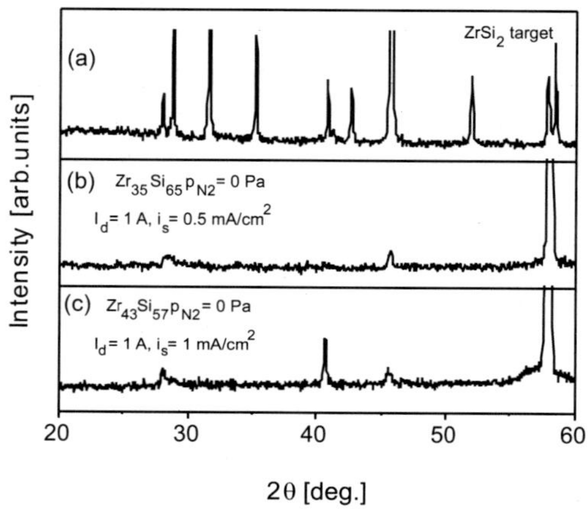

Figure 5.8 Structure of (a) an alloyed $ZrSi_2$ target and the Zr-Si films non-reactively sputter deposited on steel substrates at $I_d = 1$ A and (b) $i_s = 0.5$ mA/cm^2, and (c) $i_s = 1$ mA/cm^2. The reflections from the steel substrate were subtracted.

The variety of accessible information makes X-ray diffraction analysis a powerful tool for the structural analysis. Using this technique it was found that the structure of the Zr-Si-N films strongly depends on the elemental composition accordingly to the structure zone model proposed by Patscheider for Me-Si-N films [46]. The observed continuous transition from the crystalline microstructure to the very fine one with increasing partial pressure of nitrogen p_{N_2} completely corresponds to the assumptions of preferentially bonding in the Zr-Si-N films. Typical structure evolution of approximately 3 μm thick Zr-Si-N films sputter deposited on steel substrates at $I_d = 1$ A, $i_s = 1$ mA/cm^2, $U_s = -100$ V, and increasing partial pressure of nitrogen p_{N_2} is illustrated in Fig. 5.9. Here, the values of the film microhardness, H and the ratio N:Si are also given.

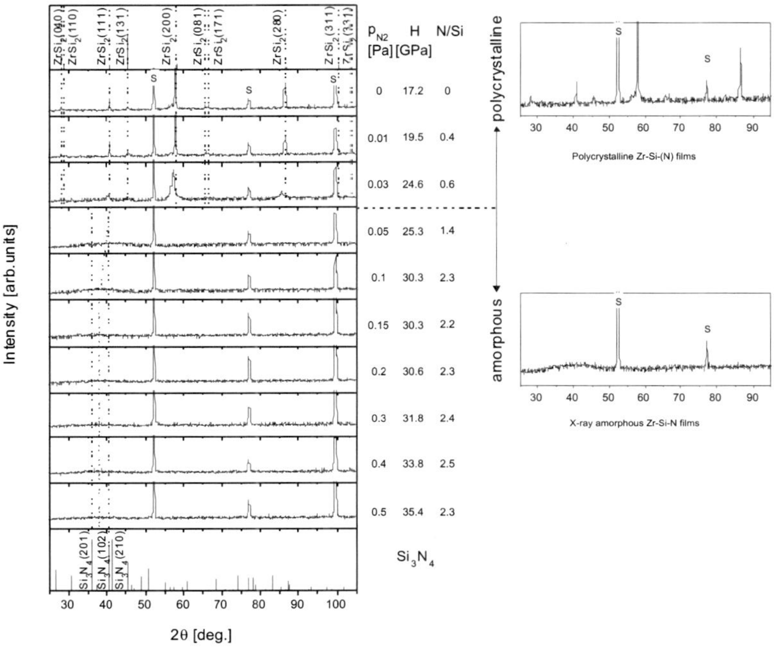

Figure 5.9 Development of the structure of the Zr-Si-N films sputter deposited at I_d = 1 A on steel substrates kept at a d.c. bias U_s = −100 V, i_s = 1 mA/cm^2, with increase in partial pressure of nitrogen p_{N_2}; here S denotes the reflection from the steel substrate. Reprinted from Ref. [28], Copyright (2005), with permission from Elsevier.

The detailed XRD scan of the non-reactively sputter deposited single-phase Zr-Si film (p_{N_2} = 0 Pa) shows strong narrow reflection lines attributed to the $ZrSi_2$ grains grown in (200) and (280) directions. A high intensity of these narrow diffraction peaks suggests a high amount of relatively large grains. The mean value of the crystallite size, calculated from the corresponding peak position and full width at half maximum intensity (FWHM) according to the Scherrer formula [47], is about 30 nm. On the other hand, many of the weak reflections from $ZrSi_2$ grains with (111), (131), and (040) orientation with a low intensity were also observed. The structure of the Zr-Si film markedly changes if a reactive gas is introduced into the deposition system. The crystalline feature of the films sputtered at $p_{N_2} \leq 0.03$ Pa dramatically changes to X-ray amorphous one with increase in p_{N_2} as a consequence of the phase evolution. Since only small amount of Si is preferentially consumed with N at low p_{N_2}, the silicide phase is further formed with available Zr atoms. The presence of the $ZrSi_2$ phase in the film, together with a relatively high heat of formation of the zirconium disilicide, illustrates that this reaction dominates at low $p_{N_2} < 0.05$ Pa. As the N content increases in the film with increase in p_{N_2} and i_s (see Fig. 5.2), a higher amount of Si is bonded to N. Then, an amount of established Si_3N_4 **phase increases in** the film to the detriment of the $ZrSi_2$ one. In summary, the structure of non-reactively sputtered Zr-Si films, consisting of the crystalline $ZrSi_2$ phase, develops into the mixture of the amorphous Si_3N_4 and crystalline $ZrSi_x$ **phases at higher** p_{N_2}. The fact that the film structure obviously, alters when the N is incorporated into the growing film confirms its significantly high effect on the phase evolution of the Zr-Si-N films. In principal, it is provided through the amorphous Si3N4 phase effectively hindering the grain growth. The observed intensity decrease of the narrow high-intensive reflection (100) and (140) lines of the $ZrSi_2$ grains and their pronounced broadening serve as an indicator of the formation of a fine grained film structure. Furthermore, an obvious peak shift to the lower angles of 2θ was observed with increasing p_{N_2}. This fact is closely related to the macrostress induced as N atoms are gradually incorporated into the growing film. Increase of the film macrostress reflects directly a distortion of the $ZrSi_2$ lattice together with the formation of Si–N bonds at higher p_{N_2}. At lower p_{N_2}, the film is consisted mainly of the crystalline $ZrSi_2$ phase embedded within a thin amorphous Si_3N_4 tissue. The attendance of the Si_3N_4 matrix effectively inhibits the grain capability to growth, which finally results in the formation of a nanocomposite structure. Once the N content in the film is sufficiently high and exceeds an amount needed to form bonds with all the Si atoms, ZrN_x phase starts to be formed as well. Thus, the films deposited in the transition mode of sputtering significantly differ from the films sputtered in the metallic mode since they are consisted of two hard and stable nitrided phases: Si_3N_4 and ZrN_x. This fact is indicated by the structural changes as well as by changes of the film morphology and consequently by mechanical properties of those films.

The presence of an amorphous matrix results in hindering of the grain growth also in the case of the ZrN grains. While the Si content in the film increases up to 25 at.%, the Si_3N_4 phase becomes dominant and an amorphous-like structure characterized by one very broad, low-intensity reflection line with a maximum located at $2\theta \approx 37.5^{\circ}$ at $p_{N_2} > 0.03$ Pa is finally developed. It corresponds to the zone "C" of the Me-Si-N structure zone model [46] where the ultra-fine grained ZrN_x phase is dispersed in an amorphous matrix (see Fig. 4.1). Although ZrN is not detectable by XRD at higher p_{N_2}, it might be incorrectly interpreted as an amorphous phase. The X-ray reflections may be just below the detection limit since it reflects a small quantity of grains or ultra-fine grains dispersed in higher amount of an amorphous matrix [48]. Indeed, an amorphous feature of the ZrN_x phase has been typically observed above the film stoichiometry [44]. The tendency to amorphization might explain why no X-ray reflections were detected in the Zr-Si-N film sputtered at higher partial pressure of N. Moreover, it has been referred that no ternary Zr-Si-N phase exists [36]. It means that the films must be composed of individual phases and the components can, in principal, be in an X-ray amorphous state. Such a nanocomposite nature of the Zr-Si-N films has been revealed by detailed high-resolution transmission electron microscopy (HRTEM) investigations [14] and confirmed by Fourier transform infrared spectroscopy (FTIR) and X-ray photoelectron spectroscopy (XPS) measurements [15].

5.6 Morphology of Zr-Si-N Films

Morphology of the deposited films depends strongly on deposition conditions. The substrate temperature, composition of the sputtered flux, and working gas as well as the energy of impinging particles play an essential role in the film growth. It is generally accepted that the surface and bulk diffusion are the most determinant atomic processes of the structure evolution. This phenomenon, however, becomes more complicated if the film growth is affected by the chemical reactions of individual film forming particles. The addition of Si into transition metal nitride films has been shown to hinder the grain growth and thus, result in a decrease of grain size and weakening of film texture. Similar structure refinement effect was observed after the N is introduced into the Zr-Si films. Also in that case the structure of the growing film significantly alters through so formed Si_3N_4 phase. In order to explore, its decisive effect on the film growth in more detail, the morphology of four representative high Si-containing Zr-Si-N films was investigated by SEM. Their cross-sectional SEM images are depicted in Fig. 5.10.

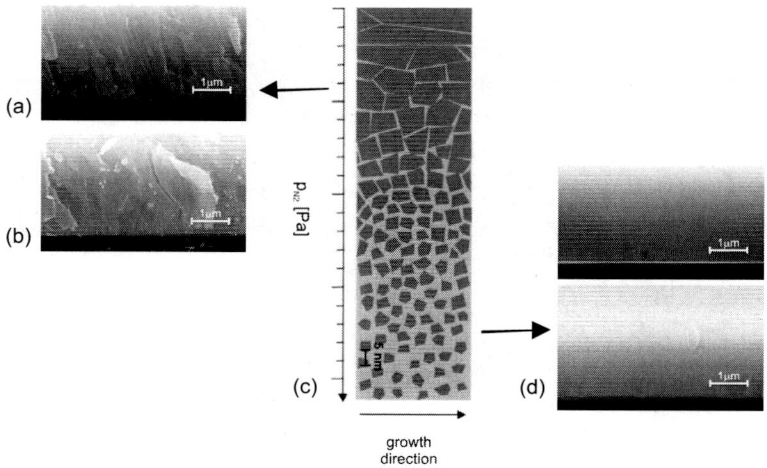

Figure 5.10 Morphology of the Zr-Si-N films in cross-section sputter deposited at $U_s = -100$ V, $I_d = 1$ A, $i_s = 1$ mA/cm^2, and various p_{N_2} (a) 0 Pa, (b) 0.03 Pa, (c) 0.1 Pa, and (d) 0.4 Pa.

According to the reported experimental results, it can be seen that the amorphous Si_3N_4 phase acts as a strong grain growth inhibitor decisively affecting the structure evolution [49]. While the non-reactively deposited films exhibit fine columnar structure corresponding to the zone "T" of the Thornton's structural model [50], vanishing of the well-developed columns was observed already at relatively low N partial pressure $p_{N_2} = 0.03$ Pa (see Fig. 5.10b). The structure is still columnar, however, the columnar feature becomes slightly blurred. The tendency of the Si_3N_4 phase to hinder the grain growth results in a refinement of the film structure, which is very pronounced at higher content of Si_3N_4 phase formed in the film (the highest volume content of Si3N_4 is formed at approximately 60 at.% of N in the film; see Fig. 5.6) and the ultra-fine grained nanocomposite structure typical for the Zr-Si-N films with a high (>25 at.%) Si content, where the ZrN grains are dispersed in a high amount of an amorphous matrix, is fully developed. A continuous addition of N into silicon-rich Zr-Si-N films reveals a formation of dense films with a morphology altering from columnar to featureless one at high N content. The glassy appearance of the films deposited at $p_{N_2} > 0.1$ Pa is clearly evident (see Fig. 5.10c and d).

Glass-like structure is characteristic for low ordered or amorphous structures with a low density of growth defects. Such films are typically characterized by low intensity broad X-ray diffraction peaks similarly to the Zr-Si-N films sputter deposited at $p_{N_2} \geq 0.05$ Pa [28,51]. Very compact cross-section morphology of those films is also associated with a high level of energetic ion bombardment during their growth. An enhancement of the surface diffusion by bombarding ions may, in addition, improve mechanical properties of sputtered films through a more compact structure formed. However, to find optimal conditions for deposition of films with acceptable properties is a complicated matter. The intensity of the ion bombardment can strongly affect not only the film morphology but also the surface roughness as well. It has been reported that the structure and surface morphology of the films deposited without the substrate bias may be rough although they contain relatively high amount of the Si_3N_4 phase [52]. Insufficient adatom mobility, which is no longer promoted by the energy of energetic ions bombarding the growing film, leads into development of porous structures with a high amount of defects and voids. On the contrary, high bias voltage may again result in a rough surface since the high ion bombardment brings out obvious resputtering effect due to high collision energy of ions [52]. The use of a bias voltage of −100 V may thus result in both very dense structure and a smooth surface.

All the experimental findings show a development of very specific type of a nanocomposite structure. The Zr-Si-N films containing a high Si content are composed of the mixture of ultra-fine grained ZrN phase and dominant amorphous Si_3N_4 if they are sputter deposited from an alloyed $ZrSi_2$ target.

5.7 Surface Roughness

Typical development of the surface roughness R_a of the Zr-Si-N films with increasing p_{N_2} is illustrated in Fig. 5.11. The decrease of R_a corresponds to the refinement of the film structure which is related to the increasing amount of N incorporated into the film. While N is bonded to Si, well-developed columnar feature of the Zr-Si-N films with a low N content vanishes and fine-grained morphology is formed. As a consequence of the impeding effect of Si_3N_4 on the grain growth, the roughness of the film surface decrease from 11 nm of the non-reactively sputtered Zr-Si film down to 8 nm corresponding to the film deposited at p_{N_2}=0.03 Pa. The Zr-Si-N films with a N content higher than 50 at.% are characterized by a glassy appearance and the roughness close to 6 nm. Especially insulating and optically transparent Zr-Si-N films sputter deposited at p_{N_2} > 0.15 Pa reveal very smooth surface due to the ultra-fine grained structure formed through the high amount of the amorphous Si_3N_4 phase.

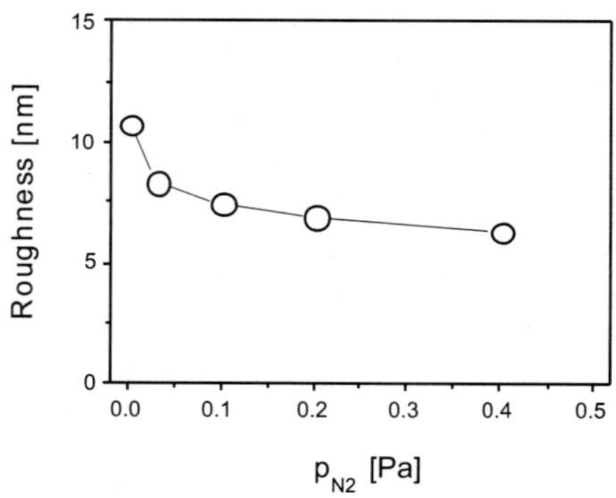

Figure 5.11 The mean surface roughness R_a of the Zr-Si-N films sputter deposited at U_s = −100 V, I_d = 1 A, and i_s = 1 mA/cm^2, as a function of p_{N_2}. The values are evaluated from corresponding images of optical profilometry.

For a quantitative analysis, the surface plots were generated from the optical profilometry measurements. Figure 5.12 shows the influence of the partial pressure of nitrogen p_{N_2} on the development of surface roughness of the Zr-Si-N films. A detailed investigation of the surface topography gives additional insights in the growth phenomenon involved during the film growth. It is well known that surface roughness directly reflects the film structure and morphology. Depending on the deposition conditions and resulting structure of the initial film, a characteristic change of the surface morphology is observed. The analysis of height distributions of rougher surfaces shows a few randomly spread craters on the film surface with a sharp peak–valley distribution. It is characteristic for the crystalline structure which is developed in the Zr-Si-N films sputtered at p_{N_2} pressures below 0.1 Pa. Here, the depth distribution was determined from the deviation of surface points with respect to the mean surface level; negative values are defined as height and positive values as depth. The rougher surface appearance of the film continuously changes into a smooth one of regular hill-like surface structure with significantly smaller feature size, as a fine-grained nanocomposite structure is formed. An essential role of the deposition conditions and the phase composition on the film structure is obviously seen. It is fully in accordance with other findings and verifies close relationship between the surface topography and the structure development.

Although the films deposited at the low N pressure regime exhibit well-developed crystalline structure with relatively large grains arranged in a columnar feature, the differences in the surface morphology are not extremely high. Very low film roughness below 10 nm of all the Zr-Si-N films is also related to relatively high ion bombardment during deposition resulting in typically very dense and compact film structure [53]. It has been observed that generally higher fluxes of ions lead to the higher nucleation density and consequently, to fine-grained morphology [54]. On the contrary, lower values of the ion current results in coarse grains with a high density of defects. Hence, the structure consisting of sub-grains and dislocation cell boundaries is usually observed [55]. The synergetic effect of the structure refinement through the formation of the nanocomposite morphology and the coarsening caused by insufficient surface mobility of adatoms should always be considered. In order to ensure the development of a dense, compact, defects-free structure of deposited films with homogeneously dispersed fine ZrN grains in an amorphous Si_3N_4 matrix, sufficient energy has to be delivered into to the growing film. Low surface mobility conditions of the deposited material at low deposition temperature and/or absence of the ion bombardment during film growth leads to the development of different phase composition and structure morphology [56]. Only the enhanced surface mobility, either by an increase in temperature or by the substrate biasing, induces the pronounced phase evolution and the formation of the ultra-fine nanocomposite structure composed of the nanocrystalline ZrN_x and the amorphous Si_3N_4 phases.

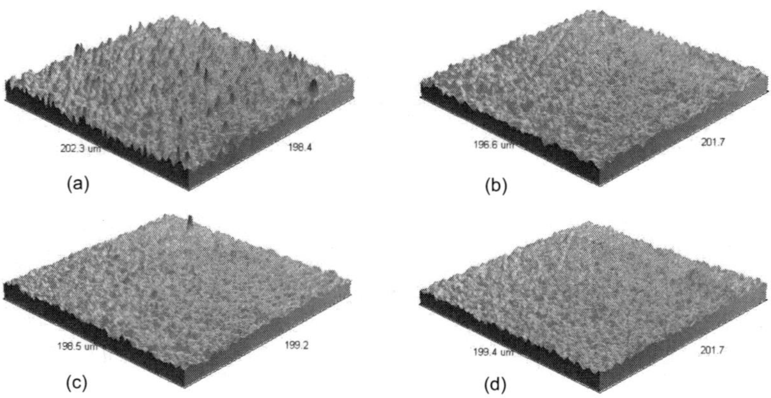

Figure 5.12 Surface topology of the Zr-Si-N films sputter deposited at U_s=−100 V, I_d = 1 A, i_s = 1 mA/cm^2, and various p_{N_2} (a) 0 Pa, (b) 0.03 Pa, (c) 0.1 Pa, and (d) 0.4 Pa, measured by optical profilometer.

From the above shown experimental results, a strong influence of the volume fraction of the Si_3N_4 phase on the film properties can be seen. The synergetic effect of the Si and N content on the phase composition is expressively reflected by the film structure and the surface morphology. The development of other film properties, particularly the film hardness and elastic properties of the Zr-Si-N films with a high (>25 at.%) Si content will be discussed in more detail in following sections.

5.8 Mechanical Properties

5.8.1 Microhardness of Zr-Si-N films

Understanding the relationships between microstructure and mechanical properties has always been one of the primary goals of the materials science. With respect to the possible usage of the Zr-Si-N films as protective coatings in load-bearing applications, the behavior under applied load has to be fully understood. Mechanical properties of thin films can be sufficiently characterized by their microhardness, elastic constants, elastic recovery, wear and abrasion resistance, friction coefficient, fatigue, adhesion, resistance to plastic deformation and crack propagation, strength, or mechanical fatigue damage. These characteristics are usually determined from indentation techniques, tribological, scratch, or surface fatique testing. Development of microhardness of the Zr-Si-N films, sputtered at different combinations of discharge current I_d and ion current density i_s with increasing partial pressure of nitrogen p_{N_2} and the film deposition rate a_D, is depicted in Fig. 5.13.

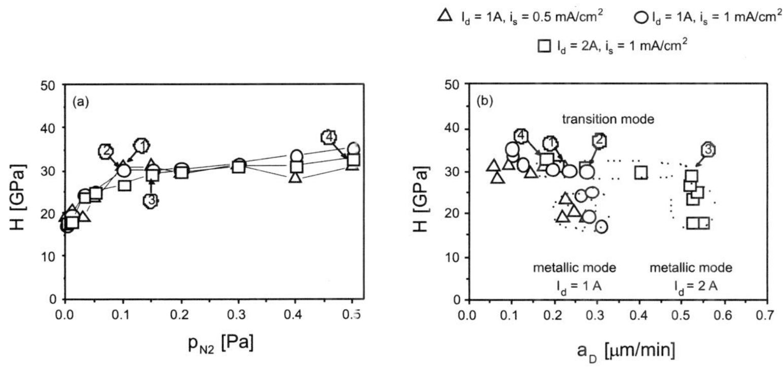

Figure 5.13 Microhardness H of the Zr-Si-N films as a function of (a) partial pressure of nitrogen p_{N_2} and (b) the film deposition rate a_D. The films were sputter deposited at $U_s = -100$ V and different combinations of discharge current I_d and ion current density i_s: (i) $I_d = 1$ A, $i_s = 0.5$ mA/cm^2; (ii) $I_d = 1$ A, $i_s = 1$ mA/cm^2; and (iii) $I_d = 2$ A, $i_s = 1$ mA/cm^2. The labels correspond to films denoted in Fig. 5.1. Reprinted from Ref. [28], Copyright (2005), with permission from Elsevier.

Figure 5.14 shows that the high Si-containing Zr-Si-(N) films can be divided into three different groups. The first one presents the non-reactively deposited $ZrSi_2$ alloy films sputtered in a pure Ar atmosphere ($p_{N_2} = 0$). They exhibit relatively high hardness of about 17 GPa, much higher than the hardness of other alloyed compounds [57]. This fact is associated to very compact and dense microstructure, developed during deposition under relatively intensive ion bombardment and high temperature, which allows the film to bear high loads. The second group of films significantly differs in terms of the phase composition and hardness due to N incorporation into the films. Addition of N into the films results in a significant change of the phase composition and consequently, in a linear increase of the film hardness. It is attributed to the formation of the nanocomposite structure consisted of increasing amount of the amorphous Si_3N_4 phase and the $ZrSi_2$ crystallites while the films are deposited in the metallic and transition mode of sputtering. The crack propagation and dislocation movement under applied load are then effectively hindered through the fine-grained film microstructure. This effect is even more pronounced if both hard Si_3N_4 and ZrN phases are developed in the films prepared at the nitride sputtering mode corresponding to p_{N_2} higher than 0.1 Pa and 0.15 Pa (the third group of films in Fig. 5.14). Hardness of the Zr-Si-N films is nearly two times higher than hardness of individual components, from which the films are composed ($H_{Si_3N_4}$ = 17.2 GPa and H_{ZrN} = 16 GPa [34]). It stands for the enhanced hardness of the nanocomposite Zr-Si-N films with a high (>25 at.%) Si content. Typical saturation of the hardness value to approximately 30 GPa is closely related to a maximum amount of N, which can be incorporated into the films during their reactive sputtering in a mixture of Ar and N (see Fig. 5.2). Strong bonds between Si, Zr, and N atoms and a high enthalpy of formation of the individual compounds suggest an easy formation of both amorphous Si_3N_4 and crystalline ZrN_x phases. It is, however, dependent on the efficiency of the dissociation and ionization of molecular N in the plasma discharge and its extraction and incorporation into the growing film. The experiments summarized in this chapter unambiguously proved the essential role of the magnetron discharge current I_d and ion current density i_s on the amount of the amorphous Si_3N_4 phase formed through the incorporated N content. Hardness of the Zr-Si-N films remains constant although p_{N_2} increases further up to 0.3 Pa and just slightly increases at higher values of $p_{N_2} > 0.3$ Pa. Differences in hardness of the Zr-Si-N films, sputtered at the same value of $p_{N_2} > 0.3$ Pa and different combinations of I_d and i_s, are due to different elemental and phase composition of these films (see Figs. 5.2 and 5.4). The experimental results also show that only the right combination of deposition conditions results in development of the Zr-Si-N films with a stable microstructure and high microhardness. Thus, the hard Zr-Si-N films with high hardness $H \approx 30$ GPa are formed if they contain a maximum amount of N and consequently, the highest volume content of the amorphous Si_3N_4 phase formed (see Fig. 5.14).

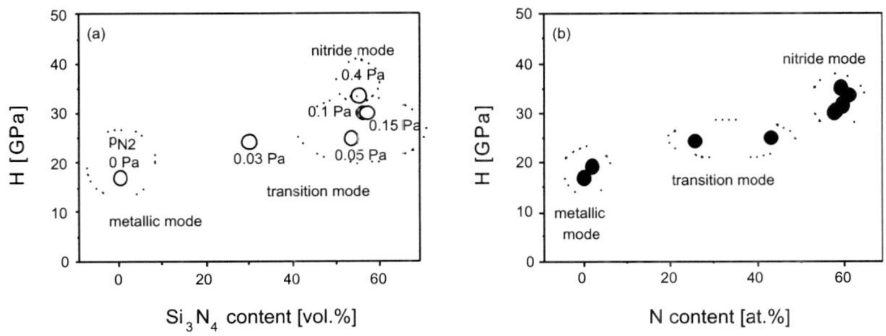

Figure 5.14 Microhardness of the Zr-Si-N films sputter deposited at the metallic, transition, and nitride modes of sputtering, as a function of the (a) Si_3N_4 and (b) N content.

For industrial applications, the deposition rate of the growing film is a highly relevant topic. Therefore, an important finding is the fact that the films with a maximum hardness can be deposited already at the end of the metallic mode of sputtering at low values of $p_{N_2} \approx 0.15$ Pa. Here, the deposition rate does not fall below 0.2 μm/min, see points 1, 2, and 3 in Fig. 5.13. From Fig. 5.13b it is seen that the hard Zr-Si-N films with a high (≥25 at.%) content of Si and microhardness $H \approx 30$ GPa can be produced at relatively high deposition rates. Here, a_D increases almost linearly with increasing I_d from $a_D \approx 0.27$ μm/min ($I_d = 1$ A) to $a_D \approx 0.52$ μm/min ($I_d = 2$ A). It is approximately four times faster than a deposition of binary nitrides of the same thickness sputtered in the nitride mode [58,59]. Hard Zr-Si-N films sputtered in the transition and nitride modes of sputtering exhibit hardness slightly higher than that of the films sputtered at the end of the metallic mode reaching the value of $H \geq 30$ GPa (see for instance the point 4 in Fig. 5.13a). The reason can be found in a change of the phase composition of the films with increasing p_{N_2}. While the Zr-Si-N film, denoted as point 1 in Fig. 5.4a, is composed of a-Si_3N_4 + (Zr,Si), the Zr-Si-N film sputtered in the nitride mode, denoted as point 4 in Fig. 5.4c, is composed of a-Si_3N_4 + $ZrN_{x>1}$. The development of the phase composition and the stoichiometry of the ZrN phase correspond to the N content incorporated into the film. It is important to note that the variation of the elemental and phase composition is directly reflected by the physical properties of the Zr-Si-N films, mainly by their thermal stability and oxidation resistance. Even if the Zr-Si-N films sputter deposited at $p_{N_2} \geq 0.1$ Pa exhibit almost constant microhardness close to 30 GPa, only the films deposited at the end of the nitride mode of sputtering exhibit the highest thermal stability and oxidation resistance (see Sections 5.11 and 5.12).

5.8.2 Young's Modulus and Elastic Recovery

Besides the film hardness H, also the knowledge on the elasticity is very important for the complex characterization of the mechanical properties of thin films. The reduced Young's modulus $E^* = E/(1-\nu^2)$ and elastic recovery W_e can be easily evaluated from the loading–unloading curves measured by nano- or micro-hardness testers [60]; here E is the Young's modulus and ν is the Poisson's ratio of the measured film. Both quantities serve for additional information on the behavior of the films under the applied load. A typical development of the reduced Young's modulus E^* and elastic recovery W_e of the Zr-Si-N films, sputtered at T_s = 500 °C, U_s = −100 V and different combinations of I_d and i_s, with increasing partial pressure of nitrogen p_{N_2} is given in Fig. 5.15.

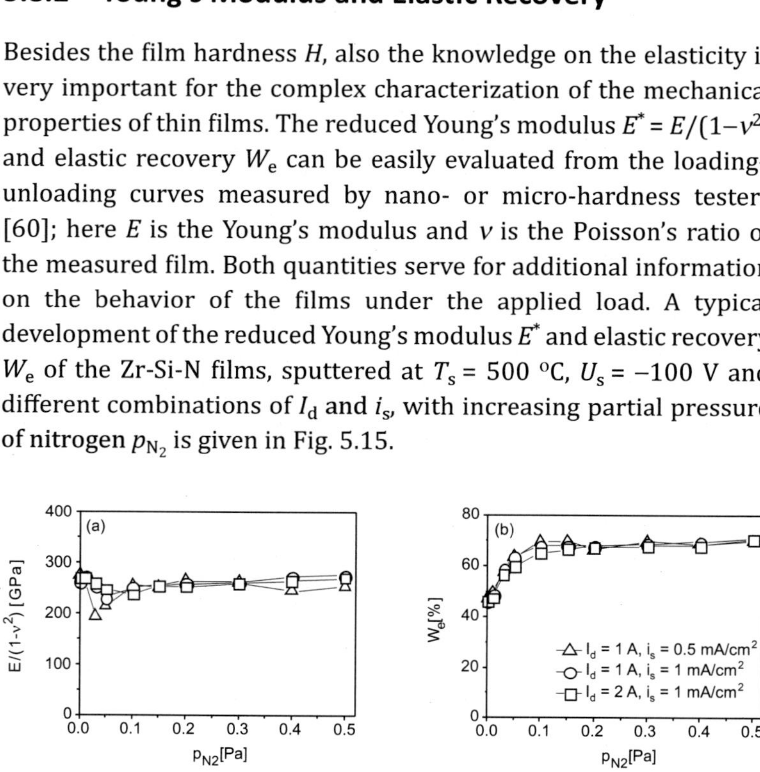

Figure 5.15 (a) Reduced Young's modulus E^* and (b) elastic recovery W_e of the Zr-Si-N films sputtered at U_s = −100 V, different combinations of I_d and i_s and increasing p_{N_2}. Reprinted from Ref. [28], Copyright (2005), with permission from Elsevier.

The values of the reduced Young's modulus E^* of the Zr-Si alloy films are comparable with those of the hardest Zr-Si-N films. Observed slight change in E^* of the Zr-Si-N films produced at $p_{N_2} < 0.1$ Pa and $p_{N_2} \geq 0.3$ Pa correlates well with the changes in elemental and phase composition, similarly as their hardness *H*. The formation of the amorphous Si_3N_4 phase results in a development of stiffer films at its intermediate and high content. However, at the very beginning of the Si_3N_4 phase formation, the Zr-Si-N films exhibit significantly lower values of E^*. It indicates that also in the case of the Zr-Si-N films with high Si content the optimum volume fraction ratio of the amorphous Si_3N_4 phase and crystalline $ZrSi_2$ or ZrN phase is required for forming hard and stiff films.

While the reduced Young's modulus remains almost constant in the whole investigated range of the partial pressure of nitrogen p_{N_2}, the elastic recovery W_e follows very similar trend as the film hardness *H* (compare Figs. 5.13a and 5.15b). The elastic recovery W_e first increases from about 0.05 Pa up to 0.1 Pa with increase in p_{N_2} and for $p_{N_2} > 0.1$ Pa is approximately constant.

5.8.3 Interrelationships between *H*, E^*, and W_e

Recently, an increasing attention is given to the relationship between the hardness *H* and the Young's modulus *E* during the development of hard films for mechanical applications [61,62]. It is well known that both quantities are closely related to the elemental and phase composition of the forming films resulting mainly from the deposition conditions. Those decisive factors determine the final film structure and corresponding mechanical properties of the films. It is generally considered that more important than to reach an exceptional high hardness is to have a good compromise between a high hardness and a low elastic modulus. Only by a right combination of both *H* and *E*, an in-service improved performance of the films can be reached [61].

The interrelationships between *H*, E^*, and W_e for the Zr-Si-N films, sputtered under different deposition conditions, are given in Fig. 5.16. The non-reactively deposited alloy $ZrSi_2$ and $ZrSi_2$(N) films with a small (<5 at.%) N content exhibit completely different behavior compared to that of the Zr-Si-N films deposited in the transition and nitride modes of sputtering. While the $ZrSi_2$ alloy films ($p_{N_2} = 0$ Pa), here represented by full symbols, exhibit high values of E^* of about 270 GPa, the formation of a nanocomposite structure in the films with an intermediate N content results in a decrease of the reduced

Young's modulus E^*. However, by exceeding the critical N content of about 20 at.%, the amorphous Si_3N_4 phase starts to dominate in the film that dramatically influences its mechanical properties. Thus, the increasing tendency of H, H^3:E^{*2}, and W_e of the Zr-Si-N films produced in the transition and nitride modes can be clearly seen if it is plotted against E^*, H, and H^3:E^{*2}, respectively (see Fig. 5.16). The H^3:E^{*2} ratio is related to the resistance of the material to plastic deformation and it is being considered that the higher is its value, the better the tribological performance of the material in a great number of wear conditions [63]. The probability of plastic deformation is significantly reduced in materials with high hardness H and low modulus E^*. It results from the fact that the stress under the applied load can be distributed over a wider area and its maximum does not exceed the critical values for the onset of the plastic deformation.

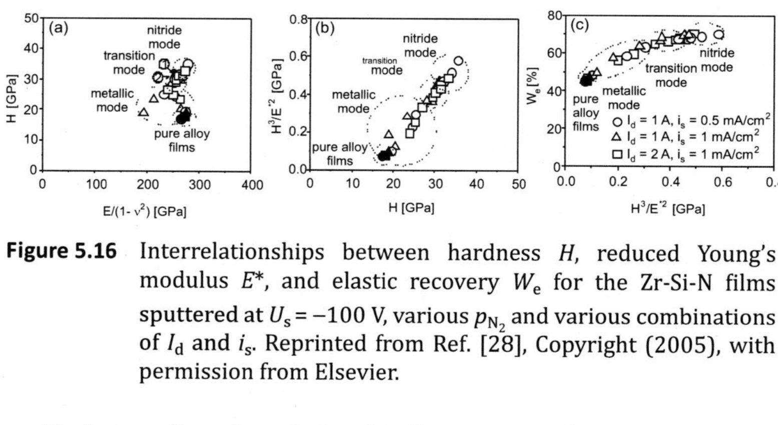

Figure 5.16 Interrelationships between hardness H, reduced Young's modulus E^*, and elastic recovery W_e for the Zr-Si-N films sputtered at $U_s = -100$ V, various p_{N_2} and various combinations of I_d and i_s. Reprinted from Ref. [28], Copyright (2005), with permission from Elsevier.

Understanding the relationship between mechanical properties and phase composition allows optimizing the conditions for films deposition with respect to their possible application. Obvious difference in the film behavior related to their structure and phase composition is evident from Fig. 5.16. The crystalline non-reactively deposited alloy $ZrSi_2$ films exhibit relatively low resistance against plastic deformation even if they are relatively stiff. The H^3:E^{*2} ratio does not exceed the value of 0.1 in that case. With higher N amount incorporated, the films bear much higher loading without failure. Besides the highest resistance to plastic deformation, indicated by a high $H^3:E^{*2}$ ratio, the Zr-Si-N films produced in the nitride mode exhibit the most elastic behavior (see Fig. 5.16c).

The values of $H^3:E^{*2}$ ranging over a wide range between 0.3 and 0.6 are considerably higher than those for Zr-Si-N films with a low (<7.5 at.%) Si [15]. The difference is given by the volume fraction ratio of the individual phases of which the film is composed. While in the Zr-Si-N films with a low (<10 at.%) Si content the ZrN phase dominates over Si_3N_4, the films with a high (>25 at.%) Si content and the ratio N:Si > 1.33 are composed of the ZrN grains embedded within dominating amorphous Si_3N_4 matrix. This fact has a positive impact on the stability of the film structure at elevated temperature and on the resistance against oxidation.

5.9 Macrostress σ in Zr-Si-N Films

It is of a great importance to understand mechanisms of the residual stress generation during deposition of thin films. The magnitude of macrostress is a decisive factor while the films are used in tool applications. If the tools are heavily loaded, complicated three-axial stress states occur, which can cause failures once critical stress values are exceeded. It has been proved by many investigations that the residual stress induced in thin films strongly depends on the deposition conditions [64]. Since the residual macrostress is ascribed to the amount of growth defects like point and line defects or stacking faults, it is directly associated to both energy of adatoms and intensity of ion bombardment applied during deposition. While in the first case, intrinsic stress arising from the film formation can be significantly reduced if mobile adatoms are arranged so they reach the minimum surface energy corresponding to the energetically favored equilibrium state, a bombardment of biased substrates by high energetic ions extracted from the plasma discharge causes an increase of the intrinsic stress through the generated growth defects [65]. The result of the competitive acting of both phenomena is reflected by a final stress level in deposited film. Relative high ion bombardment at the biased substrate of −100 V is referred as the limit to still act positively on the film growth. Under such conditions, dense and compact film structures with optimum stress level are typically developed in magnetron sputtered film systems [66]. Too high amount of growth defects generated in the films at higher bias voltages no longer provides enhanced mechanical properties. It is due to the fact that stresses induced in the film while coated tools are mechanically loaded contribute to those generated by high energetic ions. Once critical stress level in the film is exceeded, its failure typically occurs. Hence, a high stress level implies lowering of the film hardness, resistance against plastic deformation, and wear [67]. Moreover, material mismatches at the interface between substrate and film and corresponding difference in the thermal expansion coefficients have to be considered as well. For those reasons, optimization of hard coatings through the right material selection is needed to avoid the adhesion problems resulting from high stresses concentrated at the interfaces. If the stresses exceed critical limits film delamination may occur [68].

The macrostress σ in sputtered high Si-containing Zr-Si-N films has been shown to be relatively low. Its development with increasing p_{N_2} and the Si content is depicted in Fig. 5.17a and b, respectively. Crystalline alloy Zr-Si-N films deposited at very low partial pressures of N exhibit tensile stresses characteristic for pure metals. Relatively, intensive ion bombardment at substrate bias $U_s = -100$ V and corresponding ion current density $i_s = 1$ mA/cm^2 together with the deposition temperature $T_s = 500$°C result in an enhanced adatom mobility, and in consequence in a formation of films with dense and compact microstructures and low tensile stress level, not exceeding 300 MPa. Introduction of N into the plasma discharge causes a formation of the nanocomposite structure with dominating volume fraction of the Si_3N_4 phase. It consequently results in the transition from tensile to compressive stress state. The films with a mixture of a comparable amount of both the crystalline $ZrSi_2$ and amorphous Si_3N_4 phases exhibit compressive stress of about −0.5 GPa. On the contrary, the Zr-Si-N films reactively deposited at the transition and nitride mode of sputtering exhibit higher stress level. Compressive stress, indicated by a convex shape of the coated-Si wafers, slightly increases from −1.1 GPa close to −1.3 GPa with increasing N content in the film (see Fig. 5.17a). Film hardness follows the same trend in corresponding p_{N_2} range. In addition, high Si-containing Zr-S-N films exhibit very good adhesion and stability on both Si and steel substrates.

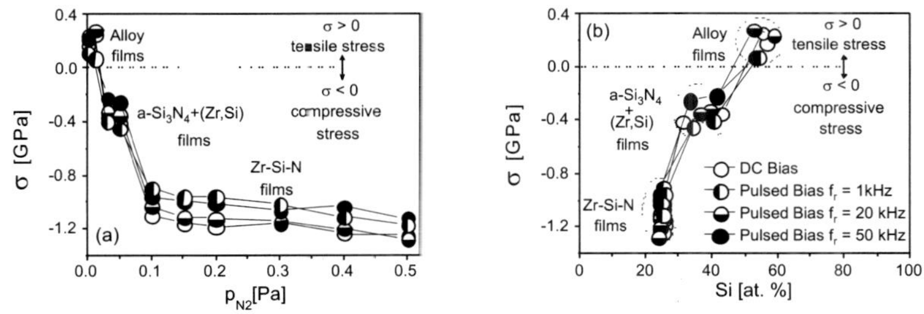

Figure 5.17 Macrostress σ in the Zr-Si-N films, sputtered on Si(100) substrate at I_d = 1 A, T_s = 500 °C and at d.c. and pulsed bias of U_s = −100 V, as a function of (a) partial pressure of nitrogen p_{N_2} and (b) Si content in the film. Reprinted from Ref. [28], Copyright (2005), with permission from Elsevier.

A decrease of the macrostress in transition metal nitride films with an increasing amount of the amorphous Si_3N_4 phase formed has been investigated in detail up to 20 at.% of the Si content [14,15,21]. It has been shown that a strong dependence of the stress level on the Si concentration in the film reflects marked changes in the phase composition and the film structure. Similar relationship between the film structure and stress evolution was observed also in the Zr-Si-N films with a high (>25 at.%) Si content. A significant stress reducing effect of Si in the Zr-Si-N films is presented in Fig. 5.17b. While the non-reactively deposited polycrystalline Zr-Si films with the highest Si content around 60 at.%, composed mainly of $ZrSi_2$ phase, exhibit a low magnitude of the tensile stress, incorporation of N into the film and a formation of the amorphous Si_3N_4 phase induces compressive stresses. It is even higher in the reactively sputtered Zr-Si-N films with the Si content of about 25 at.% with the nanocomposite structure, where ultra-fine grains are completely embedded within an amorphous Si_3N_4 tissue. It is, however, not clear yet if the low stress level in the Zr-Si-N films is ascribed to the presence of high Si content in the films or to very fine-grained structure. Therefore, a further investigation of causes, which may result in a reduction of macrostress σ, is needed.

Strong relation between macrostress and mechanical properties of Me-Si-N films is well known especially at low and intermediate Si content in the film. By exceeding the critical value of the amorphous to crystalline volume fraction ratio, the film hardness is no longer promoted by the magnitude of the film macrostress, as it is seen from Fig. 5.18. Here, a great change in the macrostress σ from 0.4 GPa to −1.2 GPa corresponds only to a small increase in film hardness H. This implies that changes in H of the hard Zr-Si-N films are not due to a compressive macrostress σ but it is very probably related to the phase evolution where an amount of the Si_3N_4 phase increases to the detriment of the crystalline one. Besides the deposition conditions, also the T_s:T_m ratio serves as a decisive factor that determines the final stress level of deposited films. It is closely attributed to material characteristics since T_m is the melting point of the deposited material and T_s is the temperature of the substrate during deposition. In the case of the Zr-Si-N film sputtered at T_s = 500 °C, the T_s:T_m ratio is relatively high, since $T_{m\ Si_3N_4}$ of dominating Si_3N_4 phase is 1900°C [68]. T_s:$T_{m\ Si3N4}$ = 0.26, at which the hard Zr-Si-N films with $H \geq 30$ GPa are produced, results in typically low values of macrostress σ due to its relaxation in the course of the continuous film growth. It is given by sufficiently high amount of the Si_3N_4 phase dominating over ZrN phase providing an automatical recovery of the compressive macrostress σ in the Zr-Si-N films sputtered at p_{N_2}> 0.15 Pa. The reduction of the macrostress σ with incorporation of Si into hard films without any effect upon the film hardness is very important from the point of view of technological applications.

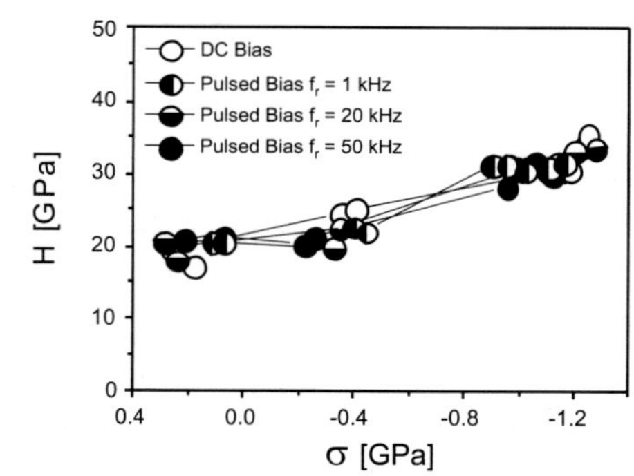

Figure 5.18 Microhardness H of the Zr-Si-N films, sputtered on Si(100) substrate at I_d = 1 A and at d.c. and pulsed bias U_s = −100 V, as a function of macrostress σ. Reprinted from Ref. [28], Copyright (2005), with permission from Elsevier.

The effect of the T_s:T_m ratio on the development of the film hardness H and macrostress σ was investigated in more detail in the Zr-Si-N film deposited at constant p_{N_2} = 0.2 Pa and varying substrate temperature T_s. The elemental and phase composition as well as the volume fraction of the amorphous Si_3N_4 phase remain obviously constant in all sputtered films which allows to vary exclusively the T_s:T_m ratio and to study its effect on the structure and mechanical properties of the film. The T_s:T_m ratio alters from 0.25 to 0.47, while T_m of the dominating Si_3N_4 phase is constant and the substrate temperature T_s ranges from 300 to 750 °C. The dependence of H and σ on the substrate temperature T_s is given in Fig. 5.19.

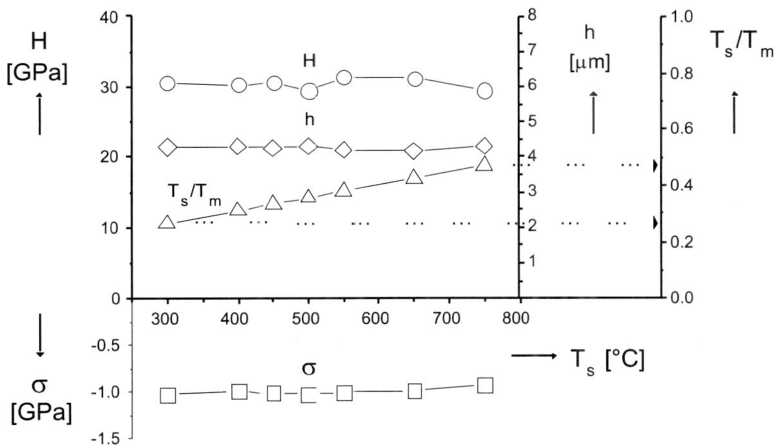

Figure 5.19 Dependences of H, σ, h, and the $T_s : T_{m\,Si_3N_4}$ ratio for the Zr-Si-N films, sputtered on pulsed biased (f_r = 50 kHz, $t_{reverse}$ = 2 μs) Si(100) substrate at I_d = 1 A, U_s = −100 V, p_{N_2} = 0.2 Pa, and p_T = 0.7 Pa, on the substrate temperature T_s. Reprinted from Ref. [28], Copyright (2005), with permission from Elsevier.

An enhanced hardness $H \approx 30$ GPa of the Zr-Si-N films, which is almost two times higher than that of Si_3N_4 and ZrN ($H_{Si_3N_4}$ = 17.2 GPa and H_{ZrN} = 16 GPa [34]), almost does not depend on the substrate temperature T_s in a wide investigated temperature range. Considering the fact that the film thickness h is also not influenced by the deposition temperature, hardness H is independent on both the

T_s:T_m ratio and the film thickness h. All the facts give a rise into the consideration that already at the lowest $T_s \approx 300°C$ corresponding to the ratio T_s:T_m = 0.27, the energy delivered to the growing film by the substrate heating is sufficient to recover a growth (intrinsic) macrostress σ [32]. It suggests that the high ratio T_s:$T_m \geq 0.27$ is a main reason why even 4 μm thick Zr-Si-N films exhibit a low macrostress $\sigma \approx -1$ GPa. This experiment also indicates that the films with a high (≥25 at.%) content of Si should exhibit a low compressive macrostress σ if they are deposited already at relatively low values of $T_s \approx 300°C$ independently on their thickness h.

5.10 Effect of Substrate Bias

5.10.1 Effect of Pulse

In magnetron systems, the fluxes of ions and electrons incident at the substrate are controlled by the substrate bias voltage. For d.c. bias applications, it is typical that the current drawn at the substrate saturates at bias voltages of the order of −100 V [69]. Here, the saturation current is an ion current, since any electrons approaching the substrate are repelled at this voltage. However, bombardment of the Zr-Si-N films by high energetic ions extracted from the plasma discharge toward the biased substrate is significantly reduced while the films are deposited at $p_{N_2} > 0.1$ Pa. Due to a high volume fraction of the amorphous Si_3N_4 phase formed, the Zr-Si-N films with a N content higher than 50 at.% become electrically insulating. Thus, the charge is accumulated on the surface of the film that grows on a d.c. biased substrate and so its surface region is no longer affected by high energetic ions. In order to positively influence the film growth and enhance, the adatom mobility through the ion bombardment also in that case, a pulsed d.c. substrate bias needs to be applied. Pulsing the substrate bias voltage in the mid-frequency range has been found to significantly increase the ion current drawn at the substrate [70]. A schematics of the pulsed voltage signal with U_s = −100 V, f_r = 50 kHz, and $t_{reverse}$ = 2 μs applied to the substrate is given in Fig. 5.20.

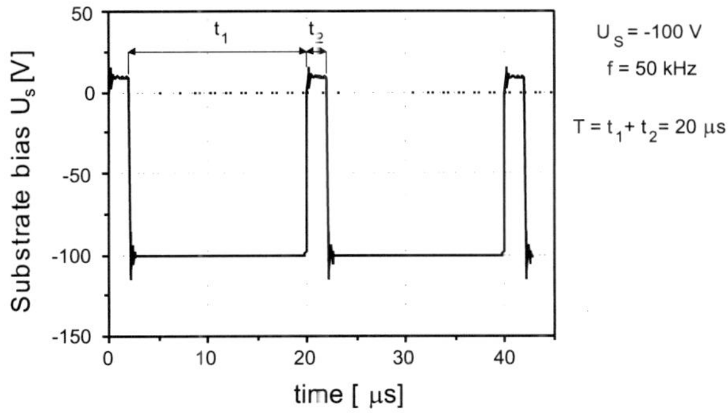

Figure 5.20 Typical 50 kHz pulsed voltage signal shown for the case of a 100 V negative d.c. biased substrate. Here, 2 μs positive excursion at 10% of the d.c. voltage and 18 μs negative excursion at 100% of the d.c. voltage are present.

The main parameter, which determines the efficiency of the pulsed d.c. substrate bias on the structure and mechanical properties of deposited films, is the ratio t_1:τ. Here, t_1 is the length of the negative pulse; $\tau = RC$ is the time required to charge the insulating film; R, the plasma resistance; $C = K\varepsilon_0 A/h$, the capacitance of the film [71]; K, the relative dielectric constant; ε_0 the permittivity of free space; A the substrate area (≈5 cm^2); and h is the film thickness (≈4 μm). Bias voltage $U_s = -100$ V giving the substrate ion current density i_s of 1 mA/cm^2 corresponds to the plasma resistance R = 20,000 Ω over the substrate area. Due to the high volume content of the Si_3N_4 phase, dielectric properties of amorphous-like electrically insulating Zr-Si-N films with a high (≥25 at.%) Si concentration can be well characterized by the dielectric constant of the amorphous bulk Si_3N_4 material $K_{Si_3N_4} = n^2 = 7.5$ [72]. In fact, the magnitude of the dielectric constant of Zr-Si-N films with X-ray amorphous structure will be lower since they are composed of two phases: Si_3N_4 and ZrN, where K_{ZrN} of the latter phase is lower than $K_{Si_3N_4}$. Based upon above assumptions, the given formulas gives C = 8.25 pF for the film capacitance and τ = 165 μs for time required for charging. The time constant τ is longer than the length of the negative pulse $t_1 < \tau$ for f_r = 50 and 20 kHz and shorter $t_1 > \tau$ than that for f_r = 1 kHz. It means that the charge can be accumulated on the surface of insulating Zr-Si-N film during the negative voltage pulse.

In order to investigate the efficiency of the d.c. pulsed bias, an ion current I_s was measured on the steel sample with the surface area of 5 cm^2. Evolution of I_s for various repetition frequencies f_r versus deposition time t_d and film thickness h is illustrated in Fig. 5.21. While at d.c. pulsed bias a rapid decrease of the amount of bombarding ions was observed already at the film thickness of 200 nm, a significant improvement of the ion bombardment efficiency is obvious at f_r = 1 kHz. Although the charging effect is not totally eliminated at low frequencies, it is sufficiently reduced at mid-frequency pulsed d.c. biasing with f_r = 50 kHz. In that case, the film is formed at the ion current density of nearly i_s = 1 mA/cm^2 during the whole deposition process. It also suggests that the charge accumulated on the film surface is greater at f_r = 1 kHz and smaller at f_r = 50 kHz, while a d.c. substrate bias cannot be used for the biasing of insulating Zr-Si-N films.

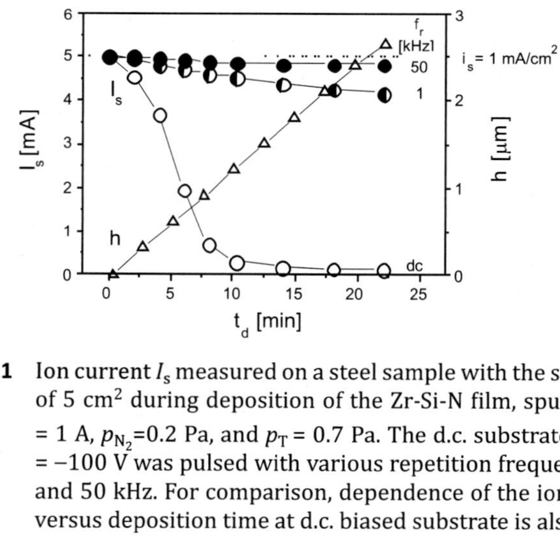

Figure 5.21 Ion current I_s measured on a steel sample with the surface area of 5 cm^2 during deposition of the Zr-Si-N film, sputtered at I_d = 1 A, p_{N_2}=0.2 Pa, and p_T = 0.7 Pa. The d.c. substrate bias of U_s = −100 V was pulsed with various repetition frequencies f_r = 1 and 50 kHz. For comparison, dependence of the ion current I_s versus deposition time at d.c. biased substrate is also given.

Comparing the results of the hardness, structure and stress measurement, one can conclude that even the pulsed d.c. substrate bias has practically no effect on the film growth and mechanical properties of the sputtered Zr-Si-N films. Almost no changes in the macrostress σ (Fig. 5.17) and the hardness H (Fig. 5.18) were observed if d.c. or pulsed bias was used. This might indicate that defects, which are induced by the energetic ion bombardment within the near-surface region of the growing film and typically influence its mechanical properties, are recovered during continuous growth when the bulk of the film is not accessible to the accelerated ions [73].

5.10.2 Effect of Bias Voltage

Strong effect of the ion current density on the elemental composition of the Zr-Si-N films is discussed in Section 5.2. By increasing the bias voltage, the resputtering effect is more pronounced due to the increasing energy of arriving ions accelerated in the substrate sheath [52]. However, the efficiency of the ion bombardment on the structure and phase composition of the Zr-Si-N films was found to be highly dependent on the volume content of the Si_3N_4 phase in the film.

The development of the elemental composition on the bias voltage is shown in Fig. 5.22 While the Zr-Si-N film sputtered at floating potential exhibit 35 at.% of Si, much lower Si content of 25 at.% is involved in the film prepared under the ion bombardment at −100 V bias voltage (see Fig. 5.22a). This effect is attributed to the weight difference between Si and Zr atoms where lighter Si is preferentially resputtered away. Thus, an amount of Zr just slightly decreases in the biased films. As it was already mentioned, a typical saturation of the ion current density has been found at relatively low substrate bias voltages where further increase in the substrate bias does not lead to a further increase in ion current [69]. It is ascribed mainly to the maximum ionization efficiency of N which is most probably reached already at −100 V. Thus, since the threshold of an enhanced incorporation of N into the films is attained, its content exhibits a constant value close to 60 at.% in a wide investigated range of the substrate bias. Significant difference in the elemental composition of the Zr-Si-N films grown on biased and non-biased substrates can be also seen from Fig. 5.22b. Here, the Si:Zr ratio of the Zr-Si-N film varying from 1.9 to 1.5 is depicted. While the composition of the non-bombarded film corresponds to the composition of the alloyed $ZrSi_2$ target with Si:Zr = 2, it is expressively lowered due to relatively strong resputtering effect at biased films. On the other hand, similarly to the N content in the Zr-Si-N films, an amount of Si and Zr incorporated into the film remains constant as well, if the bias voltage increases further. It has been reported that the metal atoms are easily resputtered away by the impinging particles with high energy if those atoms exist in a free state or in a low nitridation level. Reduction in both the Si content and the growth rate with an increase of the substrate bias voltage above −100 V has been typically observed in the films composed of dominant volume fraction of crystalline phase due to resputtering and ion bombardment phenomena. On the contrary, nitride phases fully developed in the Zr-Si-N films with a high N content result in more stable structures, which are highly resistant against an ion bombardment. It is attributed to the difference of sputtering yield of free unbounded Si and Zr atoms in elemental states compared to that of nitrided phases. It suggests that sputtered films exhibit stable structure just if they are composed of stable phases. Then, the resputtering effect is effectively hindered by strong bonds between individual elements and does not exceed a certain level. In addition, a high resistance to ion bombardment is typically associated to a high thermal stability and resistance against oxidation (see Sections 5.11 and 5.12).

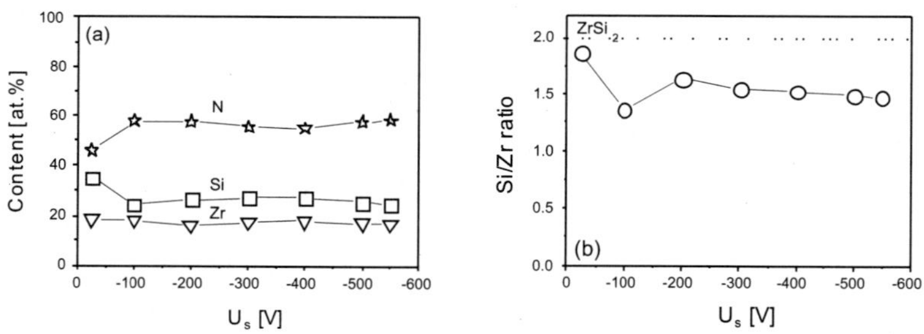

Figure 5.22 (a) Elemental composition and (b) the Si:Zr ratio of the Zr-Si-N films sputter deposited at $I_d = 1$ A, $i_s = 1$ mA/cm^2, $p_{N_2} = 0.2$ Pa, $p_T = 0.7$ Pa, and increasing negative pulsed bias voltage U_s (f_r = 50 kHz, $t_{reverse} = 2$ μs).

Figure 5.23 shows a dependence of the deposition rate on the substrate bias voltage U_s. The decrease of a_d with an increase in the substrate bias voltage from U_{fl} to −100 V is related to resputtering phenomenon and densification of the coating due to strong collision energy of ions. However, the films are sputtered at constant deposition rate at the bias voltages higher than −100 V. No significant influence of the bias voltage on the deposition rate a_d above this limit value, through possible preferential resputtering of individual elements present in the film, demonstrates a high stability of the film structure as well.

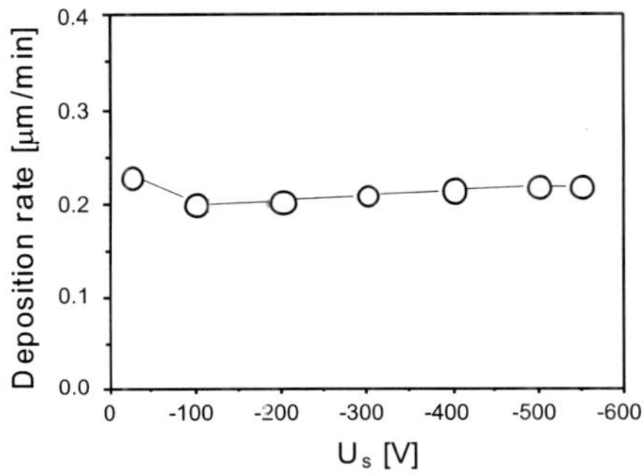

Figure 5.23 Deposition rate a_d of the Zr-Si-N films as a function of increasing negative pulsed bias voltage U_s (f_r = 50 kHz, $t_{reverse}$ = 2 μs). The films were sputter deposited at I_d = 1A, i_s = 1 mA/cm^2, p_{N_2} = 0.2 Pa, and p_T = 0.7 Pa.

If the films are affected by low energetic ions they typically differ also in a density and color compared to those which are not exposed to intensive ion bombardment during their growth. Very dense and compact Zr-Si-N films can be prepared by applying a negative bias voltage in a wide range between −100 and −550 V. The variation of the incident energy is, however, typically accompanied by changes in the elemental and phase composition, which is reflected by variation of the film color from yellow to light brown shade. Significant changes in the color is evident already in the range of a negative bias voltage from U_{fl} = −22 V to −100 V. High energy of arriving ions accelerated at the biased substrate is reflected also by a surface roughness of the deposited films. It increases in a consequence of a high ion bombardment effect on the film growth. In extreme cases, the surface roughness reaches a value of an order of magnitude higher than that of the films which are not exposed to the ion bombardment [74]. In that case, high bias voltages bring out an etching of the films due to high collision energy of accelerated ions.

The effect of the substrate bias on the mechanical properties of the Zr-Si-N films, namely microhardness H and Young's modulus E^*, is shown in Fig. 5.24.

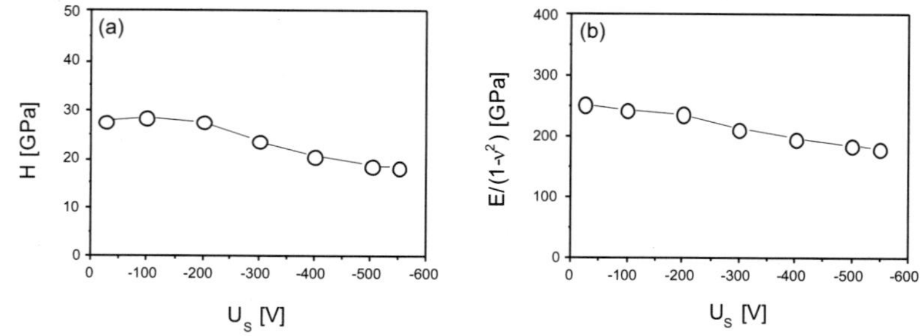

Figure 5.24 (a) Microhardness H and (b) reduced Young's modulus $E^* = E/(1-\nu^2)$ of the Zr-Si-N films sputter deposited at I_d = 1A, i_s = 1 mA/cm^2, p_{N_2} = 0.2 Pa, p_T = 0.7 Pa and increasing negative pulsed bias voltage U_s (f_r = 50 kHz, $t_{reverse}$ = 2 μs).

It is well known that the amount of the growth defects generated in the films, exposed during deposition to accelerate ions, is the higher the higher is their energy. If the critical value of the substrate bias is exceeded, the more disordered film structure is formed making probably the plastic deformation of the loaded film easier. The reduction of the film hardness, as shown in Fig. 5.24a, is thought to be attributed mainly to the synergetic effect of the changes in atomic arrangement, generation of growth defects, and reduction of the stress level. The relaxation of the residual stress has been referred as the consequence of the additional energy delivered into the film through the energetic ions [75]. As illustrated in Fig. 5.24b, the substrate bias voltage has also a distinct effect on the reduced Young's modulus. It shows a similar trend with increasing U_s to the film hardness.

5.11 Thermal Stability of Zr-Si-N Films

The usage of the binary transition metal nitride films of the Me-N type, where Me = Ti, Zr, Ta, W, Cr, etc., is strongly limited while the coated tools are heavily loaded at high temperatures. Besides good mechanical properties, the application of hard protective films in the mechanical engineering also requires a high thermal stability and a good oxidation resistance. At high temperatures, a good performance and a long lifetime of the coated tools are ensured only if the structure and mechanical properties of films are sufficiently stable [76]. Since the resistance against oxidation of transition nitride films is relatively low and does not exceed 700°C [77–81], various alloying elements have been added into the films in order to improve the stability of the structure and mechanical properties at elevated temperatures and to enhance the oxidation resistance. The addition of Si into binary nitrides of transition metals has been found to significantly improve both the thermal stability and the oxidation resistance [29,82–84]. Due to a good thermal stability, the Zr-Si-N films have been applied not only as hard protective coatings in tooling applications but also as Cu diffusion barriers in microelectronics [85].

Indeed, the Zr-Si-N films have been systematically investigated with their small and intermediate Si content (<25 at.%) only. The influence of the amount of Si on the structure, mechanical properties, and partially on the thermal behavior of the films has been reported just in this compositional range [17,85–87]; and there is only a little knowledge on the properties of the Zr-Si-N films with a high (≥25 at.%) Si content [23]. It has been shown that this kind of nanocomposite consisting of the crystalline MeN phase embedded within a massive Si_3N_4 matrix exhibit-enhanced thermal stability [88]. The investigation of the thermal stability and particularly factors which determine the crystallization temperature of the X-ray amorphous Zr-Si-N films containing a high amount of the Si_3N_4 phase is therefore of a high interest.

The significant effect of the elemental and phase composition of the high Si-containing Zr-Si-N nanocomposite films on their thermal properties is demonstrated in selected representative films with different N and Si contents. The Zr-Si-N films were sputter deposited at various partial pressures of nitrogen p_{N_2} and discharge current I_d and hence, differ in the elemental and phase composition, Si_3N_4 volume content and also mechanical, structural, and electrical properties. These diverse characteristics make them technologically interesting in the investigation of their thermal stability at different annealing temperatures, in different environment, and with or without substrate. The properties of the investigated Zr-Si-N films are summarized in Table 5.1.

Table 5.1 Properties of the investigated Zr-Si-N films in the as-deposited state

Film	p_{N_2} [Pa]	Zr [at.%]	Si [at.%]	N [at.%]	N/Si	Si/Zr	Si_3N_4 [vol.%]	H [GPa]	Phase composition	Structure	Electrical properties	Optical properties
$Zr_{43}Si_{57}$	0	43	57	0	0	1.3	0	17	$ZrSi_2$ + Zr	c	Conductive	Opaque
$Zr_{31}Si_{43}N_{26}$	0.03	31	43	26	0.6	1.4	30	25	$Si_3N_4 + ZrSi_{x<2}$	c	Conductive	Opaque
$Zr_{18}Si_{25}N_{57}$	0.1	18	25	57	2.3	1.4	55	30	$Si_3N_4 + ZrSi_{x<2} + ZrN_{x>1}$	a	Conductive	Opaque
$Zr_{15}Si_{25}N_{60}$	0.3	15	25	60	2.4	1.7	58	31	$Si_3N_4 + ZrN_{x>1}$	a	Insulating	Transparent
$Zr_{15}Si_{24}N_{61}$	0.4	15	24	61	2.5	1.6	57	32	$Si_3N_4 + ZrN_{x>1}$	a	Insulating	Transparent
Films deposited at I_d = 1 A												
$Zr_{18}Si_{29}N_{53}$	0.15	18	29	53	1.8	1.6	52	30	$Si_3N_4 + ZrN_{x<1}$	a	Conductive	Opaque
$Zr_{16}Si_{28}N_{56}$	0.2	15	24	61	2.0	1.8	55	31	$Si_3N_4 + ZrN_{x>1}$	a	Insulating	Transparent
Films deposited at I_d = 2 A												

Note: (1) The films were sputtered at $U_s = -100$ V, $i_s = 1$ mA/cm^2, $p_T = 0.7$ Pa, $T_s = 500°C$, and various partial pressure of nitrogen p_{N_2} and discharge current I_d.
(2) "c" and "a" denotes the crystalline and X-ray amorphous structure, respectively.

5.11.1 Thermal Stability of Zr-Si-N Films in Vacuum at T_a ≤700°C

It is often reported that the mismatches in thermal expansion coefficients between the film and substrate material can cause big troubles if the critical stresses at the interface are achieved. It typically results in the film delamination or film cracking [67]. Early coating failure observed in tooling applications are mainly caused by thermal overloads, either in static or dynamic conditions [76,89]. Therefore, the knowledge of the behavior of coated systems exposed to high temperatures is very important.

Thermal stability and macrostress stored in thin films are typically investigated by in situ stress–temperature measurements at different annealing temperatures T_a (Fig. 5.25) either in vacuum or in air coupled with ex situ X-ray diffraction analysis. It is well known that the biaxial macrostress σ is composed of two components, the thermal stress σ_{th} and the intrinsic stress σ_i, according to the formula:

$$\sigma = \sigma_{th} + \sigma_i \tag{5.1}$$

where

$$\sigma_{th} = \frac{E_f}{1-\nu_f}(\alpha_s - \alpha_f)(T - T_s) \tag{5.2}$$

Here, E_f and ν_f are the Young's modulus and Poisson's ratio of the film, respectively, T_s is the substrate temperature during the film deposition and T is the temperature at which σ is measured. The thermal stress σ_{th} is due to the difference in linear thermal expansion coefficients of the film and substrate and the intrinsic stress σ_i is generated in the film during its growth according to process conditions.

The development of residual stress σ in the Zr-Si-N films (i) in the as-deposited state, (ii) during the heating cycle from RT to 700°C, (iii) after annealing at T_a = 700°C for 30 min, and (iv) during the cooling from 700°C to RT with increasing p_{N_2}, that is, increasing the N content in the film, is displayed in Fig. 5.26. The macrostress σ generated in the as-deposited films changes from tensile ($\sigma > 0$, Fig. 5.26a) to compressive ($\sigma < 0$, Figs. 5.26 b–d). Indeed, its magnitude is strongly related to the phase composition of the film. Thus, the higher amount of N incorporated into the film the higher compressive stress is induced during the growth and increases up to −1.2 GPa at p_{N_2} = 0.4 Pa. During post-deposition annealing, macrostress σ in the film linearly increases with increase in annealing temperature T_a following the thermoelastic line defined by Eq. 5.2. An increase in the film stress σ is caused by higher thermal expansion coefficient α_f of the film with respect to the Si substrate. However, when the value of α_f approaches to that of Si, the slope of heating–cooling curves becomes parallel to the temperature axis. It corresponds to the constant value of the residual stress although the sample is heated. Since the thermal expansion coefficient of the Si substrate $\alpha_{Si(100)}$ = 3.55×10^{-6} K^{-1} is well known [91], the thermal expansion coefficient α_f of the Zr-Si-N films can be determined, according to Eq. 5.2, from the slope of stress–temperature heating or cooling curves shown in Fig. 5.26. The values for α_f vary between 3.6 and 8.1×10^{-6} K^{-1} depending on the elemental and phase composition of the film. The α_f decreases with increase in the N content in the Zr-Si-N film from 7.9 to 4.1×10^{-6} K^{-1} (heating curve) and from 8.1 to 3.6×10^{-6} K^{-1} (cooling curve). This slight difference in α_f during the heat treatment corresponds to the changes of residual stress σ in the film.

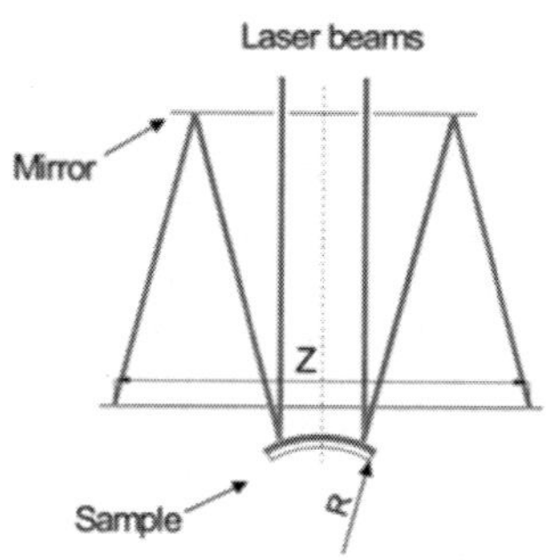

Figure 5.25 Schematic illustration of an optical beam bending method. Two parallel laser beams are reflected from the film surface and the substrate curvature is measured. Then, the macrostress is calculated using the Stoney formula [90]. The specimen can simultaneously be subjected to temperature cycles (typically between 25 and 700°C) and stress–temperature measurements can thus be performed. The advantage of the method is that no material characteristics of the investigated film are needed for the evaluation of the residual stress.

If the films are heated up to the temperature at which they were deposited, the increasing tendency of the compressive stress typically alters, see Fig. 5.26. The migration and annihilation of point defects and dislocations as well as the growth and coalescence of subgrains contributes to the recovery process and the stress relaxation [92]. Recovery is often followed by the reduction of compressive stresses mainly through annihilation of the Frenkel pairs or anti-Schottky defects. On the contrary, if the voids or vacancies are annealed-out, the volume of the film may be reduced thus resulting in increasing tensile stress [64]. Reduction of the residual stress is, however, not related just to the different type of defects in the films but also to their amount and thermal stability. Therefore, the onset temperature for recovery T_r significantly changes, as can be seen from Fig. 5.26. It is attributed to the different recovery mechanism and corresponding driving force given by the initial stress level.

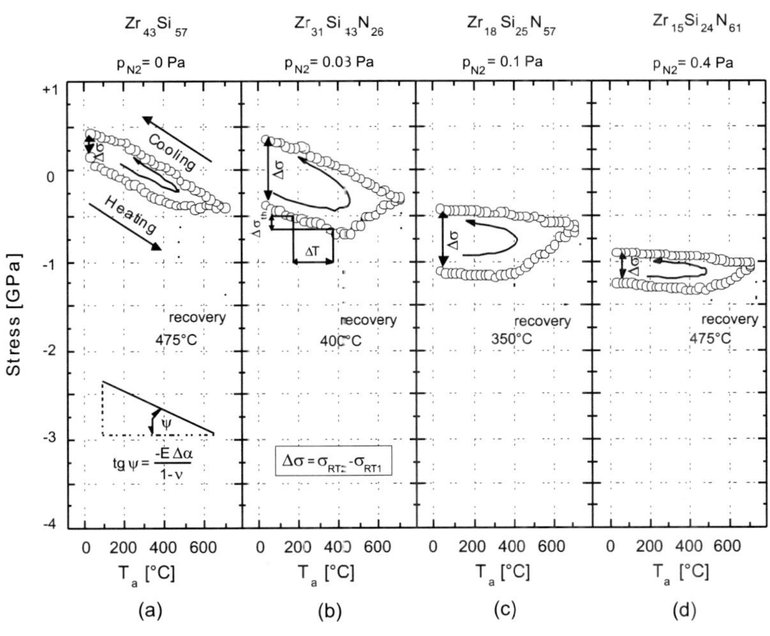

Figure 5.26 Stress–temperature curves for the Zr-Si-N films magnetron sputter deposited on Si(100) substrates. Here, σ_{RT1} and σ_{RT2} are the values of the macrostress measured at RT before and after annealing, respectively.

Indeed, there exists a strong relation between an amount of defects, stress, and thermal stability of thin films. While the stress level in the films increases with increase in amount of the film defects, the thermal stability, given mainly by the level of the stress recovery and structural changes after the annealing, consequently decreases [93]. It was found that the amount of stress relaxation is relatively low in the case of dense films with a low residual stress level and it increases with increase in stress in the as-deposited state. This fact suggests that the film stress represents a driving force for the stress recovery process—the higher the magnitude of the residual stress stored in the film, the higher the driving force for the stress recovery process. The principal is the reason why the transition metal nitride films sputter deposited under relatively high ion bombardment exhibit lower stability at elevated temperatures [6]. At higher stresses, a lower thermal activation is needed to start recovery and thus, high defect densities in the film result in its reduced thermal stability [94]. To achieve high film stability at elevated temperatures, films with stable phase boundaries have to be synthesized. It has been found that by an addition of Si into the MeN films, the nanocrystals free of dislocation are formed and the film stress and consequently, the driving force for recovery are reduced [93]. Hence, the Zr-Si-N films exhibit high hardness and improved thermal stability.

During the cooling segment, the stress–temperature curve again shows linear thermoelastic behavior (see Fig. 5.26). After the heat treatment, the magnitude of the stress recovery is determined as $\Delta\sigma = \sigma_{RT1} - \sigma_{RT2}$, where σ_{RT1} and σ_{RT2} are the values of the macrostress measured at room temperature before and after thermal annealing, respectively. Indeed, the onset temperature of the recovery and its magnitude $\Delta\sigma$ is highly important for a practical usage since it characterizes the thermal stability of the films. Both the onset temperature of the recovery T_r and $\Delta\sigma$ together with the thermal expansion coefficients of the films and the initial stress level in the films in their as-deposited state are listed in Table 5.2.

The investigation of the thermal recovery $\Delta\sigma$ yields different results for both the crystalline and amorphous-like Zr-Si-N films. For the crystalline $Zr_{43}Si_{57}$ (p_{N2} = 0 Pa) and $Zr_{31}Si_{43}N_{26}$ (p_{N_2} = 0.03 Pa) films, $\Delta\sigma$ increases with increase in stress σ generated in the film during its growth and achieves the value of 0.32 and 0.74 GPa, respectively. The onset temperature of the recovery T_r decreases with increase in N content in the film from 475°C to 400°C. On the contrary, the X-ray amorphous $Zr_{18}Si_{25}N_{57}$ (p_{N_2} = 0.1 Pa) and $Zr_{15}Si_{24}N_{61}$ (p_{N_2} = 0.4 Pa) films exhibit different behavior if they are annealed in vacuum up to 700°C. The magnitude of the stress recovery $\Delta\sigma$ decreases from 0.69 to 0.34 GPa, respectively, although the residual stress σ slightly increases further. T_r increases from 350 to 475°C in that case. In addition, the Zr-Si-N films sputtered at (i) p_{N_2}=0 and 0.4 Pa and (ii) p_{N_2}=0.03 and 0.1 Pa exhibit the same values of $\Delta\sigma$ and the same T_r even though they considerably differ in the structure, elemental composition, and macrostress. This finding clearly shows a very complex relation between the macrostress, structure, and phase composition of the nanocomposite Me-Si-N films.

Table 5.2 Properties of the investigated Zr-Si-N films after annealing in vacuum to 700°C for 30 min. The films were sputtered at U_s = −100 V, I_d = 1 A, i_s = 1 mA/cm^2, p_T = 0.7 Pa, T_s = 500°C, and various partial pressure of nitrogen p_{N_2}

Film	p_{N_2}[Pa]	h_f [μm]	σ [GPa]	$\|\Delta\sigma\|$ [GPa]	ΔH [GPa]	α_{f1} [10^{-6} K^{-1}]	α_{f2} [10^{-6} K^{-1}]	T_r [°C]
$Zr_{43}Si_{57}$	0	3.3	+0.2	0.32	−2	7.9	8.1	475
$Zr_{31}Si_{43}N_{26}$	0.03	3.4	−0.4	0.74	1	6.2	6.9	400
$Zr_{18}Si_{25}N_{57}$	0.1	3.8	−1.1	0.69	1	4.1	3.7	350
$Zr_{15}Si_{24}N_{61}$	0.4	3.6	−1.2	0.34	−1	4.2	3.6	475

Note: $|\Delta\sigma|$ is the magnitude of the stress recovery, $\Delta H = H - H_{post}$ is the difference in microhardness before and after annealing.

Besides the stress recovery, the recrystallization process plays an important role in the thermal stability of the films as well. Thermally induced grain growth may alter the volume fraction ratio between the crystalline and amorphous phase, which typically results in lower hardness. In situ stress–temperature measurement using the bending method, illustrated in Fig. 5.25, enables to determine stability of films intended for thermally loaded tools through the investigation of the residual stress development in the films during annealing. Combined with the structure and hardness investigations, it may clarify mechanisms which contribute to the changes of film properties while the films exposed to high temperatures.

Evolution of the film structure and microhardness after vacuum annealing to 700°C depicted in Figs. 5.27 and 5.28 does not show any dramatic changes. The structure of the films remains crystalline, characterized by strong (110), (131), and (002) reflections from $ZrSi_2$ grains at p_{N_2} = 0 Pa and by a preferred $ZrSi_2$(111) orientation at p_{N_2} = 0.03 Pa. A slight intensity increase of the $ZrSi_2$ diffraction patterns and their shift to higher values of 2θ is attributed to a slight crystallinity enhancement and a change of the macrostress σ. It is, however, not reflected by any change of the mechanical properties. Moreover, also X-ray amorphous Zr-Si-N films sputtered at $p_{N_2} \geq 0.1$ Pa and characterized by one very broad, low-intensity reflection line are fairly thermally stable to 700°C.

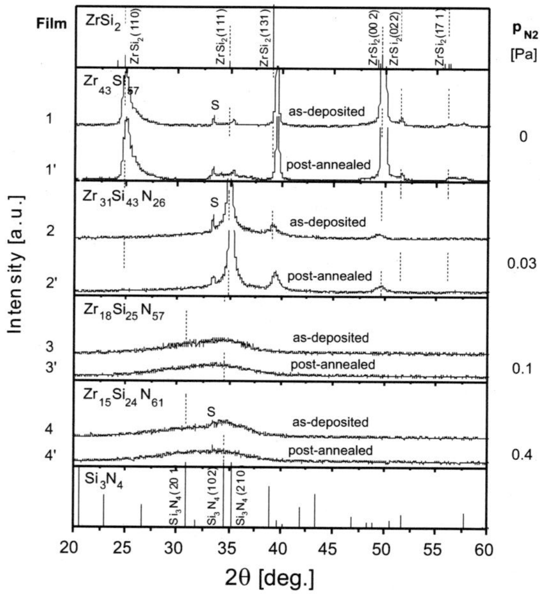

Figure 5.27 XRD patterns of the Zr-Si-N films in the as-deposited state and after annealing in vacuum to T_a = 700°C. Just a slight change of the diffraction patterns reveals a thermally stable structure of the films. Reprinted from Ref. [40], Copyright (2006), with permission from Elsevier.

Microhardness H of the Zr-Si-N films is closely related to the developed phase composition [39,46,95]. The transition from crystalline phase-dominated films prepared at $p_{N_2} \leq 0.03$ Pa to amorphous phase-rich films sputtered at higher p_{N_2} is accompanied with an increase in the film hardness. Since neither the phase composition nor the crystalline to amorphous volume fraction ratio changes when the films are annealed to 700°C, the hardness of the films remains constant as well. Also the effect of the stress level in the film is negligible. Although curves $H = f(\sigma)$ in Fig. 5.28b for the as-deposited films and those after annealing are clearly separated, there is almost no difference in microhardness. It means that a relatively large stress relaxation $|\Delta\sigma|$ from 0.3 to 0.7 GPa together with a change from compressive to tensile stress leads to a slight decrease in H about 1 GPa only (Fig. 5.28a). The change of the microhardness of the Zr-Si-N films after annealing is represented by the hardness difference $\Delta H = H - H_{post}$ given in Table 5.2. It again indicates that the high hardness value H of the Zr-Si-N films is not affected by the stress σ generated in the film during its growth and thus, the elemental and phase compositions must play more important role.

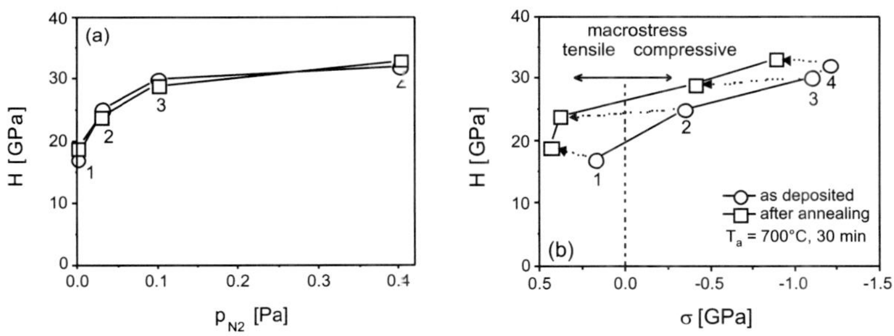

Figure 5.28 Microhardness H of the Zr-Si-N films depicted in the as-deposited state and after vacuum annealing at T_a = 700°C for 30 min versus (a) partial pressure of nitrogen p_{N_2} and (b) residual macrostress σ. The labels 1, 2, 3, and 4 denote films sputtered at various p_{N_2} = 0, 0.03, 0.1, and 0.4 Pa, respectively.

5.11.2 Thermal Stability of Zr-Si-N Films in Vacuum at T_a > 700°C

In cutting applications-coated tools are typically exposed to temperatures up to 1000°C. Under such conditions the mechanical properties of films may significantly change due to the crystallization from an amorphous phase or grain coarsening phenomenon [83,96]. Therefore, it is very important to specify the critical temperature at which the films lose their enhanced properties.

Annealing experiments of amorphous-like Zr-Si-N films with various Si and N content at temperatures exceeding 1100°C revealed significant importance of the phase composition of these nanocomposite films. Although all Zr-Si-N films sputtered at $p_{N_2} \geq$ 0.1 Pa exhibit similar elemental composition, they significantly differ in the electrical and optical properties. While the $Zr_{18}Si_{25}N_{57}$ film sputtered at p_{N_2} = 0.1 Pa is electrical conductive and optically opaque, Zr-Si-N films deposited at higher p_{N_2} are electrically insulating and optically transparent (see Table 5.1). The electrical conductivity of these films obviously originates from non-nitrided phases present in the films, for example, $ZrSi_x$ phase or free unbounded Si or Zr atoms, or stoichiometric ZrN. Hence, electrical properties of the Zr-Si-N films expressively change with a higher amount of N incorporated into the film increasing the nitrification level (see Fig. 5.7). This difference in physical properties of the Zr-Si-N films suggests that even if the films are sputtered at a relatively narrow p_{N_2} range, they greatly differ in the phase composition. Any change in electrical or optical properties serves as an indicator of the changes in the crystalline to amorphous phase volume fraction ratio and nitrification level of the present phases that is obviously reflected also in the thermal properties of the Zr-Si-N films.

The phases appearing after crystallization can be interpreted in the light of the evolution in the phase composition of the films in their as-deposited state. The phase composition was shown to dramatically differ with increase in the partial pressure of N during deposition (see Section 5.3). Figure 5.29a shows a structure development of annealed $Zr_{18}Si_{25}N_{57}$ film (p_{N_2} = 0.1 Pa). It can be seen that the film exhibits an X-ray amorphous structure up to approximately 800°C, at $T_a >$ 800°C the following phases crystallize: $Zr_5Si_3N_{1-x}$ at $T_a \approx$ 850°C, ZrN + $ZrSi_2$ at $T_a \approx$ 1000°C, and Si_3N_4 + ZrO_2 at $T_a \approx$ 1100°C. The ternary $Zr_5Si_3N_{1-x}$ phase is the high-temperature Zr_5Si_3 phase stabilized by N to lower temperatures [36]. A decrease of the reflections from the crystallized phases with increasing T_a is attributed to their decomposition and conversion to other phases. From that reason, the reflection from the Zr_5Si_3 grains almost disappears at $T_a \approx 950°C$. The $Zr_5Si_3N_{1-x}$ phase is probably converted into $ZrSi_2$ phase at $T_a \approx 1000°C$ as a consequence of an enrichment of the film with Si from the Si(100) substrate. On the other hand, the reflections from the ZrN phase are strongly reduced at $T_a \approx 1150°C$ due to the conversion of ZrN to ZrO_2. A higher intensity of the ZrO_2 reflections can be associated with a worse base pressure ($\approx 10^{-2}$ Pa) at a higher T_a. Increase in intensity of the $ZrSi_2$ reflections is probably associated with an enhanced interdiffusion of Si from the substrate to the film. This phenomenon is characteristic if the coated system is exposed to a high heating load for a relatively long time. The crystallization of the Si_3N_4 phase was observed at temperatures T_a higher than 1000°C which is fully in an agreement with the findings of other authors [97].

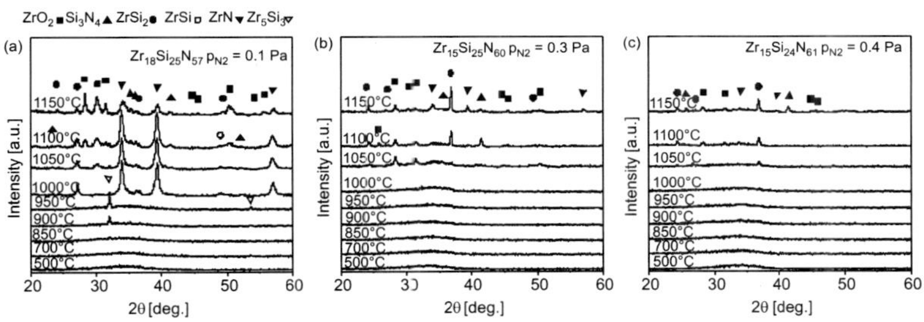

Figure 5.29 Structure development of three X-ray amorphous Zr-Si-N films with different phase composition and physical properties annealed up to 1150°C in 50°C step and with a dwell time of 30 min. Reprinted from Ref. [40], Copyright (2006), with permission from Elsevier.

The $Zr_{15}Si_{25}N_{60}$ film (p_{N_2} = 0.3 Pa) exhibits a considerably higher thermal stability compared to the $Zr_{18}Si_{25}N_{57}$ film sputtered at p_{N_2} = 0.1 Pa. Its X-ray amorphous structure is thermally stable up to 1000°C. At higher annealing temperatures T_a, the crystallization route can be described as follows: ZrO_2 + $ZrSi_2$ develop at $T_a \approx 1050°C$, ZrO_2 + $ZrSi_2$ + Si_3N_4 crystallize at $T_a \approx 1100°C$, and ZrO_2 + $ZrSi_2$ + Si_3N_4 + ZrN at $T_a \approx 1150°C$ (see Fig. 5.29b). A preferential formation of the ZrO_2 phase at $T_a \approx 1050°C$ instead of the $ZrSi_2$ can be explained by a higher affinity of Zr to O_2 ($\Delta H_{f\ ZrO2}$ = −1101.3 kJ/mol) than to Si ($\Delta H_{f\ ZrSi2}$ = −159 kJ/mol) [35]. Comparing both films, the mechanism of the formation of the crystalline phases in the $Zr_{15}Si_{25}N_{60}$ film is different from that sputtered at p_{N_2} = 0.1 Pa. Higher nitrification level of Zr and Si results in delayed crystallization which occurs first at $T_a \approx 1050°C$. It is ascribed to higher thermal stability of the film if it is composed of the overstoichiometric $ZrN_{x>1}$ phase. The ZrN phase starts to crystallize at approximately 1150°C, together with the Si_3N_4 phase. Higher thermal stability of the $Zr_{15}Si_{25}N_{60}$ film results in a higher temperature T_a, which is necessary for decomposition or desorbtion of N from the $ZrN_{x>1}$ phase. Hence, in this way produced free Zr reacts with O_2 from a base atmosphere ($\approx 10^{-2}$ Pa) at considerably higher temperatures. A strong tendency of the ZrO_2 phase to be formed to the detriment of the ZrN phase is ascribed to a high reactivity of Zr with oxygen (compare $\Delta H_{fZrO_2} = -1101.3$ kJ/mol with $\Delta H_{f\,ZrN} = -365.5$ kJ/mol) [35]. The addition of a small amount of oxygen to the film during annealing induces a production of ionic metal–oxygen bonds in a matrix of covalent metal–nitrogen bonds. Subsequently, oxide phases are readily formed since the oxidation process is markedly thermally activated.

Frequently observed atomic diffusion between the substrate and the film at higher annealing temperatures is known to be extremely high along porous grain boundaries [98]. Although the amorphous Si_3N_4 phase tends to lock diffusion paths in Zr-Si-N films through the developed dense and compact microstructure without grain boundaries [6], this effect to stabilize the structure at elevated temperatures greatly depends on the volume content of the Si_3N_4 phase. The denser and more amorphized structure of the Zr-Si-N films is formed with increasing amount of Si_3N_4 phase, the higher crystallization onset temperature of $ZrSi_x$ phase was observed corresponding to the enrichment of the film with Si diffused from the substrate. It indicates that the passivating effect of Si_3N_4 is limited by specific critical temperature. Further, improvement of the thermal stability of the Zr-Si-N films can be thus achieved using diffusion barriers deposited on the substrate.

The development of the film structure depicted in Fig. 5.29c reveals even higher thermal stability of the $Zr_{15}Si_{24}N_{61}$ film (p_{N_2} = 0.4 Pa) compared to that sputtered at p_{N_2} = 0.3 Pa. The crystallization temperature was observed at 1050°C. At this temperature the $ZrSi_2$ phase starts to crystallize. At higher annealing temperatures the following crystalline phases appear: $ZrSi_2 + ZrO_2$ at $T_a \approx 1100°C$ and $ZrSi_2 + ZrO_2 + ZrN + Si_3N_4$ at $T_a \approx 1150°C$. The reaction with oxygen occurs at evidently higher temperature suggesting higher resistance of the film against the oxidation. The XRD analysis also shows that the crystallization of the ZrN phase is very low and starts almost simultaneously with that of the Si_3N_4 phase at $T_a \approx 1150°C$. Moreover, the crystallization of Si_3N_4 occurs at higher T_a than in the case of the films sputtered at lower p_{N_2}. Experimental data are summarized in Table 5.3. In order to understand the relationship between the phase composition and the thermal stability of the Zr-Si-N films in different annealing atmospheres, more detailed study was performed and is discussed further in Section 5.12.

The study of mechanical properties of the Zr-Si-N films annealed in vacuum at $T_a > 700°C$ completes the information on the thermal stability of this investigated coating system. Besides the structure stability also the stability of the mechanical properties is very important for high-temperature applications. The loss of the enhanced hardness or the wear resistance associated to the structural changes at elevated temperatures may definitively result in film destruction and unexpected damages of the coated tools. In order to find the critical temperature T_c, when the films no longer exhibit their enhanced mechanical properties, microindentation measurements were performed after every annealing cycle at the room temperature. The development of the film microhardness as a function of the annealing temperature is presented in Fig. 5.30.

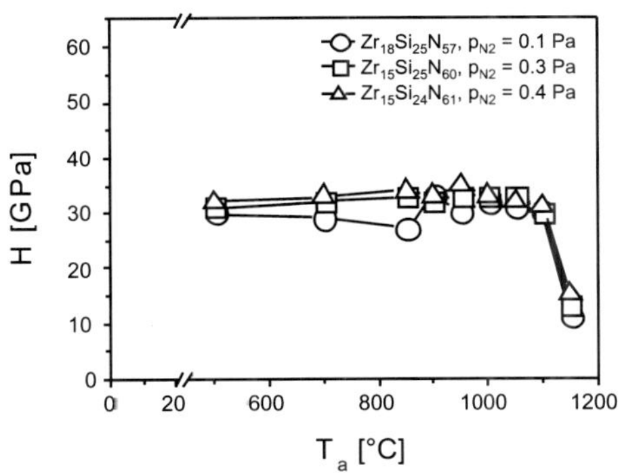

Figure 5.30 Microhardness H of the X-ray amorphous Zr-Si-N films as a function of the annealing temperature T_a. Reprinted from Ref. [40], Copyright (2006), with permission from Elsevier.

Table 5.3 Crystallization temperature T_{cr} and phase composition of amorphous-like Zr-Si-N films

Film	Phase composition of as-deposited film	T_{cr1} [°C]	First c-phase	T_{cr2} [°C]	Next c-phases	T_{cr3} [°C]	Next c-phases
$Zr_{18}Si_{25}N_{57}$	$Si_3N_4 + ZrSi_{x<2} + ZrN_{x>1}$	850	Zr_5Si_3	1000	$ZrN + Si_3N_4 + ZrO_2$	1150	
$Zr_{15}Si_{25}N_{60}$	$Si_3N_4 + ZrN_{x>1}$	~1050	ZrO_2	1100	$ZrSi_2 + Si_3N_4$	1150	ZrN
$Zr_{15}Si_{24}N_{61}$	$Si_3N_4 + ZrN_{x>1}$	~1050	$ZrSi_2$	1100	ZrO_2	1150	$ZrN + Si_3N_4$

Note: "c-" is the crystalline phase, T_{cr1}, T_{cr2}, and T_{cr3} are the crystallization temperature of the first phase and next phases, respectively.

In this investigation, a very high stability of the mechanical properties of all Zr-Si-N films was observed. The microhardness $H \approx 30$ GPa of the Zr-Si-N films remains constant up to the critical temperature T_c. It was found to be approximately 1100°C, which is almost two times higher than that of pure nitrides [99,100]. This fact suggests that the high temperature stability is provided if the films are composed of both co-existing Si_3N_4 + ZrN_x phases. However, once the critical annealing temperature is exceeded, the film hardness H suddenly drops down at T_a> 1100°C. The loss of enhanced mechanical properties seems to be related to the crystallization onset, mainly of oxide phases. A development of the ZrO_2 phase was observed right at this temperature range (Fig 5.29).

5.11.3 Thermal Stability of Zr-Si-N Film System in Air and Argon at T_a = 1300°C

From the perspective of tooling applications, the vacuum annealing experiment is an idealized case of the operating conditions of the sample. In real, the structural evolution of an annealed film is strongly affected not only by the phase composition of the film but also by the atmosphere in which a thermal annealing is carried out. This fact is demonstrated by thermal annealing of the $Zr_{18}Si_{25}N_{57}$ film in flowing air and Ar to T_a = 1300°C (Fig. 5.31).

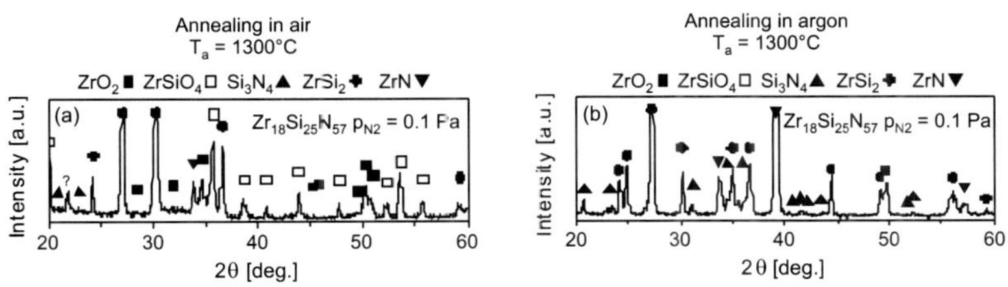

Figure 5.31 Structure of the opaque and electrically conductive $Zr_{18}Si_{25}N_{57}$ film sputtered at p_{N_2} = 0.1 Pa after annealing in (a) air and (b) argon to T_a = 1300°C. Reprinted from Ref. [40], Copyright (2006), with permission from Elsevier.

The $Zr_{18}Si_{25}N_{57}$ film (p_{N_2} = 0.1 Pa) exhibits an X-ray amorphous structure in the as-deposited state. It is composed of three phases:

amorphous Si_3N_4 phase and two ultra-fine grained $ZrSi_x$ + ZrN_x phases. This specific combination of all the phases makes the film electrical conductive and optically opaque. The annealing to T_a = 1300°C in both atmospheres results in a strong crystallization of the present phases as it is shown in Fig. 5.31. However, it can be seen that there is a huge difference in the structure of the annealed films depending on the atmosphere. The annealing in Ar induces a crystallization mainly of the ZrN and $ZrSi_2$ grains that is reflected by their strong reflections in the corresponding diffractogram (see Fig. 5.31b). It means that the film is composed of the ZrN and many of the $ZrSi_2$ grains of different crystallographic orientations after the annealing. Moreover, no amount of the ZrO_2 phase was detected. Low reflections from the Si_3N_4 grains after annealing in both Ar and air atmospheres indicate that the crystallization of the amorphous Si_3N_4 phase is strongly suppressed in both cases. Also the development of the ZrN phase was found to be hindered, while the film is annealed in air, and $ZrSi_2$, $ZrSiO_4$, and ZrO_2 phases dominate instead. Strong reaction of O_2 with the $Zr_{18}Si_{25}N_{57}$ film is closely associated to its phase composition. The oxidation process as a function of the volume fraction of the Si_3N_4 phase in the films is described in Section 5.12 [28,101,102].

5.11.4 Thermal Stability of Zr-Si-N Film System in Vacuum and Argon at T_a = 1100°C

In annealing experiments, a difference in the structural development is obvious when the films are investigated under different conditions. A strong effect of the annealing atmosphere on the film crystallization is demonstrated in the case of the $Zr_{15}Si_{25}N_{60}$ film sputter deposited at p_{N_2} = 0.3 Pa. It is composed of two different phases; a high amount of the amorphous Si_3N_4 phase and a smaller content of the fine-grained amorphous-like ZrN_x phase. In Fig. 5.32 the structure of the $Zr_{15}Si_{25}N_{60}$ film annealed in vacuum (10^{-2} Pa) and Ar (atmospheric pressure) at T_a = 1100°C for 30 min is compared.

While the film annealed in Ar exhibits a very weak crystallization of the $ZrSi_2$ and Si_3N_4 grains (Fig. 5.32b), the film annealed at the same temperature T_a in vacuum is well crystallized (Fig. 5.32a).

Particularly, $ZrSi_2$, Si_3N_4, and ZrO_2 reflections are quite strong. This means that even a very small amount of O_2 present in a residual atmosphere (vacuum of 10^{-2} Pa) is sufficient to strongly influence a process of crystallization.

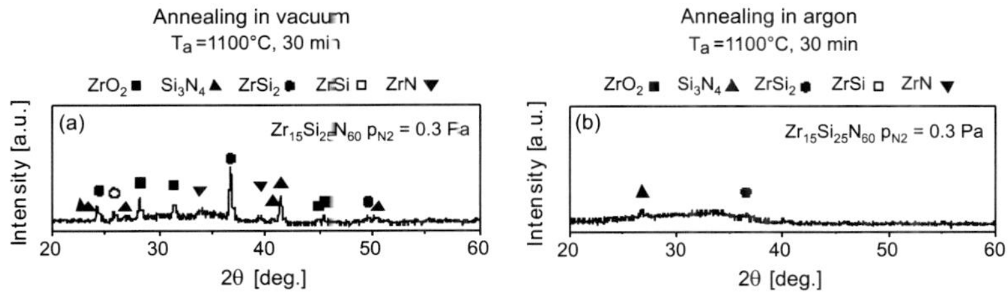

Figure 5.32 Structure of the optically transparent and electrically insulating $Zr_{15}Si_{25}N_{60}$ film sputtered at p_{N_2} = 0.3 Pa after annealing in (a) vacuum and (b) Ar to T_a = 1100°C for 30 min. Reprinted from Ref. [40], Copyright (2006), with permission from Elsevier.

5.11.5 Thermal Stability of Zr-Si-N Films in Flowing Argon at T_a = 1600°C

In high-temperature applications, the effect of the atomic diffusion between the substrate and the film has to be also taken into the consideration. The changes in the structure of the film annealed at high temperatures may be strongly affected by the interdiffusion of substrate elements into the film. This phenomenon varies with different type of films and is stronger in the crystalline films with open void boundaries compared to those with a dense and compact fine-grained structure [6,98]. The effect of the diffusion process between the Si substrate and the Zr-Si-N film can be investigated in more detail if the films, separated from the substrate by chemical etching, are annealed for comparison by using the differential scanning calorimetry (DSC). This procedure completely excludes the influence of substrate elements on the evolution of the film structure with increasing T_a.

Results of DSC measurements of two X-ray amorphous Zr-Si-N films: electrically conductive and optically opaque $Zr_{18}Si_{29}N_{53}$ film and electrically insulating and optically transparent $Zr_{16}Si_{28}N_{56}$ film, are shown in Fig. 5.33 Although both films sputtered at I_d = 2 A and p_{N_2} = 0.15 and 0.2 Pa, respectively, exhibit an X-ray amorphous structure in the as-deposited state, they differ in their phase composition. This is also indicated by different crystallization temperatures as determined by exothermic peaks on the DSC curves.

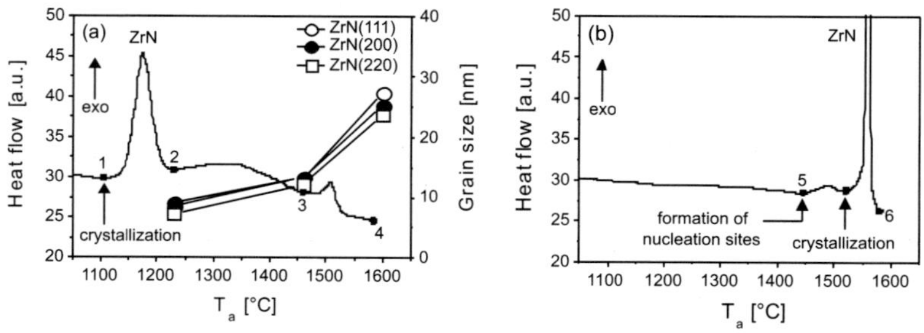

Figure 5.33 DSC curves of two X-ray amorphous (a) optically opaque $Zr_{18}Si_{29}N_{53}$ and (b) optically transparent $Zr_{16}Si_{28}N_{56}$ films with different phase composition, removed from the Cu substrate. The numbers 1–6 denote the values of T_a at which XRD patterns given in Fig. 5.34 were recorded. Reprinted from Ref. [40], Copyright (2006), with permission from Elsevier.

A strong dependence of the crystallization of the amorphous-like phase on the film composition is demonstrated by comparing the crystallization temperature T_{cr} of both Zr-Si-N films. While the opaque film starts to crystallize at $T_{cr} \approx 1130°C$, the transparent one is X-ray amorphous up to $T_{cr} \approx 1530°C$. A huge difference in the thermal stability of those films is probably caused by a different amount of nucleation sites in the as-deposited state. This fact is strongly attributed to their different phase composition. The $Zr_{18}Si_{29}N_{53}$ film is composed of two Si_3N_4 + $ZrN_{x<1}$ phases converting into a mixture of three Si_3N_4 + ZrN + Zr phases when T_a increases. A relative high amount of the formed nucleation Zr sites consequently favors crystallization at lower temperatures. The structural evolution of the annealed films was investigated using an X-ray diffraction analysis. The XRD pattern corresponding to the $Zr_{18}Si_{29}N_{53}$ film reveals that it is composed of the ZrN and Si_3N_4 grains already after annealing to T_a = 1230°C (Fig. 5.34a). The reflections from ZrN grains significantly increase and become narrow with increasing annealing temperature T_a. It is typically associated to an increase of the grain size, here, from approximately 8 nm at T_a = 1230°C to about 25 nm at T_a = 1490°C, as determined from a broadening of the ZrN reflection lines using the Scherrer formula [47]. A very low intensity of the X-ray reflections from Si_3N_4 grains indicates considerably weaker crystallization of the amorphous Si_3N_4 phase compared to ZrN.

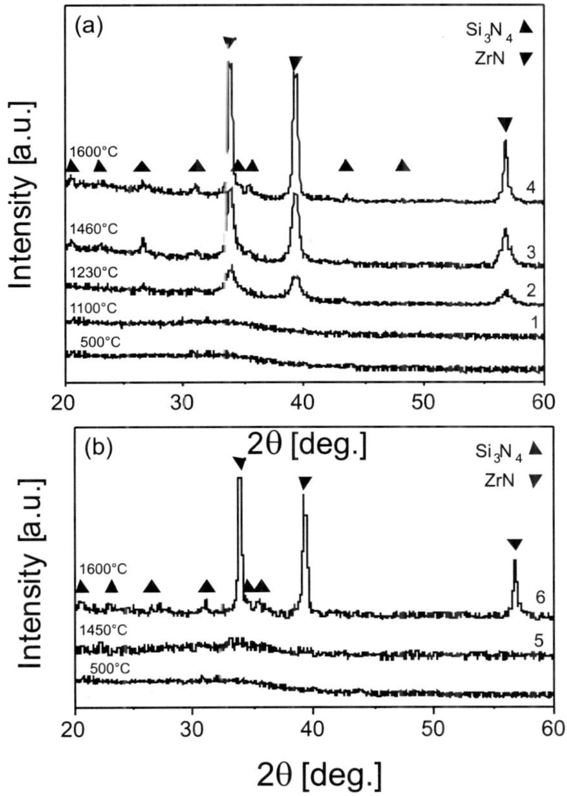

Figure 5.34 Structure of the X-ray amorphous (a) optically opaque $Zr_{18}Si_{29}N_{53}$ and (b) optically transparent $Zr_{16}Si_{28}N_{56}$ films with different phase composition after the DSC measurements. Reprinted from Ref. [40], Copyright (2006), with permission from Elsevier.

Probably in more amorphized $Zr_{16}Si_{28}N_{56}$ film, higher T_a is needed to form nucleation sites and thus the onset of crystallization is shifted to higher annealing temperatures (Fig. 5.33b). A relatively broad exothermic peak measured between 1450 and 1510°C indicates a formation of nucleation sites necessary for crystallization from the amorphous-like structure at T_{cr} = 1530°C. Thermal annealing at $T_a > T_{cr}$ results in a strong crystallization of the ZrN grains, although a very weak crystallization of the Si_3N_4 grains was observed even at T_a = 1600°C, see the XRD patterns given in Fig. 5.34b. It indicates an extreme thermal stability of the amorphous Si_3N_4 phase present in the film.

5.11.6 Summary

The above described experimental results show that the thermal stability of the Zr-Si-N films is affected by many factors. A huge dependence of the crystallization temperature and finally developed film structure on the composition of the Zr-Si-N films and the annealing atmosphere were clearly demonstrated. From the thermal annealing experiments follow that the N content and consequently, the nitrification level of present metal atoms are determinant factors for the thermal stability of the Zr-Si-N films. Firstly, it was shown that the thermal stability can be significantly enhanced through the developed nanocomposite structure consisted of nanograins of the ZrN enrobed by the amorphous Si_3N_4 phase. The presence of a high volume of the thermally stable Si_3N_4 tissue inhibits the growth of the ZrN_x nanograins justifying very high thermal stability of the films. Thus, the crystallization and the grain growth of both phases are detected at very high annealing temperatures. On the contrary, in the films without or just with a low amount of N, the increase of the annealing temperature T_a is accompanied by an increase in the grain size, as usually observed in traditional materials. Secondly, the thermal stability of the films is ascribed to the stability of all the phases from which the films are composed. If one of the phases is not thermally stable, the crystallization and oxidation processes occur at lower temperatures [102]. It has been also observed that the thermal stability of the Zr-Si-N films continuously increases as the overstoichiometric $ZrN_{x>1}$ phase is developed with an increasing amount of N incorporated into the film. It suggests its increasing thermal stability. Thus, the highest crystallization temperature is achieved in the films sputtered at the highest partial pressure of nitrogen p_{N_2}.

A strong effect of the annealing atmosphere and the interdiffusion of the elements from the substrate into the film on the thermal stability are also obvious. It was shown that the crystallization and oxidation processes occur at lower temperatures if the phases present in the film are not stable and easily decomposes and react with oxygen from the annealing atmosphere. Also a high amount of the nucleation sites present in the films in their as-deposited state or formed during annealing by a decomposition of individual phases or by an enrichment of the film with substrate elements, lowers the crystallization temperature. All the facts give a rise into the consideration that one of the possible ways how to achieve the highest thermal stability of the Zr-Si-N films is to synthesize films which exhibit very dense structure composed of thermally stable components. It can be achieved through the ultra-fine grained or even well-amorphized nanocomposite structure consisting of a high amount of the Si_3N_4 phase and the overstoichiometric $ZrN_{x>1}$. Then, formation of nucleation sites for further crystallization is effectively suppressed and the diffusion of oxygen from the annealing atmosphere is effectively limited. These conditions are satisfied in the Zr-Si-N films sputtered at high N partial pressure $p_{N_2} > 0.3$ Pa showing the highest thermal stability. Moreover, thermal stability of the Zr-Si-N films can be enhanced if the films are separated from the substrate by a barrier interlayer. The interdiffusion of substrate elements into the film can be avoided in this way together with undesirable and uncontrolled changes of its phase composition, which significantly influence the crystallization.

5.12 Oxidation Resistance of Zr-Si-N Films

The oxidation resistance of Zr-Si-N films with a Si content ranging from 5 to 15 at.% has been previously investigated [17,103] (see Section 4.3). It has been reported that the oxidation resistance can be significantly pronounced by the addition of a higher amount of Si into the films. The Zr-Si-N films already with 10 at.% have exhibited an onset temperature of oxidation T_{ox} around 600°C [17], that can be even increased if a higher amount of Si and N is incorporated into the film. The basic oxidation mechanism of the Zr-Si-N films has been shown as a formation of a ZrO_2 oxide layer on the film surface [103].

Although the Zr-Si-N coating system with a low and intermediate Si content is well known, just poor attention is devoted to the characterization of the oxidation behavior of films with a Si content higher than 25 at.%. Though the hardness of the Me-Si-N films with a high Si content does not exceed 35 GPa, the films exhibit excellent oxidation resistance [28,43,101,102,104–108]. This fact makes them more interesting for the tools applications than superhard but less thermally stable films ($H > 40$ GPa).

In order to find the relationship between the phase composition and the oxidation resistance of the Zr-Si-N films with various elemental composition, a detailed study was performed on the Zr-Si-N films sputter deposited in a wide range of the partial pressure of nitrogen $p_{N_2} = 0, 0.03, 0.05, 0.1, 0.15$, and 0.4 Pa. The films strongly differ in their phase composition varying with p_{N_2} from the crystalline $ZrSi_2$ through the mixture of a-Si_3N_4 + $ZrSi_{x<2}$ and a-Si_3N_4 + $ZrSi_x$ + ZrN_x phases further to a-Si_3N_4 + $ZrN_{x>1}$. The oxidation resistance of the Zr-Si-N films, demonstrated by thermogravimetric curves, is given in Fig. 5.35. From this figure it is clearly seen that the oxidation resistance, which is determined as the onset of the film mass increase Δm, strongly depends on the phase composition of the Zr-Si-N films. It is relatively low for the crystalline alloy films and a-Si_3N_4 + (Zr,Si) composite films with a low content of N, although also in this case no mass gain Δm corresponding to oxidation is observed up to 800°C. The oxidation resistance, however, strongly increases with the increase in content of the Si_3N_4 phase in the film. Just the amount of the Si_3N_4 phase plays an essential role in the thermal stability and oxidation resistance of the Zr-Si-N films. Its efficiency to protect the $ZrSi_x$ and further ZrN_x grains at a higher N content highly depends on the coverage level of crystallites, which is the higher the higher volume content of a Si_3N_4 tissue is formed. Although Si_3N_4 is very thermally stable phase, a difference in thermal stability and oxidation resistance of Me-Si-N films is affected mainly by a diverse stability of other phases of which these films are composed. For instance, if the MeN_x phases readily decompose at elevated temperatures, inward oxygen diffusion proceeds through channels remaining after the N release [102]. Due to a high reactivity of oxygen, a rapid oxidation process results in a disturbed microstructure since the amorphous Si_3N_4 matrix can no longer act as an efficient barrier hindering the oxidation process. The oxidation resistance is also lower, if the Zr-

Si-N films in the as-deposited state consist of free metal atoms (Zr or Si), which readily oxidize at relatively low temperatures (see the $Zr_{26}Si_{31}N_{43}$ film in Fig. 5.35). Compared to other Me-Si-N films, the Zr-Si-N films sputtered at $p_{N_2} > 0.1$ Pa exhibit much higher oxidation resistance [43,102]. This fact is ascribed to a high nitrification level of the ZrN_x phase, which makes it more stable [44,102,109]. Hence, the highest oxidation resistance exhibit films composed of a high amount of a-Si_3N_4 and overstoichiometric $ZrN_{x>1}$. In this case, no increase in mass (Δm = 0) is detected up to 1300°C.

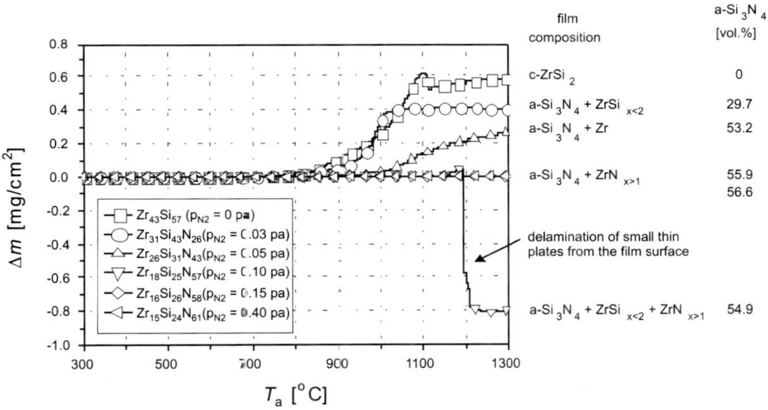

Figure 5.35 Thermogravimetric curves of the Zr-Si-N films with different phase compositions dynamically heated in flowing air as a function of annealing temperature T_a. Reprinted from Ref. [28], Copyright (2005), with permission from Elsevier.

X-ray diffraction measurements of the samples after the annealing experiments reveal significant structural changes of the oxidized films (Fig. 5.36). Although all Zr-Si-N films sputter deposited at $p_{N_2} \geq$ 0.05 Pa exhibited an X-ray amorphous structure in the as-deposited state, the phases developed upon high-temperature annealing in air show again a strong relation to the phase composition of the films.

While the orthorhombic $ZrSi_2$ phase was observed in all the films, other phases developed after annealing obviously differ. Oxidation of the Zr-Si-N films deposited at $p_{N_2} < 0.15$ Pa is accompanied by a formation of the tetragonal $ZrSiO_4$ and monoclinic ZrO_2 phases.

Moreover, these films exhibit an opaque surface and a partial film spallation or delamination after annealing. On the other hand, additional crystallization only of the Si_3N_4 phase was observed in the films sputtered at $p_{N_2} \geq 0.15$ Pa. Although, the diffraction peaks around $2\theta \approx 35°$ might also fit to the positions of the tetragonal ZrO_2 standard, crystallization of the Si_3N_4 phase is more probable. The formation of the crystalline ZrO_2 phase is typically accompanied by a volume expansion with subsequent film spallation, which was, however, not observed. On the contrary, the films remain smoothed and dense. Other peaks observed in the XRD patterns can be probably attributed to the ternary $Zr_5Si_3N_{1-x}$ phase, which is the high-temperature Zr_5Si_3 phase stabilized by N to lower temperatures [36].

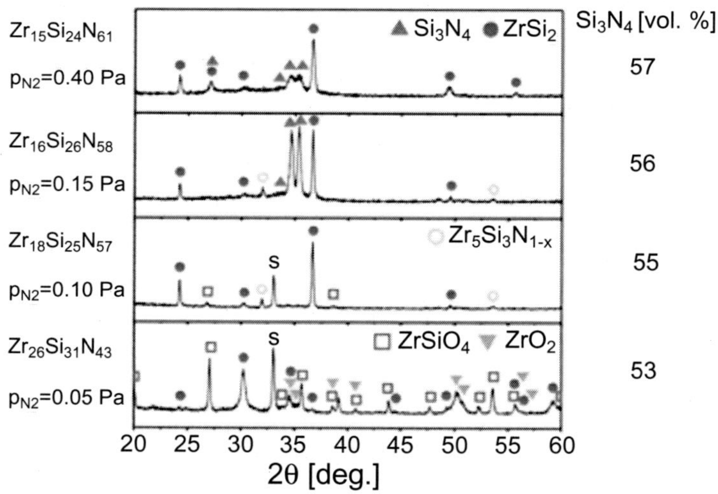

Figure 5.36 XRD patterns of the Zr-Si-N films sputtered at various p_{N_2} after annealing in air to 1300°C; here S denotes the reflection from the Si(100) substrate.

The plane-view surface images of the representative X-ray amorphous $Zr_{15}Si_{24}N_{61}$ film in the as-deposited state and after annealing in air to 1300°C taken by light optical microscopy (LOM) (Fig. 5.37) and atomic force microscopy (AFM) (Fig. 5.38) together with cross-sectional scanning electron micrographs (Fig. 5.39) serve for better understanding of the oxidation mechanisms.

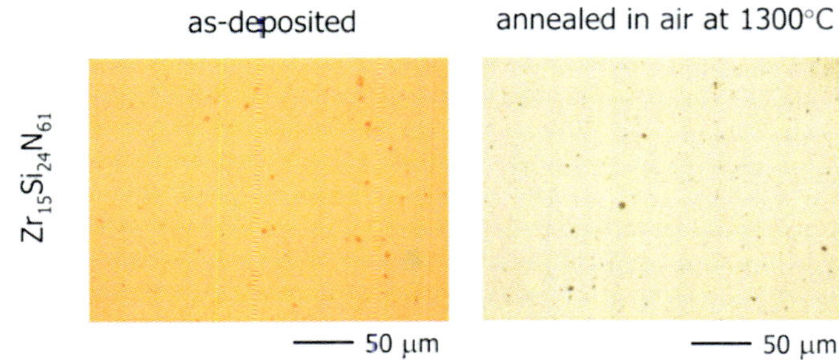

Figure 5.37 LOM surfaces of the as-deposited and annealed $Zr_{15}Si_{24}N_{61}$ film exhibiting the best oxidation resistance.

The film composed of $Si_3N_4 + ZrN_{x>1}$ exhibits the highest oxidation resistance. Its surface is smooth with the mean roughness R_a below 2 nm remaining without any defects also after annealing. Negligible increase in Δm and a slight change of the surface color indicate a formation of very thin transparent interference layer on the film surface [104]. The surface oxide layer is compact and dense with a low roughness (Fig. 5.38).

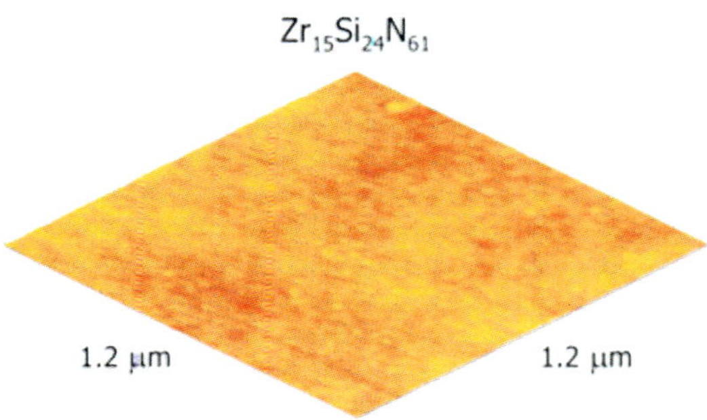

Figure 5.38 AFM surface of the annealed $Zr_{15}Si_{24}N_{61}$ film exhibiting the best oxidation resistance.

The detailed investigation of the thin oxide layer formed on the film surface after annealing in air to 1300°C by using scanning electron microscopy reveals a very thin oxide layer formed on the film surface characterized by many small globular particles (Fig. 5.39). The thickness of this layer is 115 nm, which is only about 3% of the film thickness (h_f = 3600 nm). The bulk of the film remains very dense and glass-like. It enables to measure the microhardness of the film and to characterize the stability of mechanical properties after annealing. The microhardness of the annealed film is 3 GPa lower than that of the as-deposited film. However, when the indenter penetrated deeper into the film at higher loads, the microhardness approaches to that one of the as-deposited Zr-Si-N film. This fact reveals that the bulk of the annealed Zr-Si-N film beneath the oxide layer remains unchanged.

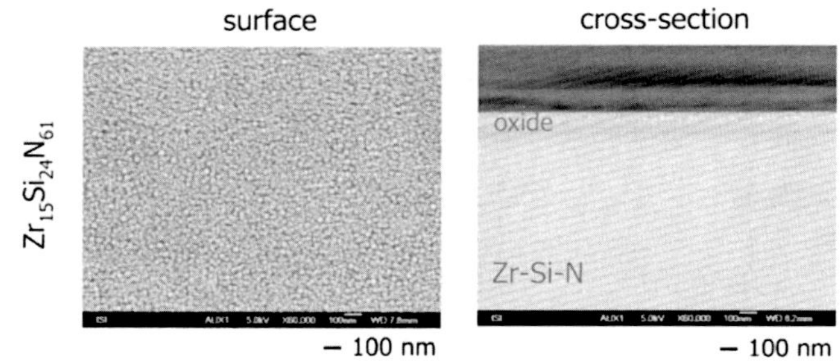

Figure 5.39 SEM plane-view and cross-sectional images of the annealed $Zr_{15}Si_{24}N_{61}$ film exhibiting the best oxidation resistance.

The above results show that the high-temperature oxidation resistance of Me-Si-N films can be achieved if the film is composed of thermally stable phases which form dense microstructure without grain boundaries. The synergetic effect of a high amount of Si and N in the film results in the formation of an X-ray amorphous microstructure where ultra-fine grains are dispersed in a high volume of the Si_3N_4 phase. In this way, the inward oxygen diffusion is efficiently prevented that promotes a high protective capability of the film against oxidation. While in the crystalline materials the oxygen diffusion proceeds readily through the open voids, defects and other diffusion paths, glass-like morphology of the Zr-Si-N films with a high (≥25 at.%) Si content hinders oxygen transport further into the film and so prevents the oxidation process. The nitrification level of Si and Zr is shown to be also very important since the bonded atoms cannot diffuse and segregate toward surface. Thus, the diffused free Si or Zr atoms would more easily form SiO_2 or ZrO_2 at the surface area compared to Si_3N_4 or ZrN_x [108]. In addition, the nanocomposite nature of the Zr-Si-N films ensures that thin oxide layer is formed on the surface separated by a sharp interface from the remaining film. This surface layer acts as a diffusion barrier against oxygen. Since the oxide layer formed on the surface of the Zr-Si-N film is probably not fully crystallized and grain boundaries are filled with amorphous SiO_2 or SiO_xN_y phases, the oxidation rate is very slow even at very high temperatures. The Zr-Si-N films with a high Si content (≥20 at.%) belongs to a new class of advanced protective coatings keeping good mechanical properties even at temperatures over 1000°C in oxidizing environment. A huge protecting effect of the Si_3N_4 phase together with a high stability of the overstoichiometric ZrN_x is evident from this experiment performed on the Zr-Si-N films sputtered at $p_{N_2} \geq 0.3$ Pa.

References

1. R. Makabe, S. Nakajima, O. Tabata, M. Aoki (1986), Dependence of the hardness of titanium nitride prepared by plasma chemical vapour deposition on the gas flow rate and the r.f. power, *Thin Solid Films* 137, 49–50.
2. W.D. Sproul (1983), Very high rate reactive sputtering of TiN, ZrN and HfN, *Thin Solid Films* 107, 141–147.
3. A. Cavaleiro, C. Louro, J.V. Fernandes, C.M.A. Brett (1999), Use of ultramicroindentation to evaluate the degradation of sputtered coatings, *Vacuum* 52, 157–162.
4. C. Subramanian, K.N. Strafford (1993), Review of multicomponent and multilayer coatings for tribological applications, *Wear* 165, 85–95.
5. C. Louro, A. Cavaleiro (1997), Thermal oxidation of tungsten-based sputtered coatings, *J. Electrochem. Soc.* 144, 259–266.
6. L. Hultman (2000), Thermal stability of nitride thin films, *Vacuum* 57, 1–30.
7. P.J. Martin, A. Bendavid, J.M. Cairney, M. Hoffman (2005), Nanocomposite Ti-Si-N, Zr-Si-N, Ti-Al-Si-N, Ti-Al-V-Si-N thin film coatings deposited by vacuum arc deposition, *Surf. Coat. Technol.* 200, 2228–2235.
8. J.G. Han, H.S. Myung, H.M. Lee, L.R. Shaginyan (2003), Microstructure and mechanical properties of Ti-Ag-N and Ti-Cr-N superhard nanostructured coatings, *Surf. Coat. Technol.* 174–175, 738–743.
9. S.Veprek (1999), The search for novel superhard materials, *J. Vac. Sci. Technol.* A17, 2401–2420.
10. E. Martinez, R. Sanjinés, O. Banakh, F. Lévy (2004), Electrical, optical and mechanical properties of sputtered CrN_y and $Cr_{1-x}Si_xN_{1.02}$ thin films, *Thin Solid Films* 447–448, 332–336.

11. M. Benkahoul, C.S. Sandu, N. Tabet, M. Parlinska-Wojtan, A. Karimi, F. Lévy (2004), Effect of Si incorporation on the properties of niobium nitride films deposited by DC reactive magnetron sputtering, *Surf. Coat. Technol.* 188–189, 435–439.
12. P. Torri, A. Mahiout, J. Koskinen, J.-P. Hirvonen, L.-S. Johansson (2000), Corrosion properties of amorphous Mo-Si-N and nanolayered Mo-Si-Nn/SiC coatings, *Scr. Mater.* 42, 609–613.
13. O. Knacke, O. Kubaschewski, K. Hesselmann (1991), *Thermochemical Properties of Inorganic Substances II*. Berlin: Springer-Verlag.
14. M. Nose, W.A. Chiou, M. Zhou, T. Mae, M. Meshii (2002), Microstructure and mechanical properties of Zr-Si-N films prepared by rf-reactive sputtering, *J. Vac. Sci. Technol.* A20, 823–828.
15. D. Pilloud, J.F. Pierson, A.P. Marques, A. Cavaleiro (2004), Structural changes in Zr-Si-N films vs. their silicon content, *Surf. Coat. Technol.* 180–181, 352–356.
16. A. Winkelmann, J.M. Cairney, M.J. Hoffman, P.J Martin, A. Bendavid (2005), Zr-Si-N films fabricated using hybrid cathodic arc and chemical vapour deposition: Structure vs. properties, *Surf. Coat. Technol.* 200, 4213–4219.
17. M. Nose, M. Zhou, T. Nagae, T. Mae, M. Yokota, S. Saji (2000), Properties of Zr-Si-N coatings prepared by RF reactive sputtering, *Surf. Coat. Technol.* 132, 163–168.
18. W.-T. Li, D.R. McKenzie, W.D. McFall, Q.-C. Zhang (2001), Effect of sputtering-gas pressure on properties of silicon nitride films produced by helicon plasma sputtering, *Thin Solid Films* 384, 46–52.
19. S.K. Habib, A. Rizk, I.A. Mousa (1998), Physical parameters affecting deposition rates of binary alloys in a magnetron sputtering system, *Vacuum* 49, 153–160.
20. T. Hurkmans, D.B. Lewis, H. Paritong, J.S. Brooks, W.D. Münz (1999), Influence of ion bombardment on structure and properties of unbalanced magnetron grown CrN_x coatings, *Surf. Coat. Technol.* 114, 52–59.
21. T. Mae, M. Nose, M. Zhou, T. Nagae, K. Shimamura (2001), The effects of Si addition on the structure and mechanical properties of ZrN thin films deposited by an r.f. reactive sputtering method, *Surf. Coat. Technol.* 142–144, 954–958.
22. B.-S. Yau, J.-L. Huang (2004), Effects of nitrogen flow on r.f. reactive magnetron sputtered silicon nitride films on high speed steel, *Surf. Coat. Technol.* 176, 290–295.

23. K.S. Fancey (1995), An investigation into dissociative mechanisms in nitrogenous glow discharges by optical emission spectroscopy, *Vacuum* 46, 695–700.

24. C.H. Shon, J.K. Lee, H.J. Lee, Y.K. Shin, Y. Yang, T.H. Chung (1998), Velocity distribution in magnetron sputter, *IEEE Trans. Plasma Sci.* 26, 1635–1644.

25. B. Fritsche, T. Chevolleau, J. Kourtev, A. Kolitsch, W. Möller (2003), Plasma diagnostic of an RF magnetron Ar/N_2 discharge, *Vacuum* 69, 139–145.

26. J.-W. Lee, S.C.N. Cheng (2000), Development of SiN_x and AlN_x passivation layers, *Thin Solid Films* 358, 215–222.

27. M. Diserens, J. Patscheider, F. Lévy (1998), Improving the properties of titanium nitride by incorporation of silicon, *Surf. Coat. Technol.* 108–109, 241–246.

28. J. Musil, R. Daniel, O. Takai (2005), Structure and mechanical properties of magnetron sputtered Zr-Si-N films with a high (≥25 at.%) Si content, *Thin Solid Films* 478, 238–247.

29. I. Safi (2000), Recent aspects concerning DC reactive magnetron sputtering of thin films: A review, *Surf. Coat. Technol.* 127, 203–218.

30. S. Schiller, K. Goedicke, J. Reschke, V. Kirchoff, S. Schneider, F. Milde (1993), Pulsed magnetron sputter technology, *Surf. Coat. Technol.* 61, 331–337.

31. M. Scherer, I. Schmitt, R. Latz, M. Schanz (1992), Reactive alternating current magnetron sputtering of dielectric layers, *J. Vac. Sci. Technol.* A10, 1772–1776.

32. C. Louro, A. Cavaleiro, F. Montemor (2001), How is the chemical bonding of W-Si-N sputtered coatings? *Surf. Coat. Technol.* 142–144, 964–970.

33. A. Vomiero, E. Bosolo Marchi, G. Mariotto, G. Della Mea, A. Scandurra, O. Puglisi (2005), The role of N in the re-sputtering inhibition of Si in W-Si-N reactively sputtered thin layers, *J. Appl. Phys.* 97, 349051–349057.

34. E.A. Brandes (1992), *Smithells Metals Reference Book*. Oxford: Butterworth-Heinemann .

35. J.A. Dean (1999), *Lange's Handbook Of Chemistry*. New York: McGraw-Hill, Inc..

36. P. Rogl, J.C. Schuster (1992), *Phase Diagrams of Ternary Boron Nitride and Silicon Nitride Systems*. Materials Park, OH: ASM International.

37. W. Spengler, R. Kaiser (1976), First and second order Raman scattering in transition metal compounds, *Solid State Commun.* 18, 881–884.

38. G. Beshkov, S. Lei, V. Lazarova, N. Nedev, S.S. Georgiev (2003), IR and Raman absorption spectroscopic studies of APCVD, LPCVD and PECVD thin SiN films, *Vacuum* 69, 301–305.

39. C.H. Zhang, Z.-J. Liu, K.Y. Li, Y.G. Shen, J.B. Luo (2004), Microstructure, surface morphology and mechanical properties of nanocrystalline TiN/amorphous Si_3N_4 composite films synthesized by ion beam assisted deposition, *J. Appl. Phys.* 95, 1460–1467.

40. R. Daniel, J. Musil, P. Zeman, C. Mitterer (2006), Thermal stability of magnetron sputtered Zr-Si-N films, *Surf. Coat. Technol.* 201, 3368–3376.

41. S. Veprek, A. Niederhofer, K. Moto, T. Bolom, H.-D. Männling, P. Nesládek, G. Dollinger, A. Bergmaier (2000), Composition, nanostructure and origin of the ultrahardness in nc-TiN/a-Si_3N_4/a- and nc-$TiSi_2$ nanocomposites with HV = 80 to ≥105 GPa, *Surf. Coat. Technol.* 133–134, 152–159.

42. H.O. Pierson (1996), *Handbook of Refractory Carbides and Nitrides.* New Jersey: Noyes Publications.

43. H. Zeman, J. Musil, P. Zeman (2004), Physical and mechanical properties of sputtered Ta-Si-N films with a high (≥40 at.%) content of Si, *J. Vac. Sci. Technol.* A22, 646–649.

44. H.M. Benia, M. Guemmaz, G. Schmerber, A. Mosser, J.-C. Parlebas, Investigations on non-stoichiometric zirconium nitrides, *Appl. Surf. Sci.* 200 (2002) 231–238.

45. M. Del Re, R. Gouttebaron, J.-P. Dauchot, P. Leclère, G. Terwagne, M. Hecq (2003), Study of ZrN layers deposited by reactive magnetron sputtering, *Surf. Coat. Technol.* 174–175, 240–245.

46. J. Patscheider, T. Zehnder, M. Diserens (2001), Structure-performance relations in nanocomposite coatings, *Surf. Coat. Technol.* 146–147, 201–208.

47. J.I. Langford (1980), National Bureau of Standards. In: *Accuracy in powder diffraction, NBS special publication no. 567.* S. Block, C.R. Hubbard (Eds.), Washington DC: National Bureau of Standards.

48. P.B. Barna, M. Adamik, J. Labar, L. Köver, J. Toth, A. Dévényi, R. Manaila (2000), Formation of polycrystalline and microcrystalline composite thin films by codeposition and surface chemical reaction, *Surf. Coat. Technol.* 125, 147–150.

49. P.B. Barna, M. Adamik (1998), Fundamental structure forming phenomena of polycrystalline films and the structure zone models, *Thin Solid Films* 317, 27–33.

50. J.A. Thornton, D.W. Hoffman (1989), Stress-related effects in thin films, *Thin Solid Films* 171, 5–31.

51. A.P. Marques, A. Cavaleiro (2003), Structural and mechanical properties of amorphous W-Si-N sputtered films after thermal annealing, *Thin Solid Films* 441 150–160.

52. S.R. Choi, I.-W. Park, J.H. Park, K.H. Kim (2004), Influence of substrate bias voltage on deposition behavior and microindentation hardness of Ti-Si-N coatings by a hybrid coating system of arc ion plating and sputtering techniques, *Surf. Coat. Technol.* 179, 89–94.

53. S.R. Choi, I.-W. Park, S.H. Kim, K.H. Kim (2004), Effects of bias voltage and temperature on mechanical properties of Ti-Si-N coatings deposited by a hybrid system of arc ion plating and sputtering techniques, *Thin Solid Films*, 447–448, 371–376.

54. L. Hultman, U. Helmersson, S.A. Barnett, J.-E. Sundgren, J.E. Greene (1987), Low-energy ion irradiation during film growth for reducing defect densities in epitaxial TiN(100) films deposited by reactive-magnetron sputtering, *J. Appl. Phys.* 61, 552–555.

55. I. Petrov, L. Hultman, U. Helmersson, J.-E. Sundgren, J.E. Greene (1989), Microstructure modification of TiN by ion bombardment during reactive sputter deposition, *Thin Solid Films* 169, 299–314.

56. F. Vaz, L. Rebouta, P. Goudeau, T. Girardeau, J. Pacaud, J.P. Riviere, A. Traverse (2001), Structural transitions in hard Si-based TiN coatings: The effect of bias voltage and temperature, *Surf. Coat. Technol.* 146–147, 274–279.

57. F.J. Cherne, M.I. Baskes, R.B. Schwarz (2003), Atomistic simulations of the phase stability and elastic properties of nickel-zirconium alloys, *J. Non-Crystalline Solids* 317, 45–51.

58. M. Kawamura, K. Kumagai, Y. Abe, K. Sasaki, H. Yanagisawa (1998), Characterization of TiN films prepared by r.f. sputtering using metal and compound targets, *Vacuum* 51, 377–380.

59. T. Banno, H. Kutsuna, S. Minami (2006), Application of high-density plasma to sputtering and reactive sputtering processes, *Vacuum* 80, 667–670.

60. W.C. Oliver, G.M. Pharr (1992), An improved technique for determining hardness and elastic modulus using load and displacement sensing indentation experiments, *J. Mater. Res.* 7, 1564–1583.

61. A. Leyland, A. Matthews (2000), On the significance of the H/E ratio in wear control: A nanocomposite coating approach to optimised tribological behavior, *Wear* 246, 1–11.

62. J. Musil, F. Kunc, H. Zeman, H. Poláková (2002), Relationships between hardness, Young's modulus and elastic recovery in hard nanocomposite coatings, *Surf. Coat. Technol.* 154, 304–313.

63. T.Y. Tsui, G.M. Pharr, W.C. Oliver, C.S. Bhatia, R.L. White, S. Anders, A. Anders, I.G. Brown (1995), Nanoindentation and nanoscratching of hard carbon coatings for magnetic disks, *Mater. Res. Soc. Symp. Proc.* 383, 447–452.

64. H. Oettel, R. Wiedemann (1995), Residual stresses in PVD hard coatings, *Surf. Coat. Technol.* 76–77, 265–273.

65. D.M. Mattox (1989), Particle bombardment effects on thin-film deposition: A review, *J. Vac. Sci. Technol.* A7, 1105–1114.

66. F.M. d'Heurle, J.M.E. Harper (1989), Note on the origin of intrinsic stresses in films deposited via evaporation and sputtering, *Thin Solid Films* 171, 81–92.

67. V. Teixeira (2002), Residual stress and cracking in thin PVD coatings, *Vacuum* 64, 393–399.

68. H. Holleck (1986), Material selection for hard coatings, *J. Vac. Sci. Technol.* A4, 2661–2669.

69. P.J. Kelly, R.D. Arnell (1998), Development of a novel structure zone model relating to the closed-field unbalanced magnetron sputtering system, *J. Vac. Sci. Technol.* A16, 2858–2869.

70. P.J. Kelly, R.D. Arnell (2000), Magnetron sputtering: A review of recent developments and applications, *Vacuum* 56, 159–172.

71. P. Yashar, J. Rechner, M.S. Wong, W.D. Sproul, S.A. Barnett (1997), High-rate reactive sputtering of yttria-stabilized zirconia using pulsed d.c. power, *Surf. Coat. Technol.* 94–95, 333–338.

72. S.P. Murarka, M. Eizenberg, A.K. Sinha (2003), *Interlayer Dielectrics for Semiconductor Technologies*. Amsterdam, Boston: Elsevier Academic Press.

73. J. Procházka, P. Karvánková, M.G.J. Vepřek-Heijman, S. Vepřek (2004), Conditions required for achieving superhardness of ≥45 GPa in nc-TiN/a-Si_3N_4 nanocomposites, *Mater. Sci. Eng.* A384, 102–116.

74. W.-J. Chou, G.-P. Yu, J.-H. Huang (2002), Bias effect of ion-plated ZrNfilm on Si(100), *Thin Solid Films* 405, 162–169.

75. J. Musil, R. Daniel (2003), Structure and mechanical properties of magnetron sputtered Zr-Ti-Cu-N films, *Surf. Coat. Technol.* 166, 243–253.

76. C. Mitterer, F. Holler, F. Üstel, D. Heim (2000), Application of hard coatings in aluminium die casting - soldering, erosion and thermal fatigue behavior, *Surf. Coat. Technol.* 125, 233–239.

77. P. Panjan, B. Navinšek, A. Cvelbar, I. Milošev, A. Zalar (1996), Oxidation of TiN, ZrN, TiZrN, CrN, TiCrN and TiN/CrN multilayer hard coatings reactively sputtered at low temperature, *Thin Solid Films* 281–282, 298–301.

78. L.A. Donohue, I.J. Smith, W.-D. Münz, I. Petrov, J.E. Greene (1997), Microstructure and oxidation-resistance of $Ti_{1-x-y-z}Al_xCr_yY_zN$ layers grown by combined steered-arc/unbalanced-magnetron-sputter deposition, *Surf. Coat. Technol.* 94–95, 226–231.

79. I. Wadsworth, I.J. Smith, L.A. Donohue, W.-D. Münz (1997), Thermal stability and oxidation resistance of TiAlN/CrN multilayer coatings, *Surf. Coat. Technol.* 94–95, 315–321.

80. I. Milošev, H.-H. Strehblow, B. Navinšek (1997), Comparison of TiN, ZrN and CrN hard nitride coatings: Electrochemical and thermal oxidation, *Thin Solid Films* 303, 246–254.

81. L. Krusin-Elbaum, W. Wittmer (1983), Oxidation kinetics of ZrN thin films, *Thin Solid Films* 107, 111–116.

82. S. Vepřek, S. Reiprich (1995), A concept for the design of novel superhard coatings, *Thin Solid Films* 268, 64–71.

83. C. Louro, A. Cavaleiro (2000), Mechanical behavior of amorphous W-Si-N sputtered films after thermal annealing at increasing temperatures, *Surf. Coat. Technol.* 123, 192–198.

84. J.W. Nah, W.S. Choi, S.K. Hwang, C.M. Lee (2000), Chemical state of (Ta,Si)N reactively sputtered coating on a high-speed steel substrate, *Surf. Coat. Technol.* 123, 1–7.

85. Y. Wang, C. Zhu, Z. Song, Y. Li (2004), High temperature stability of Zr-Si-N diffusion barrier in Cu/Si contact system, *Microelectron. Eng.* 71, 69–75.

86. Z.X. Song, K.W. Xu, H. Chen (2004), The characterization of Zr-Si-N diffusion barrier films with different sputtering bias voltage, *Thin Solid Films* 468, 203–207.

87. Z.X. Song, K.W. Xu, H. Chen (2004), The effect of nitrogen partial pressure on Zr-Si-N diffusion barrier, *Microelectron. Eng.* 71, 28–33.

88. H.-D. Männling, D.S. Patil, K. Moto, M. Jílek, S. Veprek (2001), Thermal stability of superhard nanocomposite coatings consisting of immiscible nitrides, *Surf. Coat. Technol.* 146–147, 263–267.

89. M. Stoiber, M. Panzenböck, C. Mitterer, C. Lugmair (2001), Fatigue properties of Ti-based hard coatings deposited onto cold-working tool steels, *Surf. Coat. Technol.* 142–144, 117–124.

90. J.D. Wilcock, D.S. Campbell (1969), A sensitive bending beam apparatus for measuring the stress in evaporated thin films, *Thin Solid Films* 3, 3–12.

91. H. Ebert (1971), , Langot-Börnstein Eigenschaften der Materie in ihren Aggregatzustanden. K. Schafer, G. Beggerow (Eds.), In: *Mechanish Thermische Zustandsgrossen*. Berlin: Springer-Verlag.

92. J.D. Verhoeven (1975), *Fundamentals of Physical Metallurgy*, New York: Wiley.

93. P.H. Mayrhofer, C. Mitterer (2000), High-temperature properties of nanocomposite TiB_xN_y and TiB_xC_y coatings, *Surf. Coat. Technol.* 133–134, 131–137

94. P.H. Mayrhofer, F. Kunc, J. Musil, C. Mitterer (2002), A comparative study on reactive and non-reactive unbalanced magnetron sputter deposition of TiN coatings, *Thin Solid Films* 415, 151–159.

95. S. Veprek, M.G.J. Veprek-Heijman, P. Karvánková, J. Procházka (2005), Different approaches to superhard coatings and nanocomposites, *Thin Solid Films* 476, 1–29.

96. C. Louro (2004), Hardness evaluation of W-Si-N sputtered coatings after thermal degradation, *Surf. Coat. Technol.* 180–181, 544–550.

97. Y. Hirohata, N. Shimamoto, T. Hino, T. Yamashima, K. Yabe (1994), Properties of Si3N4 films prepared by magnetron sputtering, *Thin Solid Films* 253, 425–429.

98. G.I. Grigorov, K.G. Grigorov, M. Stoyanova, J.L. Vignes, J.P. Langeron, P. Denjean (1993), Aluminium diffusion in titanium nitride films. Efficiency of TiN barrier layers, *Appl. Phys.* A57, 195–197.

99. F.-H. Lu, W.-Z. Lo (2004), Degradation of ZrN films at high temperature under controlled atmosphere, *J. Vac. Sci. Technol.* A22, 2071–2076.

100. C. Héau, R.Y. Fillit, F. Vaux, F. Pascaretti (1999), Study of thermal stability of some hard nitride coatings deposited by reactive magnetron sputtering, *Surf. Coat. Technol.* 120–121, 200–205.

101. J. Musil, R. Daniel, J. Soldán, P. Zeman (2006), Properties of reactively sputtered W-Si-N films, *Surf. Coat. Technol.* 200, 3886–3895.

102. P. Zeman, J. Musil (2006), Difference in high-temperature oxidation resistance of amorphous Zr-Si-N and W-Si-N films with a high Si content, *Appl. Surf. Sci.* 252, 8319–8325.

103. D. Pilloud, J.F. Pierson, M.C. Marco de Lucas, M. Alnot (2004), Stabilisation of tetragonal zirconia in oxidized Zr-Si-N nanocomposite coatings, *Appl. Surf. Sci.* 229, 132–139.

104. P. Zeman, J. Musil, R. Daniel (2006), High-temperature oxidation resistance of Ta-Si-N films with a high Si content, *Surf. Coat. Technol.* 200, 4091–4096.

105. J. Musil, P. Dohnal, P. Zeman (2005), Physical properties and high-temperature oxidation resistance of sputtered Si_3N_4/MoN_x nanocomposite coatings, *J. Vac. Sci. Technol.* B23, 1568–1575.

106. J.-P. Hirvonen, I. Suni, H. Kattelus, R. Lappalainen, P. Torri, H. Kung, T.R. Jervis, M. Nastasi, J.R. Tesmer (1995), Crystallization and oxidation behavior of Mo-Si-N coatings, *Surf. Coat. Technol.* 74–75, 981–985.

107. C. Louro, A. Cavaleiro (1999), The oxidation behaviour of mixed tungsten silicon sputtered coatings, *Thin Solid Films* 343–344, 51–56.

108. J.B. Choi, K. Cho, M.-H. Lee, K.H. Kim (2004), Effects of Si content and free Si on oxidation behavior of Ti-Si-N coating layers, *Thin Solid Films* 447–448, 365–370.

109. L. Pichon, T. Girardeau, A. Straboni, F. Lignou, Ph.Guérin, J. Perrière (1999), Zirconium nitrides deposited by dual ion beam sputtering: Physical properties and growth modelling, *Appl. Surf. Sci.* 150, 115–124.

Chapter 6

High Si-Containing W-Si-N Nanocomposite Films

Tungsten-based thin films have been developed as an alternative to traditional coating systems for many various applications. Sputter-deposited films have found their usage as diffusion barriers in microelectronics [1], semiconductor devices [2], interconnect technology [3], and also in the field of hard coatings for mechanical applications [4]. Although tungsten-based thin films generally exhibit excellent mechanical properties, their performance as protective coatings is markedly limited due to relatively low thermal stability and oxidation resistance of this system at high temperatures. Coatings for mechanical applications, such as cutting and forming applications or machining operations, however, perform typically under extreme conditions. Besides high hardness and resistance to fracture toughness, it also requires high thermochemical stability. An improvement of oxidation behavior and wear resistance of tungsten-based films in tooling applications has gained interest of many researchers. In last decade, many works have been devoted to investigations of the influence of the addition of various elements into these films on their physical and mechanical properties [5,6]. The main motivation of all the studies was to improve the performance of tungsten-based films in aggressive environments without changes of the mechanical properties reached during their development [7].

Novel Nanocomposite Coatings: Advances and Industrial Applications
Rostislav Daniel and Jindřich Musil

ISBN 978-981-4411-17-2 (Hardcover), 978-981-4411-18-9 (eBook)
www.panstanford.com

One of the promising candidates to improve the oxidation behavior of tungsten-based films has been found silicon as an alloying element. Selection of silicon as the third element in the ternary nitride films is based on the fact that silicon easily forms silicon nitride phase, which leads to the formation of a stable nanocomposite structure. At low silicon content in the films, thin silicon nitride tissue coexisting with the crystalline WN_x phase ensures both high hardness and enhanced oxidation resistance. Silicon nitride is known for its high thermal stability and efficiency to cover MeN grains and thus, to protect them against oxidation. Its efficiency to prevent oxidation is even more pronounced if the amorphous Si_3N_4 phase dominates in the film. In that case, fine-grained structure is developed remaining stable to high temperatures due to the crystallization temperature of Si_3N_4 considerably exceeding 1000°C. The amorphous structure is an excellent barrier for a penetration of oxygen to the substrate surface through the film and thus, very high oxidation resistance is expected to be achieved in the Me-Si-N films with a high volume content of amorphous Si_3N_4. This phenomenon is given by the fact that the fine-grained structure do not possess grain boundaries, which typically act as diffusion paths as in the case of films with well-developed columnar structure [8]. Nearly amorphous structure also provides chemical inertness and low diffusivity within the film, which is essential for these films to be used as diffusion barriers in microelectronics [9]. In addition, the absence of grain boundaries after the incorporation of silicon into MeN films has been shown to reduce the macrostress generated during the film growth and increase the thermal stability of mechanical properties of the films [10]. Indeed, the influence of annealing at high temperatures on the physical properties of Me-Si-N films is one of the main points of interest. In spite of the fact that silicon containing Me-Si-N films have been investigated in many laboratories [11–15], there is no sufficient information on properties of the films with a high (≥20 at.%) content of silicon. Although, some works have been focused on the effect of thermal annealing on mechanical properties and structural and compositional evolution of the Me-Si-N films [16], the information on the oxidation mechanisms involved in the films after exposing to oxygen at high temperatures are still missing and are not fully explained yet.

Hence, this chapter is focused on the systematic investigation of the W-Si-N films with a high silicon content consisting of fine-grained, WN_x grains dispersed in dominating silicon nitride tissue. The interrelationships between the film structure, phase composition, macrostress, mechanical properties, their thermal stability, and oxidation resistance will be discussed in detail. Besides Zr-Si-N films presented in previous chapter, the great efficiency to stabilize the film structure through metalloids, such as silicon and nitrogen is demonstrated in the case of the W-Si-N system, which might be an alternative to traditional hard coatings for high-temperature mechanical applications.

6.1 Deposition Rate a_D and Energy E_{bi} Delivered to the W-Si-N Films

Deposition rate is an important process property for producing thin films with reasonable processing times. In industrial conditions, maximizing the deposition rate is an important way of lowering process costs. One of the main advantages of d.c. magnetron sputtering is the high deposition rate, especially of metals in an inert gas atmosphere. However, while a reactive gas is introduced into the deposition system, the deposition rates of reactively sputtered nitrides even from a metallic target are about five times lower. In addition, the film synthesis is also attended, besides poor deposition rates, by poisoning and arcing of the target. The use of a strongly poisoned target leads to an unstable deposition process and thus, particle contamination and surface defects of the thin film. With respect to high deposition rates needed in many applications, the use of target voltage controlled deposition processes and composite targets may be a promising and useful way in industry production for preparation of thin film materials [17].

However, besides the deposition rate a_D determining the production costs of the sputtered films, mainly the deposition parameters, such as magnetron discharge current I_d and voltage U_d, substrate bias U_s, substrate ion current density i_s, substrate temperature T_s, substrate-to-target distance d_{s-t}, partial pressure of reactive gas p_{rg} or total pressure of sputtering gas p_T, have a crucial effect on the structure of the sputtered film and so on its physical and functional properties. Every combination of these parameters gives, however, only one discrete structure. It means that the structure of the films cannot be continuously changed by changing one process parameter in this combination. In next sections, the effect of the energy delivered to the growing film through substrate heating, particle bombardment or exothermic chemical reactions, when a heat is released during phase formation, is discussed for high Si-containing W-Si-N nanocomposite films. Although the efficiency of individual effects can expressively differ depending on the deposited material, these parameters can be relatively easily controlled and so the film properties may be changed in an exact way. Mainly the energy E_{bi} delivered to the film by ion bombardment can strongly affect the film growth. In a collisionless discharge, this energy can be calculated from three easily measured quantities, that is, the substrate bias U_s, the substrate ion current density i_s, and the deposition rate a_D of the film, according to the following formula [18]

$$E_{bi} = \frac{i_s U_s}{a_D}$$

The above formula clearly follows that the energy delivered into the film may significantly alter just in consequence of the variation in the deposition rate. Thus, even a small change in the Ar:N_2 pressure ratio, used as one of the process parameter, can greatly influence the final film structure. Indeed, the film growth is affected by the variation in the partial pressure of reactive gas in two different ways. Firstly, it determines an amount of reactive gas atoms incorporated into the film and secondly, it determines the energy, E_{bi} through reduced deposition rate. Especially in the reactive mode of sputtering this effect is usually enormously strong [19].

The effect of the target current, I_d and ion-current density, i_s on the deposition rate, a_D and the energy density E_{bi} delivered into the growing W-Si-N films by ion bombardment is illustrated in Fig. 6.1. Here, the development of a_D (Fig. 6.1a) and E_{bi} (Fig. 6.1b) at three combinations of the target current and ion current density: I_d = 1 A, i_s = 0.5 mA/cm^2, I_d = 1 A, i_s = 1 mA/cm^2, and I_d = 2 A, i_s = 1 mA/cm^2 as a function of increasing partial pressure of nitrogen p_{N_2} is displayed. Since other deposition conditions, such as U_s = −100 V, T_s = 500°C, and $p_T = p_{Ar} + p_{N_2}$ = 0.7 Pa, were kept constant, a pure effect of I_d, i_s, and p_{N_2} can be investigated.

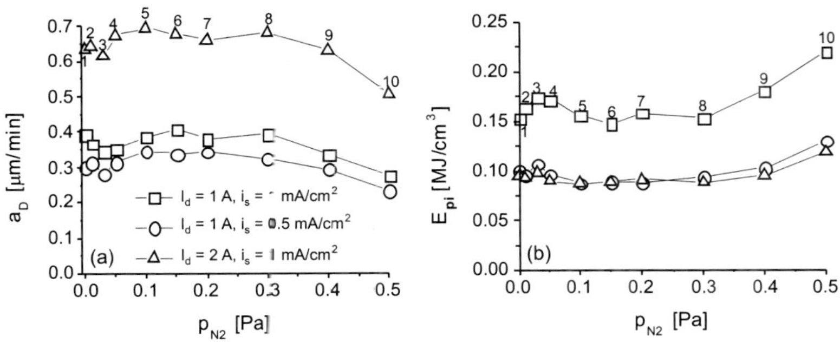

Figure 6.1 Development of (a) deposition rate a_D of W-Si-N films prepared by d.c. reactive magnetron sputtering at two values of (i) I_d = 1 and 2 A and (ii) substrate ion current density i_s = 0.5 and 1 mA/cm^2 and (b) energy E_{bi} delivered to these films during their growth as a function of partial pressure of nitrogen p_{N_2}. Reprinted from Ref. [20], Copyright (2006), with permission from Elsevier.

A decreasing tendency of the deposition rate a_D of the W-Si-N films is due to a continuous change in the mode of sputtering with increasing p_{N_2}. Indeed, two different operating modes of the reactive magnetron sputtering can be seen from Fig. 6.1, the metallic and the reactive mode. The metallic mode is characterized by a high sputter rate; therefore, a high reactive gas consumption and low partial pressure of the reactive gas. In the reactive mode, the sputter rate and the reactive gas consumption is low. Compared to other transition metal nitride films, even alloyed with aluminum, silicon, or boron, the variation in a_D is not expressively pronounced [21]. For instance, the deposition rate of the Zr-Si-N films sputtered from an alloyed $ZrSi_2$ target in the metallic mode of sputtering is about three times higher than that of films sputtered in the nitride mode [10]. Although in the reactive mode of sputtering the rate of the film growth is not significantly affected by the formation of nitrided compounds on the target surface and by a change in the sputtering efficiency, with lowering an amount of argon ions in the discharge at higher p_{N_2}, a strong relation to the deposition process parameters was found. While the ion current density i_s has only a small effect on the magnitude of a_D, the increase of I_d from 1 to 2

A results in a strong increase in the deposition rate. It is typically observed due to an enhancement of the magnetron discharge and a strong increase of the target sputtering. It can also be seen that the transition between the metallic and nitride mode shifts slightly to higher values of p_{N_2} with increase in I_d. The difference in deposition rate of films sputtered in the metallic and nitride mode at various ion current densities decreases with increase in the magnetron current. It is mainly caused by broadening of an erosion zone on the surface of sputtered target with increasing I_d. On the other hand, although the deposition rate of the films increases with increase in magnetron current I_d, it is consequently accompanied by an obvious decrease in the energy E_{bi}, according to the formula $E_{bi} = \frac{i_s U_s}{a_D}$ (Fig. 6.1b). The experiment shows further that the energy delivered into the film by ion bombardment is the lowest in the metallic mode of sputtering and continuously increases with increase in p_{N_2}. The maximum value is achieved at p_{N_2} = 0.5 Pa. The value of E_{bi} at I_d = 2 A is almost the same as that delivered to the W-Si-N films at a lower value of i_s = 0.5 mA/cm^2. It corresponds to the deposition rate approximately of 0.3 μm/min. This fact might explain some unexpected properties of the W-Si-N films sputtered under different combinations of I_d, i_s, and U_s, discussed in following sections.

6.2 Elemental Composition

Typical development of the elemental composition of the W-Si-N films, sputtered at the magnetron current I_d = 1 A, substrate bias U_s = −100 V and two values of ion current density i_s = 0.5 and 1 mA/cm^2, with increasing p_{N_2} is given in Fig. 6.2.

Similar to the Zr-Si-N films, the nitrogen content increases in the W-Si-N films with increase in p_{N_2} for both combinations of the ion current density i_s and target current I_d. This is attributed to the increasing amount of nitrogen incorporated into the growing film during the deposition process. A high affinity of nitrogen to the silicon and tungsten atoms sputtered from the target in the plasma discharge results in reaching its content of 50 at.% in the film at relatively low nitrogen partial pressure p_{N_2} = 0.2 Pa. A rapid increase in the nitrogen content, however, slows down while the metallic mode of sputtering alters into the transition and further to nitride mode of sputtering. Compared to the Zr-Si-N films [10], the maximum of the nitrogen content is achieved at the maximum partial pressure of nitrogen p_{N_2} used in the experiment, although it increases just slightly at the high-end of p_{N_2} range between 0.1 Pa and 0.5 Pa. The content of tungsten and silicon decreases to the detriment of nitrogen as the nitrided compounds are formed in the films with increase in p_{N_2}. Deposition of the W-Si-N films at higher ion current density i_s results in a remarkable increase of the amount of nitrogen incorporated into the films. It is due to the more intensive and effective dissociation and ionization of the molecular nitrogen in more intensified discharge. The acceleration of positive ions in the field of the biased sample results in impact dissociation and production of reactive atomic nitrogen, N and N_2^+ ions and thus higher amount of nitrogen is subsequently incorporated into the growing film. However, the difference in the nitrogen content in the films prepared at i_s = 0.5 and 1 mA/cm^2 is more pronounced at lower p_{N_2} in the range between 0.01 and 0.2 Pa. At higher p_{N_2}, the difference in the nitrogen concentration is smaller. It should be noted that the effect of the more intensified discharge on the elemental composition of the W-Si-N films is not so expressive compared to the Zr-Si-N films, probably due to the differences in the affinity of nitrogen to tungsten (ΔH_{fW_2N} = −71.9 kJ/mol) in contrast to Zr (ΔH_{fZrN} = −365.5 kJ/mol) [22,23].

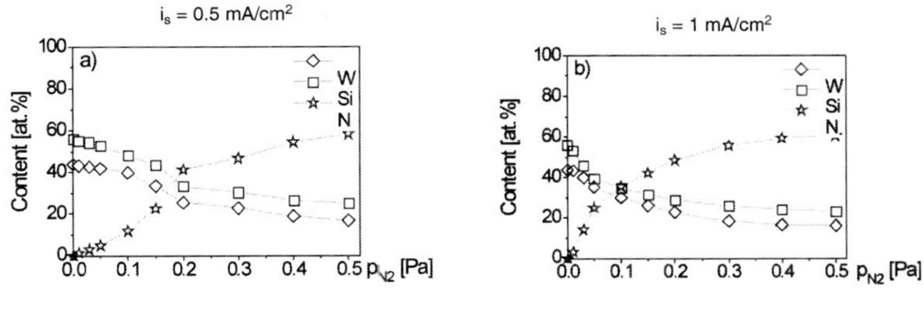

Figure 6.2 Elemental composition of the W-Si-N films reactively sputtered from the WSi_2 target at U_s = −100 V, I_d = 1 A and (a) i_s = 0.5 mA/cm^2 and (b) i_s = 1 mA/cm^2, as a function of partial pressure of nitrogen p_{N_2}. Reprinted from Ref. [20], Copyright (2006), with permission from Elsevier.

The development of the Si:W atomic ratio of the W-Si-N films shows its strong dependence on the ion current density i_s and the amount of nitrogen incorporated into the film, see Fig. 6.3. It is attributed to the resputtering phenomenon resulting in a variation of the silicon and tungsten content. The amount of silicon and tungsten in the film linearly decreases with increase in the nitrogen content. However, the content of these atoms is very sensitive to both the bombardment of low-energetic ions and the formation of nitrided compounds in the film as well as on the target surface. In Fig. 6.3, the Si:W ratio of the films sputtered at two different values of the ion current density i_s = 0.5 mA/cm^2 and 1 mA/cm^2 is displayed.

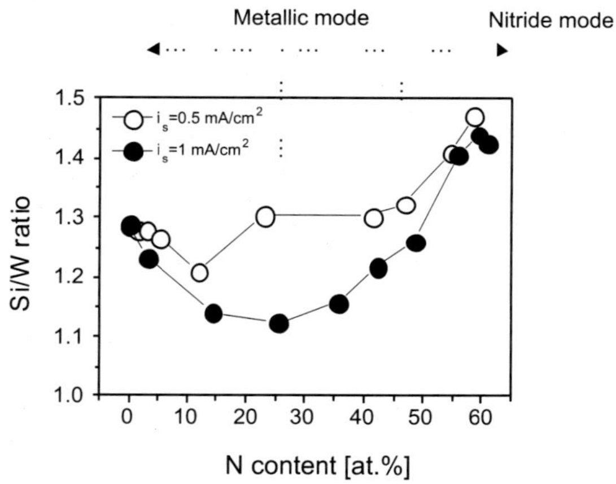

Figure 6.3 Development of the Si:W ratio of the W-Si-N films, sputter deposited at I_d = 1 A, U_s = −100 V and two values of the ion current density i_s = 0.5 mA/cm^2 and 1 mA/cm^2, as a function of the nitrogen content in the film.

Although the deposition of the non-reactively sputtered W-Si films is not affected by reactive gas in the deposition system, the elemental composition of those films reveals an obvious difference compared to the composition of the alloyed WSi_2 target. It is well known that the progressive change in the target surface composition due to the different sputtering yields of individual components alters the sputtered flux. After a sputtering of a few hundred atom layers, a steady-state transfer of atoms from the bulk target to the plasma is, however, reached, what typically results in the film stoichiometry which is the same as the original target composition. Significantly, lower value of Si:W = 1.3 of both films sputtered at i_s = 0.5 and 1 mA/cm^2, compared to the Si:W ratio of the target equal to 2, indicates a strong effect of ion bombardment on the elemental composition of the growing film. The negative substrates bias $U_s = -100$ V ensures relative intensive bombardment of the film by ions with a high impact energy extracted from the plasma discharge. With respect to the large weight difference of silicon (m_{Si} = 28 amu) and tungsten (m_W = 184 amu), the lighter silicon is preferentially resputtered away from the film even if it is bonded to tungsten forming WSi_2 or W_5Si_3 compound. The binding energies corresponding to the heat of formation of both tungsten silicides are similar ($\Delta H_{fWSi_2} = -91.9$ kJ/mol, $\Delta H_{fW_5Si_3} = -125$ kJ/mol [24]). Although ion bombardment seriously alters the film composition, there is just a slight effect of the ion current density in the range from 0.5 mA/cm^2 to 1 mA/cm^2.

In reactive atmosphere, nitrided compounds are formed in the growing film while silicon and tungsten atoms are bonded to nitrogen. Nitrogen is transported toward the substrate from the plasma discharge where it readily reacts in its atomic form with the metal atoms. Strong bonds between metal atoms and nitrogen make resputtering of individual components more difficult since the sputtering yield of nitrides is much higher than that of pure metals [25]. This phenomenon is also reflected in the Si:W ratio evolution while the films are sputtered in a N_2 atmosphere, see Fig. 6.3. Here, the decrease in the Si:W ratio to 27 at.% of nitrogen content in the W-Si-N film is attributed mainly to the resputtering of silicon induced by the ion bombardment ($U_s = -100$ V). Moreover, its content in the film is also reduced due to its lower sputter efficiency from the target. A formation of the Si_3N_4 compound on the target surface reduces significantly the amount of silicon in the sputtering flux. However, with respect to relatively low nitrogen content in the film up to p_{N_2} = 0.05 Pa, there is enough free or weakly bonded silicon in the film, which can be resputtered away. Once all silicon is bonded to nitrogen forming stable silicon nitride phase, its content just slightly decreases together with an amount of tungsten. This decrease on behalf of nitrogen content, as a consequence of higher nitrogen amount incorporated into the film with increase in p_{N_2}, causes that the Si:W ratio remains constant in this narrow p_{N_2} range. An opposite trend where the Si:W ratio increases with increase in nitrogen content can be observed for lower nitrogen concentrations in the films of lower silicon content. It is due to the complete consumption of silicon with nitrogen at lower p_{N_2} [26]. When the nitrogen content in the film is higher than that needed to form the stoichiometric Si_3N_4 phase, WN also starts to form. With respect to high tungsten content, free unbounded tungsten may be still preferentially resputtered in relation to silicon. As a consequence, the Si:W ratio increases. The W-Si-N films sputter deposited at $p_{N_2} \geq 0.3$ Pa and containing the nitrogen content higher than 55 at.% are already fully nitrided. Therefore, the Si:W ratio remains constant with increase in p_{N_2} since the resputtering of both metals are expressively attenuated due to the strong bonds developed among silicon, tungsten, and nitrogen atoms. Understanding of the resputtering phenomenon and the effect of the target poisoning is important for the control of the film composition and consequently the film structure. It is shown that while the elemental composition of the W-Si-N films is minimally affected through the ion bombardment, it can be easily controlled by the variation in the partial pressure of nitrogen.

6.3 Chemical Bonding and Phase Composition

Taking into account in previous research studies, where the preferential type of bonding between the elements was investigated, the phase composition route of the W-Si-N films can be described in several steps depending on the nitrogen content in the film as follows: Non-reactively sputtered W-Si films are expected to be composed of the silicides phases only. From the ternary phase diagram of the W-Si-N system, it is known that the W_5Si_3 or more probably WSi_2 are the phases to be formed. The heat of formation of individual phases are stated as $\Delta H_{f\,WSi_2} = -91.9$ kJ/mol for WSi_2 and $\Delta H_{f\,W_5Si_2} = -125$ kJ/mol for W_5Si_3 [24]. The ternary phase diagram of the W-Si-N system is depicted in Fig. 6.4.

The W-Si-N films with very low nitrogen content are still consisted of the tungsten silicides; however, the developed Si–N bonds should also be expected. It is due to the fact that the affinity of silicon for nitrogen is much higher in comparison to tungsten for nitrogen. Therefore, Si–N bonding is preferentially established in the film forming a phase of the Si_3N_4 type. The heat of formation of silicon nitride is $\Delta H_{f\,Si_3N_4} = -745.1$ kJ/mol [27]. As the nitrogen content is low, only a small part of silicon is consumed in the bonding leaving enough silicon content to form silicides with the tungsten available. When the nitrogen content increases in the W-Si-N film, higher amount of the silicon nitride phase is formed to the detriment of silicides. Since just a small percentage of nitrogen is left to form silicides at higher nitrogen content, the Si_3N_4 phase become dominant in the film. Finally, if some nitrogen is still available after total consumption of silicon, tungsten can be bonded to nitrogen and form the WN phase. Typically, the W_2N phase ($\Delta H_{fW_2N} = -71.9$ kJ/mol [23]) is formed at relatively low nitrogen content in the film, altering in WN ($\Delta H_{fWN} = -24$ kJ/mol [23]) at higher amount of nitrogen incorporated into the film, see Fig. 6.4. According to this idea the phase composition of the W-Si-N films develops with increasing p_{N_2} from WSi_2+W ($p_{N_2} = 0$) through Si_3N_4 + WSi_2 + W first to Si_3N_4 + $WN_{x<1}$ and then to Si_3N_4 + $WN_{x>1}$, see Fig. 6.5 and Table 6.1.

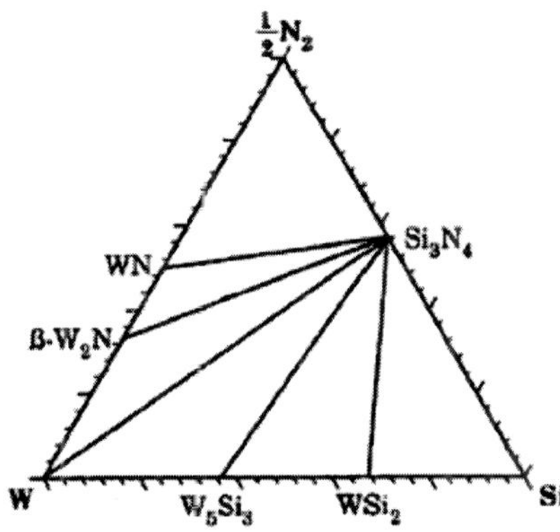

Figure 6.4 Isothermal cut through the W-Si-N ternary phase diagram showing the compositions accessible by magnetron sputtering from an alloyed WSi_2 target [24].

This experiment reveals that changes in the deposition conditions, particularly in the partial pressure of nitrogen p_{N_2} and the ion current density i_s, strongly influence development of the phase composition. It is caused mainly by changes in the elemental composition of the sputtered films. Indeed, a minimum value of $p_{N_{2min}}$, in which the amount of nitrogen incorporated into the growing film exceeds that needed to form the Si_3N_4 and $WN_{x=1}$ phases, shifts toward lower values if the films are prepared with increasing i_s

(Fig. 6.5). Moreover, when I_d increases at i_s = constant, the value of $p_{N_{2min}}$ shifts to higher values.

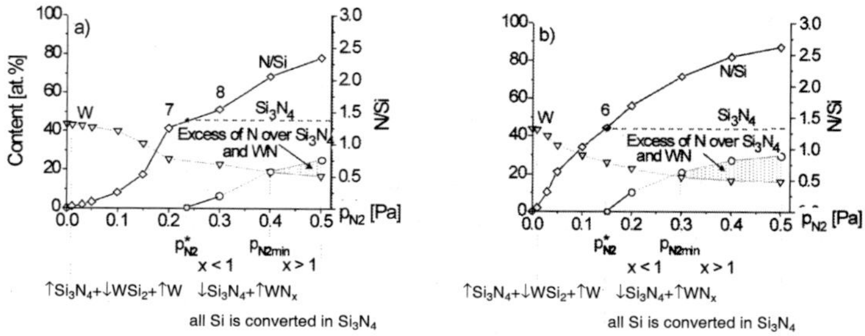

Figure 6.5 Development of the N:Si atomic ratio, content of tungsten and excess of nitrogen over that needed for the formation of the Si_3N_4 and WN phases in the W-Si-N films as a function of partial pressure of nitrogen p_{N_2}. The films were sputtered at $U_s = -100$ V, I_d = 1 A and two values of the ion current density i_s: (a) i_s = 0.5 mA/cm^2 and (b) i_s = 1 mA/cm^2. Reprinted from Ref. [20], Copyright (2006), with permission from Elsevier.

The above-mentioned hypothesis on the phase formation in the W-Si-N films has been confirmed by various experiments, such as X-ray photoelectron spectroscopy (XPS). XPS measurement reveals the bonding contributions of Si–Si(W) and W–W(Si) at lower nitrogen content and Si–N and W–N at higher p_{N_2}, that is, silicon and nitrogen establish preferential bonds and only when one of them is fully consumed, the remaining content of the other will be bonded to tungsten [26].

Table 6.1 Development of the phase composition in the W-Si-N films with increasing partial pressure of nitrogen p_{N_2}

Partial Pressure of Nitrogen	Phase Composition
$p_{N_2} = 0$	h-WSi_2 + W
$0 < p_{N_2} \leq p^*_{N_2}$	Si_3N_4 + WSi_2 + W
$p^*_{N_2} < p_{N_2} \leq p_{N_{2min}}$	Si_3N_4 + $WN_{x<1}$
$p_{N_2} > p_{N_{2min}}$	Si_3N_4 + $WN_{x>1}$

Note: (1) h is the hexagonal phase (2) $p^*_{N_2}$ labeled in Fig. 6.5.

It has been shown that Me-Si-N films with a high volume content of the amorphous silicon nitride phase exhibit unique properties. Besides a high hardness and low residual stresses, the films are thermally stable and resistant against oxidation [28]. This is related to the presence of the Si_3N_4 phase in the W-Si-N films, which volume content is predominantly given by the nitrogen content (Fig. 6.6). Although, Si_3N_4 is not detected by XRD in the W-Si-N films [16,29], its existence is revealed by XPS [26]. Thus, it must be present in the W-Si-N films in an amorphous state. It is, in fact, a typical form of silicon nitride in this type of nanocomposite films [8,30,31].

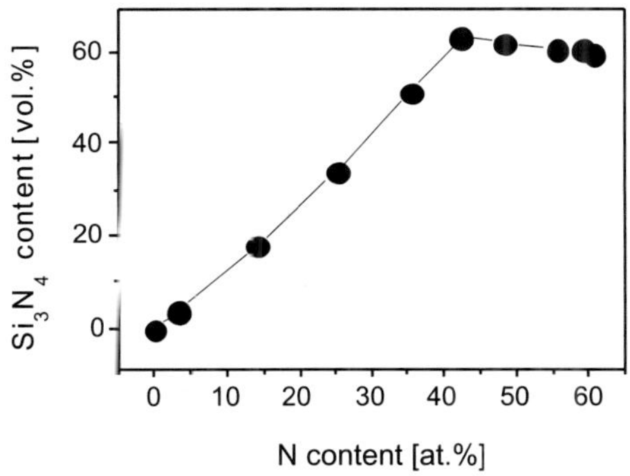

Figure 6.6 Development of the volume content of the amorphous Si_3N_4 phase as a function of the nitrogen content in the film. The W-Si-N films were sputter deposited at U_s = −100 V, I_d = 1 A, and i_s = 1 mA/cm^2.

It is clearly seen that the amount of Si_3N_4 rapidly increases in the film with increase in the nitrogen content. The fact that, the maximum volume content is achieved already at p_{N_2} = 0.1 Pa, suggests a very high affinity of nitrogen to silicon. Indeed, the silicon nitride phase is formed in the film to the detriment of the crystalline WSi_x phase what substantially alters the volume fraction ratio between the crystalline and amorphous phase in the W-Si-N films. By exceeding the critical value of p_{N_2} = 0.1 Pa, the amount of Si_3N_4 saturates having approximately 60 vol.%. The saturation of the volume content of silicon nitride is ascribed to the limited amount of silicon in the film which can be bonded to nitrogen. The silicon content even decreases in the film as a result of a lower sputtering yield of nitrided compounds formed on the target surface. It is reflected by a slight decrease of Si_3N_4 volume content in the W-Si-N film at p_{N_2} >

0.1 Pa (Fig. 6.6). On the contrary, the higher amount of nitrogen is bonded to tungsten in this p_{N_2} range, forming overstoichiometric $WN_{x>1}$ (Table 6.1).

6.4 Structure of W-Si-N Films

It has been shown that a formation of the nanocomposite structure is strongly related to the phases established in the films during their growth [32]. Although readiness of silicon, nitrogen, and tungsten to form compounds expressed in terms of the enthalpy of formation is very high. Some additional prerequisites, such as relatively high energy of impinging adatoms and their sufficiently high surface mobility, have to be satisfied to form nanocomposite structure composed of the crystalline MeN phase embedded within an amorphous Si_3N_4 matrix [33]. The phase formation is principally given by an amount of incorporated silicon and nitrogen atoms into the films and their chemical reactivity. A synergetic effect of increasing silicon and nitrogen content on the phase formation in the W-Si-N films, as revealed by structural analysis (Figs. 6.7 and 6.8) is fully in agreement with the morphological zone model proposed by Patscheider [32], and also with findings of other scientists [34].

Non-reactively sputter-deposited W-Si films exhibit a crystalline structure composed of the h-WSi_2 grains, which are mainly oriented in (111) direction. Although the stoichiometry of the film typically reflects the composition of the sputtered target source when it is deposited by magnetron sputtering in a pure argon atmosphere, the stoichiometry of the alloy W-Si films markedly differs from that of WSi_2 target. It can be explained by a variation in elemental composition of the films, which is together with the film structure markedly affected by the low-energy ion bombardment. The substrate bias of −100 V is one of the potential ways to influence the film properties through a formation of a dense and compact structure, although it is often ascribed to the resputtering effect of free or weakly bonded elements [35]. In contrast to the strong effect of the ion current density i_s on the structure of the Zr-Si films, the W-Si films sputtered at i_s in the range of 0.5–1 mA/cm^2 differs in the structure just moderately. Also the elemental composition of both films remains similar though the films are sputtered under differently intense ion bombardment.

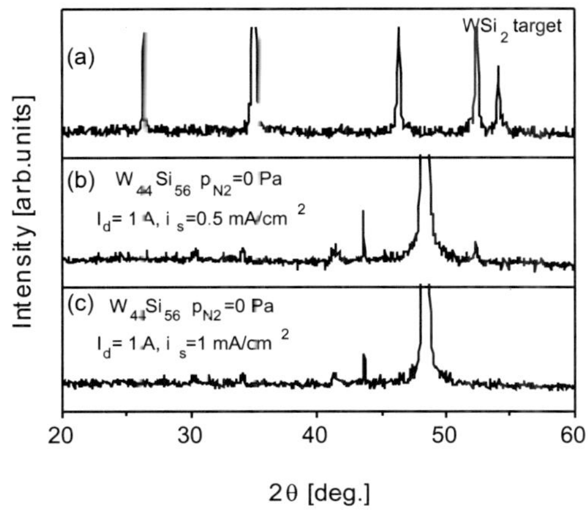

Figure 6.7 The structure of (a) the alloyed WSi_2 target and the W-Si films non-reactively sputter deposited on steel substrates at $I_d = 1$ A (b) $i_s = 0.5$ mA/cm^2, and (c) $i_s = 1$ mA/cm^2.

When the films are deposited with increase in nitrogen content in the transition and further in the nitrogen mode of sputtering, the changes in the phase composition are clearly reflected by the evolution of the film structure. The development of XRD patterns of the W-Si-N films with increasing p_{N_2} (Fig. 6.8) reveals that a change in the structure is the more pronounced the higher amount of nitrogen is incorporated into the growing film. The films sputtered at $p_{N_2} \leq 0.03$ Pa are crystalline, characterized by one broad, low-intensity reflection. More detailed analysis of the X-ray patterns reveals that this reflection is superimposed by several narrow, high-intensity reflections. With respect to the formation of the Si_3N_4 phase in this compositional range, the structure of the W-Si-N films with intermediate nitrogen content corresponds well to the nanocomposite one, where the WSi_2 grains are enrobed by an amorphous silicon nitride tissue. The presence of the amorphous phase inhibits the growth of the grains what is reflected by the broadening of the X-ray diffraction patterns. From the point of view of the preferential bonding phenomenon in the Me-Si-N films, it is expected that at higher p_{N_2} the WN phase is established to the detriment of WSi_2. Hence, no silicide phases were observed at $p_{N_2} > 0.15$ Pa since all the silicon atoms are already preferentially bonded to nitrogen. Also in this case the growth of WN grains is markedly hindered by an amorphous silicon nitride matrix. In addition, a grain size decrease is more pronounced at higher amount of Si_3N_4 when this effect becomes stronger. It results in the ultra-fine grained X-ray amorphous structure of the W-Si-N films composed of WN grains dispersed in a high amount of silicon nitride matrix. The nanocomposite structure of the W-Si-N films with a high volume content of Si_3N_4 is characterized by one very broad, low-intensity reflection (Fig. 6.8). It has been shown that mainly this characteristic of Me-Si-N films, where a high volume content of an amorphous silicon nitride is formed, is responsible for the high thermal stability of the films. Also at very high temperatures, the grain growth is efficiently prevented keeping the structure and thus mechanical properties of the films almost unaffected. The assumption that Si_3N_4 is formed as an amorphous phase is indicated by missing reflections from crystalline silicon nitride. The difference between the structure of fully crystalline non-reactively sputtered W-Si film and the nanocomposite W-Si-N film composed of the WN grains dispersed in the amorphous silicon nitride phase is illustrated in more detail in Fig. 6.9.

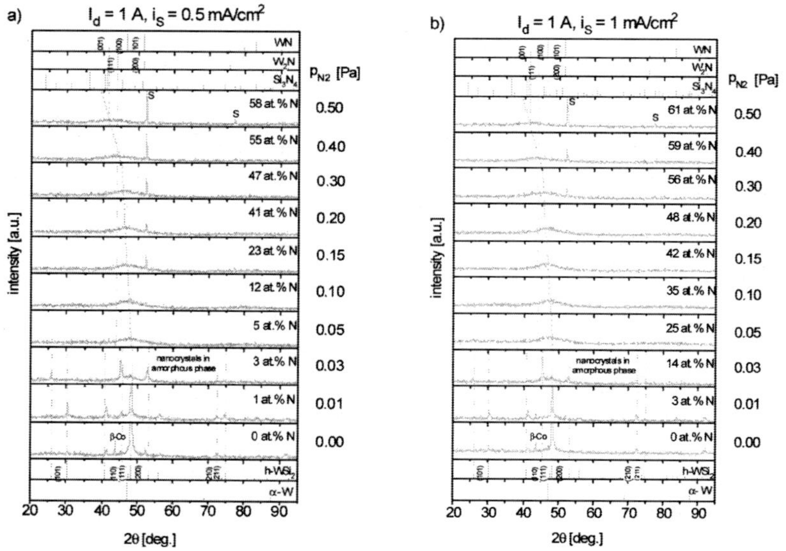

Figure 6.8 Development of the structure of the W-Si-N films sputtered on steel substrates at I_d = 1 A, U_s = −100 V and (a) i_s = 0.5 and (b) 1 mA/cm^2, with increasing partial pressure of nitrogen p_{N_2}. Reprinted from Ref. [20], Copyright (2006), with permission from Elsevier.

The addition of nitrogen into the W-Si films results in a loss of texture and decrease of the grain size. The mean value of the crystallite size of W-Si films, calculated using Scherrer's formula [36], is around 12 nm while it decreases with increasing nitrogen content below the detection limit of the XRD analysis. This behavior is attributed to an encapsulation of the growing grains by an amorphous silicon nitride tissue and thus preventing their further development. This strong efficiency of the Si_3N_4 phase to hinder the grain growth results in the formation of films with an ultra-fine microstructure exhibiting unique properties, particularly thermal stability and resistance against oxidation [10,37,38].

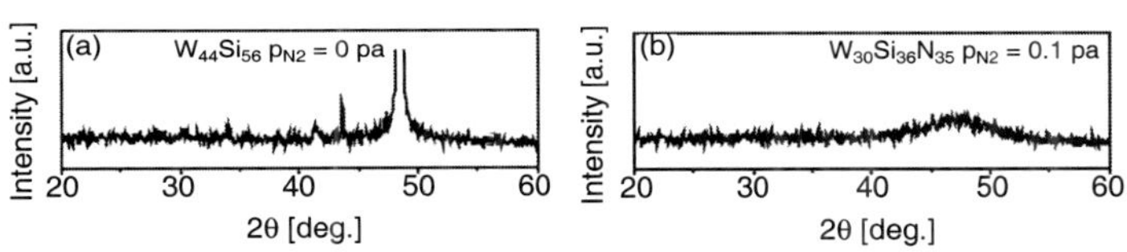

Figure 6.9 Structure of (a) the crystalline W-Si and (b) the X-ray amorphous nanocomposite W-Si-N film sputtered at $I_d = 1$ A, $i_s = 1$ mA/cm^2, and $p_{N_2} = 0$ Pa and 0.1 Pa, respectively.

6.5 Morphology of W-Si-N Films

Scanning electron microscopy (SEM) investigations performed on the W-Si-N films sputter deposited at increasing partial pressure of nitrogen show the same microstructural evolution as revealed by XRD analysis. A decrease of the grain size promoted by the amorphous Si_3N_4 phase and consequent changes of the film morphology are evident in Fig. 6.10.

The SEM cross-sectional images of four representative W-Si-N films with various elemental compositions illustrate the tendency of decreasing the column width and weakening of film texture as the nitrogen content is increased in the films. The non-reactively sputtered W-Si film exhibits columnar structure corresponding to the zone "T" of the Thornton's structural model (Fig. 6.10a) [39]. A crystalline feature of this film, however, suddenly vanishes with increasing p_{N_2} (Fig. 6.10b). It corresponds to the development of the microstructure where the crystallites are penetrated by an amorphous phase. The impeding effect of the amorphous Si_3N_4 tissue on the grain growth results in reduction of the grain size and development of a dense structure with a smooth fracture in cross-section. This refinement of the film microstructure is reflected by X-ray peak broadening as shown in Fig. 6.8. More detailed study of the W-Si-N film deposited at p_{N_2} = 0.03 Pa reveals that the crystalline grains are not distributed homogeneously in an amorphous matrix although the Si_3N_4 content achieved 20 vol.%. A dramatic change in the film morphology is observed at $p_{N_2} \geq 0.05$ Pa. While the nitrogen content in the W-Si-N films exceeds approximately 30 at.%, the size of the MeN grains decreases below 10 nm. This strong refinement effect of the silicon nitride results in a glassy appearance of the film morphology (Fig. 6.10c,d). Such an amorphous-like glassy feature of the films in cross-section corresponds to the nanocomposite structure consisting of ultra-fine grains embedded in an amorphous matrix, typically developed in Me-Si-N films with a high (≥50 vol.%) Si_3N_4 volume content.

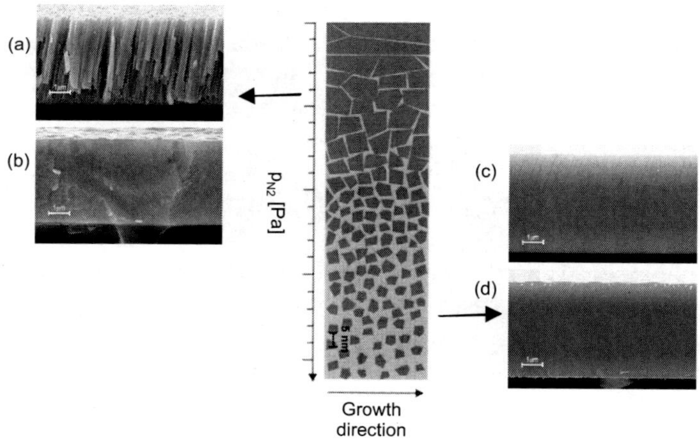

Figure 6.10 Cross-sectional micrographs of the W-Si-N films with various elemental composition obtained using a secondary electrons (SE) detector (a) $W_{44}Si_{56}$ (p_{N_2} = 0 Pa), (b) $W_{40}Si_{46}N_{14}$ (p_{N_2} = 0.03 Pa), (c) $W_{30}Si_{36}N_{35}$ (p_{N_2} = 0.1 Pa), and (d) $W_{17}Si_{24}N_{59}$ (p_{N_2} = 0.4 Pa), sputtered at $U_s = -100$ V, I_d = 1 A, and i_s = 1 mA/cm^2.

6.6 Surface Roughness

The transitions between microcrystalline and X-ray amorphous structure are also accompanied by the changes of the surface roughness. The decreasing tendency of the surface roughness as a function of increasing partial pressure of nitrogen p_{N_2} used during deposition is shown in Fig. 6.11. A rapid roughness decrease from 37 nm to 14 nm corresponds well to the phase composition changes induced in the W-Si-N films with increasing amount of nitrogen incorporated into the films. Since the Si–N bonds are established in the film forming the amorphous Si_3N_4 phase, well-developed columnar structure becomes more compact composed of smaller grains. The surface smoothening effect is ascribed to the structural refinement related to the amorphous tissue acting as a grain growth inhibitor. As the grain size decreases under the detection limit of XRD analysis, the surface roughness approaches the value of 8 nm (p_{N_2} = 0.2 Pa) which even decreases to 6 nm at p_{N_2} = 0.4 Pa.

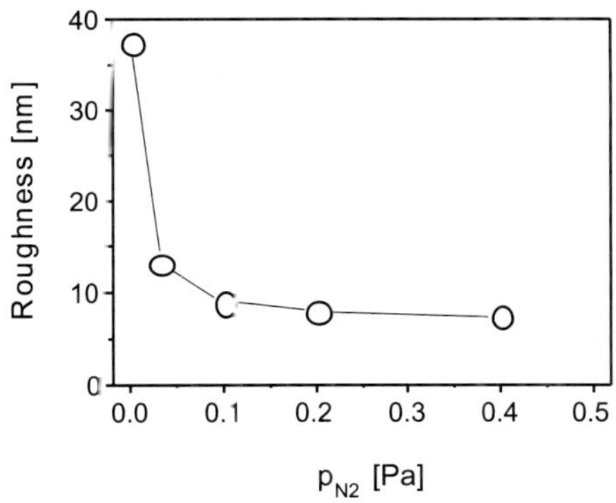

Figure 6.11 Surface roughness evolution of the W-Si-N films sputter deposited at U_s = −100 V, I_d = 1 A, and i_s = 1 mA/cm^2, as a function of p_{N_2}.

The morphological changes in the films related to the changes in their phase composition are illustrated in Fig. 6.12. The films significantly vary in their elemental composition and consequently in the crystalline-to-amorphous phase volume fraction ratio. The $W_{44}Si_{56}$ film deposited in the metallic mode of sputtering (p_{N_2} = 0 Pa) exhibits a crystalline structure consisting of well-developed large columns grown perpendicular to the substrate surface (Fig. 6.10a). The columnar structure is characteristic for magnetron sputtered films where the condensing adatoms have relatively high surface energy. During the film growth the columns experience both lateral extension and vertical stretch and thus the film surfaces become rougher with deposition time. Hence, the film surface exhibits large valleys and peaks characterizing its rough feature. While the crystalline films are characterized by rough surfaces, the films containing higher nitrogen content exhibit low surface roughness. The surface smoothening effect is associated to the less pronounced columnar nature of the films resulting from the hindering of the grain growth. Since it is provided by the silicon nitride tissue, it is very strong when the Si_3N_4 phase becomes dominant in the film (Fig. 6.12d). The transition between crystalline to fine-grained nanocomposite structures is clearly evident from both Figs. 6.11 and 6.12 as a continuous decrease of the surface roughness from very rough into a smooth one.

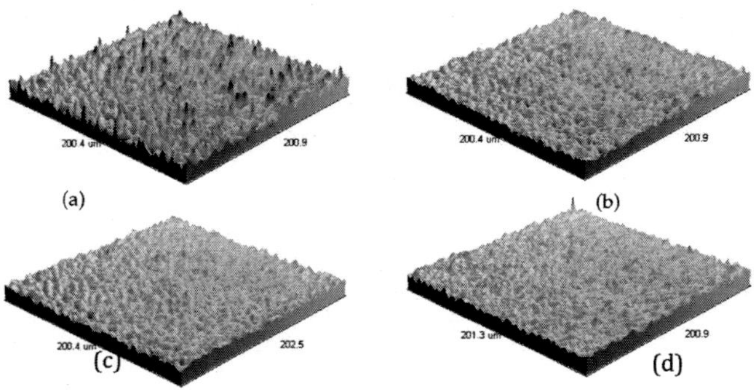

Figure 6.12 Surface topology of 3 μm thick W-Si-N films sputter deposited at $U_s = -100$ V, $I_d = 1$ A, $i_s = 1$ mA/cm^2, and various p_{N_2} (a) 0 Pa, (b) 0.03 Pa, (c) 0.1 Pa, and (d) 0.4 Pa, illustrating the transition from crystalline to X-ray amorphous structure.

Although the W-Si-N films are very attractive for usage in tooling applications as hard protective coatings, they are also often exploited as micromachining elements or optical switching. However, the major field of application is in microelectronic devices as diffusion barriers in copper-based interconnections [40,41]. From a compositional and structural point of view, the capability of the W-Si-N films in microelectronic devices increases with increasing compactness of the film structure. It significantly improves the compatibility with the electronic elements and thus the efficiency of the deposited barriers. It has been shown that the surface roughness significantly affects the transmittance as well as the electrical resistivity of the films [3].

The geometrical properties of a given surface topography, however, also influence the tribological performance of the films, particularly the friction and wear behavior. It in turn affects the functional properties in actual tribological application and thus the performance and reliability of the coated component [42].

6.7 Mechanical Properties of W-Si-N Films

6.7.1 Microhardness

Microhardness is very important intrinsic property of thin films. Since it determines performance of coated tools in many industrial applications, an optimization of mechanical properties of thin films has attracted a great interest in last decade [43–47]. It is well known that the film hardness is closely related to the microstructure. Thus, the hardness of nitride films composed of either two phases of different materials [48, 49] or a mixture of grains of the same material but with different crystallographic orientations [22,50,51] can be markedly higher than that of the bulk binary nitrides. It is due to the different deformation mechanisms, which occur when the films are mechanically loaded. In materials with the columnar structure, cracks propagate mainly along the weak boundaries under indentation load and thus high hardness cannot be obtained. This fact results from the existence of defects and voids within the boundaries [29]. On the contrary, in the nanocrystalline scale the deformation of materials does not occur through the traditional dislocation mechanism but follows completely different mechanisms [52]. Thin films with a nanocomposite structure exhibit superhardness ($H \geq 40$ GPa), which is provided through the hindering the multiplication and mobility of dislocations generated under the applied load. Thus, when the film structure is constructed as ~5 nm transition metal nitride crystallites embedded in thin amorphous matrix, the grain boundary sliding and the incoherence stress relaxation are effectively suppressed at the decreased grain size [53]. Indeed, enhanced film properties and the protective capability of films seem to arise mainly through optimized structure–property relations. However, hardness is not the only characteristic that ensures good tribological performance of the coated tools. The film resistance against wear and abrasion is mainly given by an optimal combination of hardness and elastic properties of the films [54].

Determination of the mechanical properties of thin films by indentation is very difficult if the measured properties are influenced by the substrate [55–57]. Both combinations of a hard film on a soft substrate or a soft film on a hard substrate make troubles to obtain accurate values of hardness and elastic modulus of thin films [58]. It is because the indentation response of a thin film on a substrate is a complex function of the elastic and plastic properties of both the film and substrate. In order to study "film-only" properties, the indentation depth should be reduced to less than 10% of the film thickness [59]. This commonly accepted rule was used also in the hardness measurement of the W-Si-N films. While at low loads of the Vicker's indenter the hardness values are substantially over-estimated due to the strong effect of the surface roughness and imperfection of the diamond indenter, they are under-estimated at higher loads due to the effect of the softer substrate (Fig. 6.13). The hardness should thus be measured within certain load range where the values are constant.

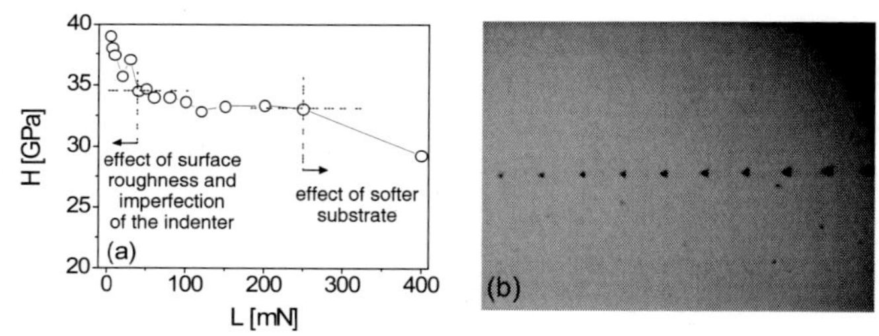

Figure 6.13 (a) Dependence of the microhardness H of 4 μm thick W-Si-N film on the indenter load L. The film was prepared at I_d = 1A, U_s = −100 V, i_s = 1 mA/cm^2, T_s = 500°C, p_{N_2} = 0.3 Pa and p_T = 0.7 Pa. (b) Residual imprints on the sample surface correspond to the increasing indenter load L used in the experiment.

The W-Si-N films substantially differ in their hardness depending on the elemental and phase composition. While the crystalline W-Si-N films with low silicon content are very hard reaching in some cases values higher than 40 GPa, the X-ray amorphous films with a high silicon and nitrogen contents exhibit lower hardness [34]. This tendency has been already observed for sputtered films of other systems, such as Cr-Ni-N [60], W-Ni-N [61], or Zr-Si-N [10]. It is principally caused by the increasing amount of the amorphous silicon nitride phase in the W-Si-N films significantly influencing the size of grains [32]. If the grain size decreases below ~4 nm, the film hardness is markedly reduced following the "reverse" Hall–Petch relation [62]. It is also characteristic for the W-Si-N films with a high (≥25 at.%) silicon content sputtered from an alloyed WSi_2 target, which exhibit fine-grained nanocomposite structure consisted of the WN phase embedded in a high volume fraction of an amorphous matrix. In spite of their X-ray amorphous structure, the W-Si-N films have a high hardness of around 30 GPa, which is the hardness almost two times higher than that of individual components, from which the films are composed (H_{WN} = 18 GPa, $H_{fSi_3N_4}H_{Si_3N_4}$ = 17.2 GPa [27]). It is given by the nanocomposite nature with a higher short-range order, characteristic for this type of structure, which makes these W-Si-N films harder compared to fully amorphous films [14]. Low hardness of fully amorphous films results mainly from the structure: Disordered network with low density cannot sustain high loading during indentation. Moreover, the lower hardness values are also typically related to the low compressive stress state of the films.

Strong relationship between the phase composition, and thus the structure, and microhardness of the W-Si-N films is illustrated in Fig. 6.14. Here, the development of the film hardness as a function of deposition conditions, that is, partial pressure of nitrogen p_{N_2} and energy delivered to the film during its growth, is depicted. It is shown that the film hardness increases with increase in p_{N_2} as a consequence of the phase evolution.

The crystalline non-reactively sputtered $W_{44}Si_{56}$ film exhibits relatively low hardness of about 10 GPa. However, it rapidly increases with the nitrification of metal silicon and further tungsten atoms. Already at p_{N_2} = 0.03 Pa, an amount of the amorphous silicon nitride tissue is sufficiently high to cover the h-WSi_2 grains, hinder their growth and form a nanocomposite structure. XRD pattern of this film reveals that the grains are present in the film with different crystallographic orientations (Fig. 6.8b). This fact is most probably responsible for the enhanced hardness of the $W_{40}Si_{46}N_{14}$ film labeled by the number 3 in Fig. 6.14a. With increasing amount of silicon nitride and formation of W–N bonds, hardness of the films increases further. While the nanocomposite structure composed of an amorphous Si_3N_4 matrix and WN_x grains is formed above 0.15 Pa of p_{N_2}, the film hardness achieves its maximum values. It is also ascribed to an increase of the stoichiometry x of the WN_x phase. Indeed, the W-Si-N film with a maximum microhardness H_{max} is composed of the overstoichiometric $WN_{x>1}$ phase and Si_3N_4 phase, which most probably co-exist separately.

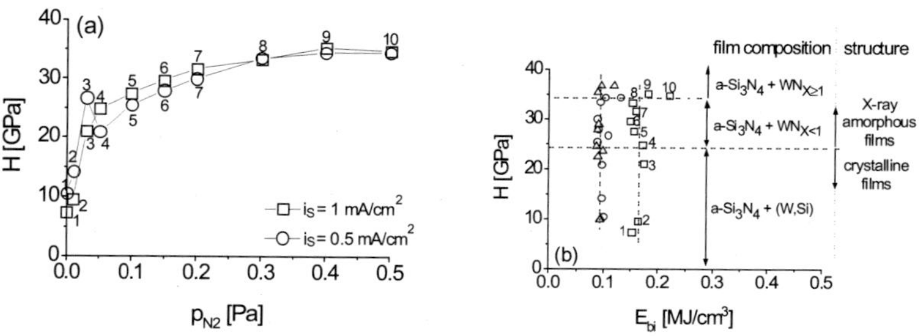

Figure 6.14 Hardness H of the W-Si-N films, magnetron sputtered at I_d = 1 A, U_s = −100 V and two values of i_s = 0.5 mA/cm^2 and 1 mA/cm^2, as a function of (a) partial pressure of nitrogen p_{N_2} and (b) the energy density E_{bi} delivered to the growing film by bombarding ions. The numbers 1–10 are the same as those defined in Fig. 6.1. Reprinted from Ref. [20], Copyright (2006), with permission from Elsevier.

The relation between the film hardness H and the energy density E_{bi} delivered to the W-Si-N film during its growth by bombarding ions is displayed in Fig. 6.14b. It is evident that the hardness almost does not depend on E_{bi}, although H alters with increasing p_{N_2} in a wide range from approximately 10 GPa to 35 GPa. Indeed, not even the maximum hardness H_{max} is affected by the magnitude of E_{bi}. Its value must be for a given combination of process parameters (U_s, i_s, and I_d) higher than a minimum energy density $E_{bi\ min}$≈0.1 MJ/cm^3; however, already at $E_{bi}>E_{bi\ min}$ the maximum hardness H_{max} remains constant. This fact indicates that the hardness of the W-Si-N film is determined mainly by its phase composition. If the phase composition of the films does not alter with increasing E_{bi}, H_{max} does not increase as well. Thus, the maximum hardness H_{max} is achieved in the W-Si-N films, which are composed of stable phases, that is, of $WN_{x>1}$ and amorphous Si_3N_4.

The effect of the deposition temperature T_s and the film thickness h on hardness H of the W-Si-N films, sputtered at p_{N_2} = 0.3 Pa, is shown in Figs. 6.15a and b, respectively. In these W-Si-N films, all silicon is completely converted into the Si_3N_4 phase and its volume content dominates over that of the WN_x phase. In fact, mainly such a high volume content of silicon nitride is essential to provide a constant hardness value of the films since its contribution to formation of a stable nanocomposite structure with the WN_x phase is very strong. A slight increase in hardness of the film with increase in T_s can be explained by an increase of stoichiometry x = N/W of the WN_x phase, see Fig. 6.15b. At higher deposition temperatures, N_2 molecules are more effectively dissociated at the surface of growing film, which results in a higher amount of nitrogen incorporated into the film.

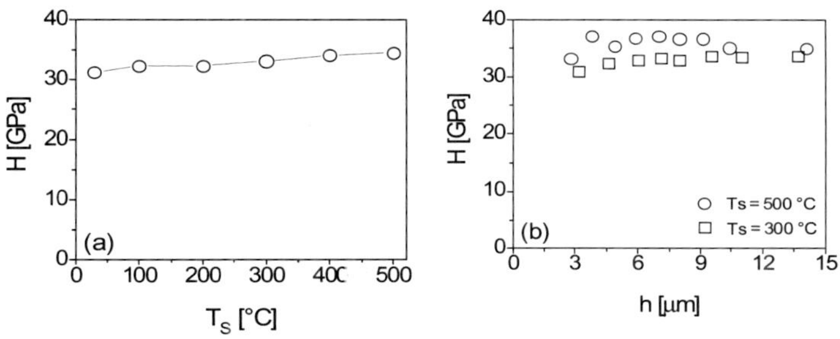

Figure 6.15 Hardness H of the W-Si-N film, sputtered at I_d = 1 A, U_s = −100 V, i_s = 1 A/cm^2 and p_{N_2} = 0.3 Pa as a function of (a) substrate temperature T_s (h = 5 μm) and (b) film thickness h for two values of T_s = 300°C and 500°C.

6.7.2 Young's Modulus and Elastic Recovery

The microindentation test has been developed from conventional hardness testing to provide mainly hardness of materials. However, since a continuous measure of load and displacement is obtained, it is possible to determine other properties, such as elastic modulus and recovery. The elasticity is one of the important film properties characterizing the response of coated systems to indenter load and together with hardness assesses the penetration resistance of the material. The reduced Young's modulus $E^* = E/(1-\nu^2)$ and elastic recovery W_e of the W-Si-N films, sputtered at T_s = 500°C, U_s = −100 V and different combinations of I_d and i_s, evaluated from loading–unloading curves as a function of the partial pressure of nitrogen p_{N_2} is given in Fig. 6.16.

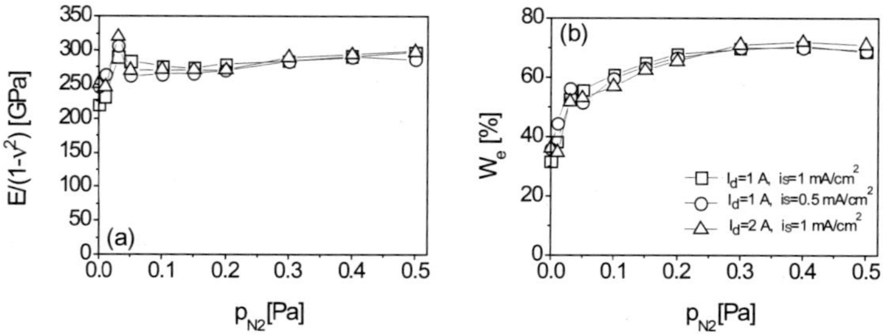

Figure 6.16 Development of (a) reduced Young's modulus E^* and (b) elastic recovery W_e of the W-Si-N films, sputtered at U_s = −100 V and different combinations of I_d and i_s, as a function of p_{N_2}.

It is evident from Fig. 6.16 that also elastic properties of the W-Si-N films are affected by the development of various phases in the films with an increasing amount of nitrogen incorporated. While the reduced Young's modulus of the crystalline alloy $W_{44}Si_{56}$ film (p_{N_2} = 0 Pa) is relatively low, it increases from about 230 GPa to 320 GPa at p_{N_2} = 0.03 Pa. This rapid increase of E^* corresponds well to the encapsulation of WSi_2 grains by an amorphous silicon nitride tissue and formation of a nanocomposite structure. Although the $W_{40}Si_{46}N_{14}$ film sputtered at p_{N_2} = 0.03 Pa exhibits the maximum value of Young's modulus, its hardness is higher just in comparison with the W-Si-N films with the nitrogen content lower than 25 at.%. In the p_{N_2} range between 0.05 and 0.2 Pa, the value of E^* = 270 GPa remains constant and just slightly increases with p_{N_2} up to 300 GPa at p_{N_2} = 0.5 Pa. On the contrary, the hardness of the W-Si-N films continuously increases in the same p_{N_2} range. It is related to the formation of the amorphous silicon nitride phase, which ensures both high hardness and stiffness through the developed nanocrystalline structure. However, preparation of films with the highest E^* requires an optimum in the crystalline-to-amorphous phase volume fraction ratio. Similarly to the film hardness, the elastic recovery W_e increases with increase in p_{N_2} and achieves its maximum at the maximum nitrogen content in the film at p_{N_2} = 0.5 Pa. This seems to be attributed to the increase of the WN_x stoichiometry

6.7.3 Interrelationships between H, E^*, W_e, and H^3/E^{*2}

During past years, scientific research has been focused toward the achievement of ultra-high hardness with associated high elastic modulus. Mainly hardness has been regarded as an indicator and a primary material property which defines wear resistance of the films. However, many present research studies have shown that the magnitude of the elastic modulus strongly affects wear behavior of the film as well [54]. Hence, the H:E ratio, called the elastic strain to failure, and H^3:E^{*2}, called the film resistance against plastic deformation, have been established as more suitable parameters for predicting the wear resistance of materials. For instance, the nanocomposite coatings with high hardness and low elastic modulus exhibit improved toughness preventing crack propagation even if the film is heavily mechanically loaded [63]. Wear improvement through the optimized H:E ratio is, however, given by phase composition of deposited films, which determines the final film properties. Since magnetron sputtering allows close control over process parameters, it is one of the potential techniques for the preparation of new generations of wear resistant films with unique, non-equilibrium compositions and structures. The interrelationships between H, E^*, W_e, and the film resistance against plastic deformation expressed by the H^3:E^{*2} ratio are given in Fig. 6.17 for the W-Si-N films prepared under various deposition conditions.

With respect to mechanical properties, the W-Si-N films can be divided into three different groups related to metallic, transition, and nitride modes of sputtering. While the films deposited in the metallic and nitride modes of sputtering show low and high hardness and stiffness, respectively, films synthesized in the transition mode exhibit values lying between. Especially, the alloy W-Si-N films with a low nitrogen content exhibit poor resistance against plastic deformation and wear. These films are characterized by the H^3:E^{*2} ratio lower than 0.05, that is characteristic for soft pure metals and/or alloy films. Moreover, their tribological performance is even lowered by a high surface roughness resulting from the crystalline nature of the films.

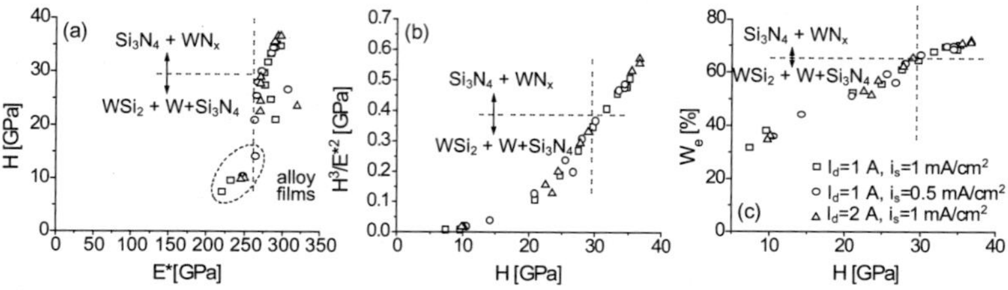

Figure 6.17 Interrelationships between hardness H, reduced Young's modulus E^*, elastic recovery W_e, and the H^3:E^{*2} ratio of the W-Si-N films reactively sputtered at $U_s = -100$ V and different combinations of I_d, i_s, T_s, and p_{N_2}. Reprinted from Ref. [20], Copyright (2006), with permission from Elsevier.

However, just a small amount of nitrogen incorporated into the films leads in a significant enhancement of mechanical properties of the W-Si-N films. That is attributed to several effects, such as a decrease of the grain size and formation of compact and very dense structure. All these facts lead to the enhancement of the film hardness and resistance against plastic deformation and wear, which is even more pronounced with increasing nitrification level of the films at higher p_{N_2}. This continuous improvement of the film properties can be clearly seen from Fig. 6.17 as an increasing tendency of H, H^3:E^{*2}, and W_e of the W-Si-N films plotted versus E^* and H, respectively. It is typical for the W-Si-N films that the hardness rapidly increases with slight increase in elastic modulus. Indeed, it is a reason why the resistance to plastic deformation of the W-Si-N film with $H \geq 20$ GPa considerably increases with increasing H. On the contrary, the elastic recovery W_e increase relatively slowly with increasing hardness H (Fig. 6.17c). The maximum values of the H^3:E^{*2} ratio, ranging from 0.35 to 0.6, were found in the W-Si-N films characterized simultaneously by the film hardness $H > 30$ GPa and a high (≈65 vol.%) content of the Si_3N_4 phase. This result demonstrates that the enhanced microhardness and resistance against plastic deformation of the W-Si-N films are strongly ascribed to the phase evolution in the films. The mechanical properties significantly increase with the structure evolution developing from the crystalline WSi_x structure through a mixture of the WSi_2 grains with different crystallographic orientation embedded in the thin silicon nitride matrix to a nanocomposite structure. The film with the highest wear resistance represents a new generation of Me-Si-N films with enhanced mechanical properties, characteristic of a structure consisting of very fine-grained overstoichiometric $MeN_{x>1}$ grains dispersed within a high amount of dominant amorphous Si_3N_4 phase. Since the Me-Si-N films with lower silicon content typically exhibit lower values of the H^3:E^{*2} ratio, the stability and protecting capability of the Me-Si-N films against wear can be evidently improved by a high amount of Si–N bonds, that is, high volume content of the Si_3N_4 phase [64].

6.8 Macrostress σ in W-Si-N Films

Excessive residual stresses in thin films can result in severe problems in many applications. It has been shown that both tensile and compressive stresses may significantly affect the mechanical resistance and integrity of deposited films. While a high level of tensile stresses typically leads to film cracking, very high compressive stresses may cause delamination of the film from the substrate surface or formation of blister zones. A stress discontinuity at the coating–substrate interface often causes the adherence problems and consequent film delamination when the forces caused by stresses exceed the adhesion forces. The origin of the residual stresses is attributed to the difference in thermal expansion coefficients between the film and the substrate (thermal stress), or to energetic particle bombardment (intrinsic stress). Very high stresses generated in traditional transition metal nitride films can, however, be expressively reduced if the films are alloyed with silicon [29,65]. Me-Si-N nanocomposite films exhibit reasonable stress level, which is recovered already during the film growth [10]. This fact very positively affects the protecting capability of the films since the stresses concentrated at the interface usually do not exceed the critical values. Since the mechanical properties of thin films are usually related to the residual stress level generated in the film, its relaxation may cause degradation of film hardness and the resistance against plastic deformation [66]. Hence, understanding of the relationship between hardness and stresses in thin films is essential for their usage in tooling applications.

The development of the macrostress σ in the W-Si-N films indicates that the phase composition of the W-Si-N films is an essential factor determining also the stress state developed during the film growth (Fig. 6.18a). The W-Si-N films with a low nitrogen content, sputtered at $p_{N_2} < 0.03$ Pa, are characterized by tensile stress ($\sigma > 0$). Those films are composed of a mixture of the WSi_2 phase, free unbounded tungsten atoms and a small volume content of the amorphous Si_3N_4 tissue. On the contrary, while a higher nitrogen amount is incorporated into the films at $p_{N_2} > 0.05$ Pa, compressive stress ($\sigma < 0$) is consequently generated. This fact is attributed to the formation of nitrided phases in the films. It is shown that the value of the compressive stress σ increases with an increasing amount of silicon nitride and achieves its maximum value if all silicon is converted into the Si_3N_4 phase and also the WN_x phase is formed. Hence, all the W-Si-N films with an X-ray amorphous structure sputtered at $p_{N_2} > 0.2$ Pa exhibit almost constant maximum value of $\sigma \approx -1.5$ GPa. The invariant phase composition of those films is probably a reason why σ does not change with increasing p_{N_2} and remains approximately constant.

The effect of the nitrification level of the sputtered W-Si-N films on the residual stress values σ is evident from Fig. 6.18b. Here, the development of σ with increasing the N:Si r atio in the films sputtered at the same values of I_d and U_s but different values of i_s is depicted. It is clearly seen that the same value of σ can be reached at different N:Si ratios. This fact is characteristic for the W-Si-N films prepared at $p_{N_2} \leq 0.2$ Pa and various i_s. For instance, the same value of the compressive macrostress $\sigma = -0.5$ GPa is measured for the W-Si-N films with N:Si $\approx$ 0.15 and 0.6 if they are sputtered at $i_s = 0.5$ and 1 mA/cm^2, respectively (compare values labeled as 4 in Fig. 6.18). In this p_{N_2} range, the films sputtered at the same ratio N:Si $\leq$ 1.25 exhibit lower stress level if they are exposed to more intensive ion bombardment during deposition. This can be explained by a higher E_{bi} delivered to the films sputtered at $i_s = 1$ mA/cm^2, which results in a greater relaxation of the growth (intrinsic) stress and thus in lower values of σ in the W-Si-N films. Furthermore, the different phase composition has also to be taken into the consideration. While the change of σ is very strong in the films with the ratio N:Si $\leq$ 1.25, the W-Si-N films composed of two $WN_{x\geq1}$ and Si_3N_4 amorphous phases exhibit almost the same value of σ for the ratio N:Si ranging from approximately 1.25 to 2.7.

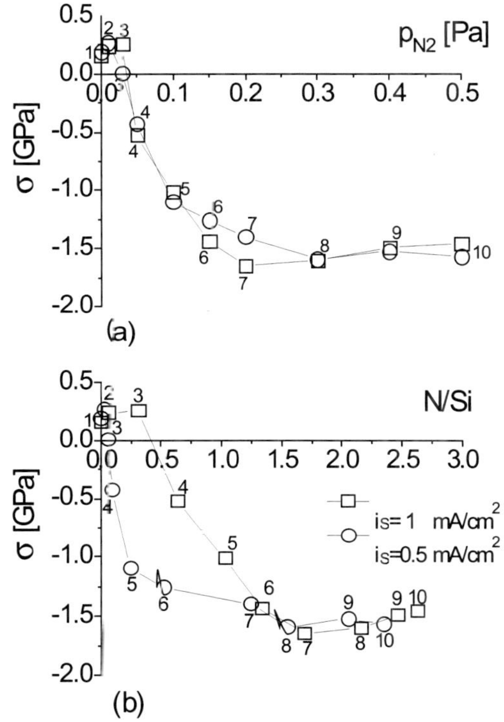

Figure 6.18 Residual stress σ in the 4 μm thick W-Si-N films, magnetron sputtered at I_d = 1 A, U_s = −100 V and two values of i_s = 0.5 and 1 mA/cm^2, as a function of (a) partial pressure of nitrogen p_{N_2} and (b) the N:Si ratio. Reprinted from Ref. [20], Copyright (2006), with permission from Elsevier.

6.8.1 Effect of Volume Content of Si_3N_4 Phase on *H* and σ

It was already shown that the amorphous silicon nitride phase substantially affects the properties of the W-Si-N films especially at its high volume content. The developed nanocomposite structure provides the enhanced mechanical properties and simultaneously reduces the residual stress, which was found to be four to five times smaller than the residual stress measured in traditional transition metal nitride films [67]. The role of an amount of the amorphous Si_3N_4 phase on the hardness values H and macrostress σ of the W-Si-N film is illustrated in Fig. 6.19. It is evident that H continuously increases with increasing volume fraction of the Si_3N_4 phase and reaches the maximum at its approximately 60 vol.%. The W-Si-N films sputtered at $p_{N_2} > 0.15$ Pa exhibit comparable microhardness and also the residual stress (Fig. 6.19). The developments of H and σ at lower partial pressure of nitrogen ($p_{N_2} \leq 0.15$ Pa), however, strongly differ when the W-Si-N films are sputtered at different values of i_s. It is related to the different energy delivered to the films during their growth.

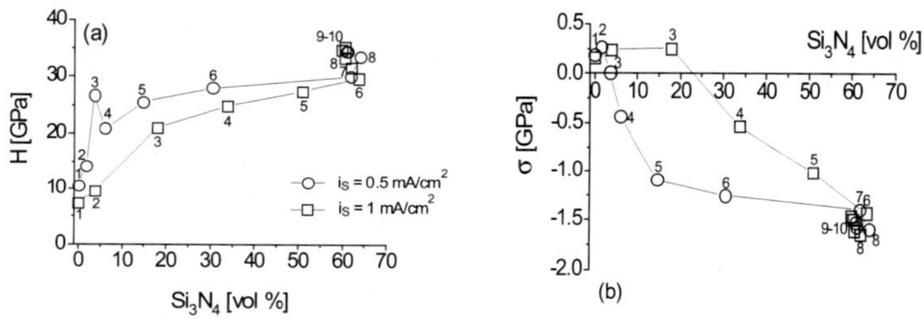

Figure 6.19 (a) Hardness H and (b) macrostress σ in the W-Si-N films, sputtered at $I_d = 1$ A, $U_s = -100$ V and two values of $i_s = 0.5$ mA/cm^2 and 1 mA/cm^2, as a function of the volume content of Si_3N_4 phase in the film. Reprinted from Ref. [20],Copyright (2006), with permission from Elsevier.

This experiment also indicates that the amount of nitrogen incorporated into the films is crucial in this case. As shown in Fig. 6.14, H and σ depend on the phases developed during deposition, particularly on the composition of the second phase, that is, on the (WSi_2 + W) and WN_x phases formed at low and high values of p_{N_2}, respectively (Table 6.1). The W-Si-N films sputtered at low p_{N_2} values are three phase composites composed of WSi_2 + W + Si_3N_4. With respect to the fact that the same amount of WSi_2 + W can be achieved at different volume combinations of both WSi_2 and tungsten components of the second phase formed at low p_{N_2}, H and σ may essentially differ in the films sputtered at $i_s = 0.5$ mA/cm^2 and 1 mA/cm^2. On the contrary, the W-Si-N films sputtered at high values of p_{N_2} are two phase composites composed of Si_3N_4 + WN_x. The hardness of those films where all silicon is converted into the Si_3N_4 phase and which contain almost the same amount of the Si_3N_4 phase (Fig. 6.19) slightly increases with increase in p_{N_2} (Fig. 6.14a). This increase in H correlates well with increased amount of nitrogen incorporated into the W-Si-N film. The macrostress σ of these films, however, remains approximately constant.

6.8.2 Effect of Substrate Temperature T_s and Film Thickness h on σ

From the technological point of view, the stability of the film structure is a very important factor, which should however also be combined with the stability of the mechanical properties. The relaxation of stresses in the films at elevated temperatures may cause a sudden film failure and thus, the lifetime shortening of coated tools if the hardness decreases with the working temperature.

The effect of the substrate temperature on the properties of the W-Si-N films is demonstrated in Fig. 6.20a. The X-ray amorphous $Wi_{18}Si_{26}N_{56}$ film (p_{N_2} = 0.3 Pa) exhibits constant compressive stress of $\sigma \approx -1.5$ GPa over the whole range of the deposition temperatures from room temperature to 500°C. This temperature range corresponds to the T_s:T_m ratio varying from 0.013 to 0.26, which represents the energy delivered into the films by the substrate heating determining the growth process phenomena and also the efficiency of the thermal energy to recover the growth macrostress in the films [68]. Here, T_m is the melting point of the deposited material and T_s is the temperature of the substrate during deposition. Since the W-Si-N films sputtered from an alloyed WSi_2 target at higher partial pressures of nitrogen $p_{N_2} > 0.15$ Pa are characterized with a dominating amount of the Si_3N_4 phase achieving approximately 64 vol.% (Fig. 6.6), the T_s:T_m ratio is mainly determined by the melting point of silicon nitride $T_{m\ Si_3N_4}$ = 1900°C [69]. The development of σ with deposition temperature suggests that the dominating Si_3N_4 phase provides spontaneous stress recovery during the film growth already at low deposition temperatures and ensures that the amorphous-like W-Si-N films with a high (≥60 vol.%) content of Si_3N_4 phase are thermally stable at a wide range of T_s. This is reflected by a relatively low and constant compressive stress level to 500°C.

Residual stress of thin solid films typically varies with their thickness, which is related to the structural evolution during film growth. Almost all the constant residual stress in a representative high silicon-containing W-Si-N film with increase in film thickness up to 14 μm (Fig. 6.15b) indicates that, the macrostress σ is independent of both T_s and h, if all silicon is converted into Si_3N_4 phase, which becomes a dominant component of the film. It means that such W-Si-N films can be prepared with a low stress level already at low substrate temperatures T_s independently on their thickness h.

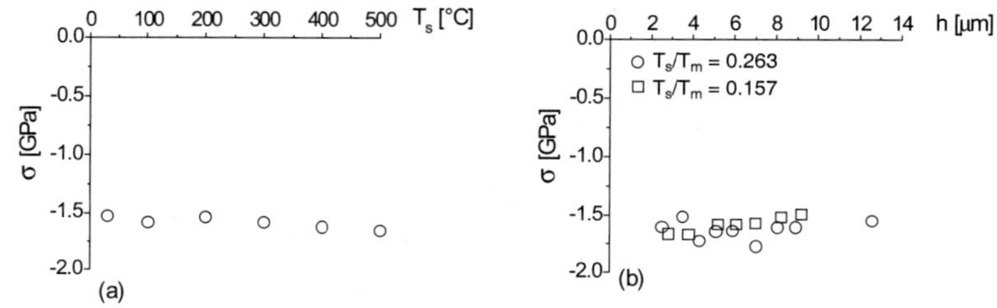

Figure 6.20 Macrostress σ in the W-Si-N films, sputtered at I_d = 1 A, U_s = −100 V, i_s = 1 mA/cm^2, and p_{N_2} = 0.3 Pa, as a function of (a) substrate temperature T_s for h = 4 μm and (b) film thickness h for two values of T_s = 300 and 500°C. Reprinted from Ref. [20], Copyright (2006), with permission from Elsevier.

Reduction of the compressive macrostress below −1.5 GPa would broaden the utilization of those films in other applications, such as electronics. However, both phases in the X-ray amorphous W-Si-N films with a high (≥60 vol.%) content of Si_3N_4 phase are very stable at $T_s \leq 500$°C and E_{bi} delivered to the growing film during deposition is probably too low to influence the stress level in the films. The results indicate that the residual stress is strongly associated to the phase composition of the W-Si-N films that in addition ensures stability of the mechanical properties and stress also at elevated temperatures. Strong dependence of the material properties on the phase composition is further demonstrated by detailed studies of thermal stability and oxidation resistance of the X-ray amorphous high Si-containing W-Si-N films (Section 6.9: Oxidation Resistance). These investigations show that the crystallization temperature of the amorphous W-Si-N films shifts to higher values with increasing volume content of the Si_3N_4 phase.

6.9 Oxidation Resistance

The procedure adopted in the past years for the development of hard coatings for mechanical applications was based upon two main scopes. First, a synthesis of films with excellent mechanical properties, such as very high hardness and very high fracture toughness, has been a main driving force for the development studies. However, besides outstanding mechanical properties, the thermal stability of films used in tool applications has been found to be relevant as well. An effort to prepare very hard films stable at high temperatures has turn the focus of the researchers on the addition of alloying elements into the transition metal nitride films. A special attention has been paid to the elements, which are known to stabilize the film structure with respect to its contribution to the thermal behavior of the film system. The addition of alloying elements, such as aluminum or chromium, is one of the possible solutions of how to reach simultaneously very high hardness and excellent behavior at high temperatures. Since the oxidation resistance of films is one of very important factors affecting the durability of coated tools working at high temperatures, new possibilities to avoid a rapid decrease in wear protection and a loss of adherence of the films that are exposed to aggressive environments has recently gained much attention. One of the studied elements, showing a strong potentiality to protect thin films against high-temperature oxidation, is silicon. Although silicon has been used in many coating systems, the best results have been obtained for its highest content in the films [37,38,70]. Although the films have exhibited mostly amorphous-like structure, the film hardness has been higher than the hardness of corresponding conventional transition nitrides. The presence of the amorphous phase has been shown to inhibit the growth of the nanograins justifying the very high thermal stability of these films. Hence, after film annealing no significant grain growth is typically detected in contrast to traditional materials [70].

The oxidation resistance of the W-Si-N films with a high silicon content has been shown to be strongly dependent on their phase composition, namely on the volume content of the Si_3N_4 phase. The role of individual phases, in which the film is composed, on the resistance of the films against oxidation is typically investigated by means of a symmetrical high-resolution thermogravimetry, where the onset of oxidation is determined as the change of the film mass Δm, combined with structural analysis of the films after annealing experiments by means of X-ray diffraction and electron microscopy.

Experimental results of four representative W-Si-N films with an X-ray amorphous structure but various phase composition annealed to 1300°C in synthetic air are displayed in Fig. 6.21. Although all films contain volume content of Si_3N_4, which is higher than 50 vol.%, they differ in composition of other phases.

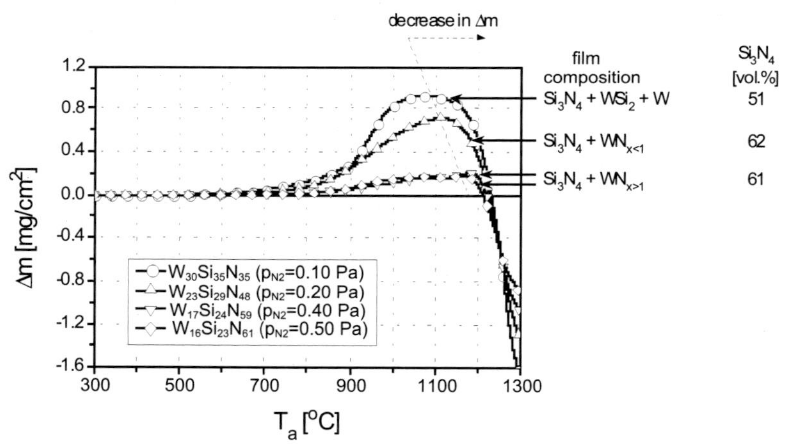

Figure 6.21 Thermogravimetric curves of W-Si-N films with different phase compositions showing the mass gain Δm observed during annealing in flowing air as a function of annealing temperature T_a. The films were sputtered at following deposition conditions: I_d = 1 A, U_s = −100 V, i_s = 1 mA/cm^2, T_s = 500°C and different values of p_{N_2} = 0.1 ($Wi_{30}Si_{35}N_{35}$), 0.2 ($Wi_{23}Si_{29}N_{48}$), 0.4 ($Wi_{17}Si_{24}N_{59}$), and 0.5 Pa ($Wi_{16}Si_{23}N_{61}$). Reprinted from Ref. [20], Copyright (2006), with permission from Elsevier.

It has been already shown that the silicon-containing WN films exhibit much higher oxidation resistance than tungsten and WN films [71]. It is evident from Fig. 6.21 that the onset temperature of the oxidation strongly depends on the silicon and nitrogen content and thus, on the phase composition of the W-Si-N films in their as-deposited state. It is evident that no mass gain is observed up to 600°C for all the films and the oxidation onset markedly shifts to higher annealing temperatures when the nitrogen content in the films increases. While the tungsten wire oxidized already at approximately 600°C [72], the W-Si-N films started to oxidize in the range from 650°C to 800°C depending on their phase composition (Fig. 6.21). The lowest oxidation resistance showed the $Wi_{30}Si_{35}N_{35}$ film (p_{N_2} = 0.1 Pa) composed of Si_3N_4 + WSi_2 + W. The presence of free tungsten in this compound provides a rapid oxidation of the film resulting in an Δm increase up to 1 mg/cm^2. The films containing substoichiometric $WN_{x<1}$ nitride exhibit a similar behavior. The substoichiometric $WN_{x<1}$ is a metal (W) rich compound (WN + W) where free tungsten again leads to relatively intensive oxidation. This is a reason why the W-Si-N films composed of Si_3N_4 + WSi_2 + W and Si_3N_4 + $WN_{x<1}$ phases ($p_{N_2} \leq 0.3$ Pa) exhibit a low oxidation temperature of 650°C. The oxidation resistance can be, however, increased to approximately 800°C when the W-Si-N films contain overstoichiometric $WN_{x>1}$ nitride.

There are several points of interest to be noted in Fig. 6.21. The trend of the thermogravimetric curves can be divided into two main parts. Firstly, an increase in the mass was observed for temperatures up to 1100°C. This increase ($\Delta m > 0$) corresponds to a formation of the tungsten oxide. Secondly, a rapid decrease of the mass ($\Delta m < 0$) can be seen in the temperature range between 1100 and 1300°C. This phenomenon is related to decomposition of WN_x compound ($WN_x \rightarrow W + N$) at $T_a > 1100$°C and subsequent formation of volatile tungsten WO_x oxides, which spontaneously release from the film. Rapid oxidation of tungsten and WN slightly above 1000°C has been demonstrated by thermogravimetric measurements of a pure tungsten wire [72] and differential scanning calorimetry (DSC) measurements of a $W_{46}N_{54}$ compound in a controlled atmosphere of N_2/H_2 at annealing temperatures up to 1400°C [73].

Structural development of the high Si-containing W-Si-N films after annealing in air up to 1300°C is shown in Fig. 6.22. The microstructure of the as-deposited W-Si-N films prepared at $p_{N_2} \geq 0.1$ Pa exhibits typically fine-grained X-ray amorphous nanocomposite nature, which is characteristic for most of the Me-Si-N films with a high amount of the Si_3N_4 phase. After the thermal annealing in air to

1300°C the amorphous-like structure is converted to a crystalline triclinic WO_3 oxide one. This fact corresponds well to the stated decomposition route revealed from the thermogravimetric and DSC measurements, where the WN_x phase firstly decomposes to tungsten and nitrogen and then free unbounded tungsten reacts with oxygen after the nitrogen release. No other phases were detected, which suggests that silicon remains bounded to nitrogen in a form of the amorphous Si_3N_4 phase or reacts with oxygen forming SiO_2 or SiO_xN_y phases. Both phases exist in an amorphous state and thus are not detectable by XRD analysis.

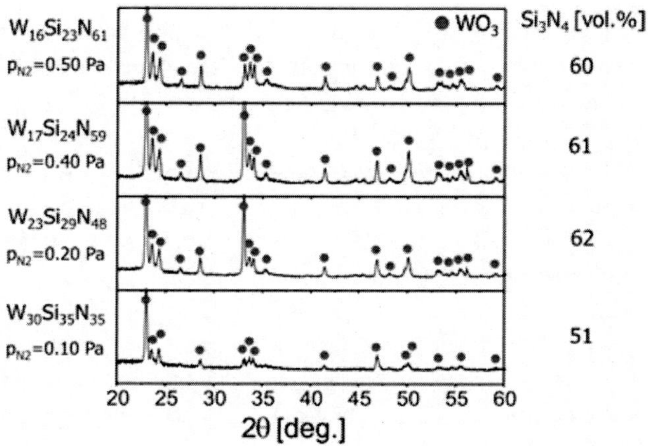

Figure 6.22 XRD patterns of the W-Si-N films with different phase composition after annealing in air to 1300°C.

The highest oxidation resistance of the W-Si-N films was achieved for the films containing the highest volume content of the Si_3N_4 phase and the overstoichiometric $WN_{x>1}$. However, in spite of the fact that the X-ray amorphous W-Si-N films exhibit approximately the same amount of silicon nitride as the Zr-Si-N films with a high (≥25 at.%) silicon content, their resistance against oxidation is much lower. While the $Zr_{15}Si_{24}N_{61}$ film (p_{N_2} = 0.4 Pa) exhibits no increase in mass (Δm = 0) up to 1300°C (Section 5.12), the $Wi_{16}Si_{23}N_{61}$ film (p_{N_2} = 0.5 Pa) reveals a strong oxidation already at 800°C. It is indeed of a great scientific interest to find the differences in the oxidation behavior and to fully understand the oxidation mechanisms involved in the X-ray amorphous Me-Si-N films with various phase compositions.

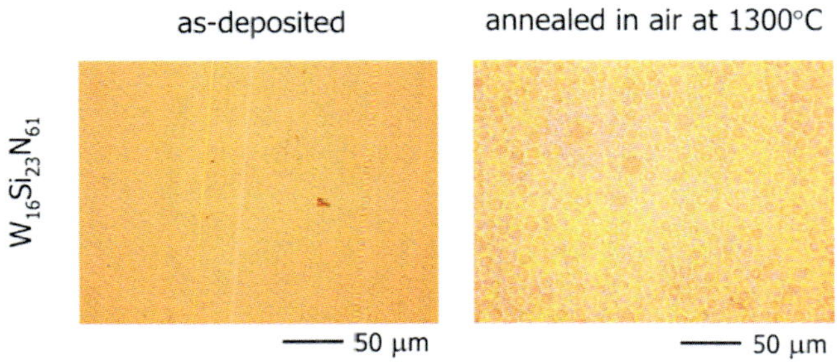

Figure 6.23 LOM surfaces of the as-deposited and annealed $Wi_{16}Si_{23}N_{61}$ film exhibiting the best oxidation resistance.

The reason for why the high Si-containing W-Si-N films behave differently at elevated temperatures with respect to the Zr-Si-N films with the same volume content of the Si_3N_4 phase is indicated in the structural development of the $Wi_{16}Si_{23}N_{61}$ film after annealing. This film consists of Si_3N_4 + $WN_{x>1}$ with 60 vol.% of Si_3N_4 and exhibits the highest oxidation resistance. The surface morphology of this film in the as-deposited state and after annealing in air to 1300°C investigated using light optical microscopy (LOM) is depicted in Fig. 6.23. The surface of the as-deposited film is very smooth without any pronounced defects with the mean roughness R_a < 2 nm, similarly to the $Zr_{15}Si_{24}N_{61}$ film. However, a dramatical change of the surface morphology has been observed after annealing at 1300°C. The surface becomes opaque with hill-valley features and a higher roughness (R_a > 100 nm). The 3D surface morphology of the film investigated using atomic force microscopy (AFM) is shown in Fig. 6.24.

The massive degradation of the film structure is related to the formation of an oxide layer, which grows in thickness as the film reacts with oxygen. The cross-sectional SEM images presented in Fig. 6.25 show in more detail the oxide layer formed during annealing. It is evident that the W-Si-N film is fully oxidized (h_{oxide} = h_f) and the structure of the oxidized film is completely disturbed as a consequence of volatilization of the tungsten oxide at T_a > 1000°C. It escapes from the film during annealing at relatively low temperatures and thus a porous sponge-like structure is fully developed after the test. The pores are also obvious on the surface of the annealed film what confirms a rapid volatilization of the tungsten oxide.

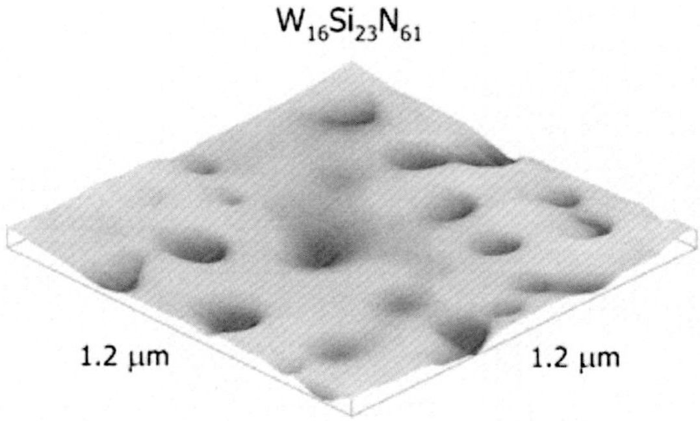

Figure 6.24 Atomic force microscopy surface of the annealed $Wi_{16}Si_{23}N_{61}$ film exhibiting the best oxidation resistance.

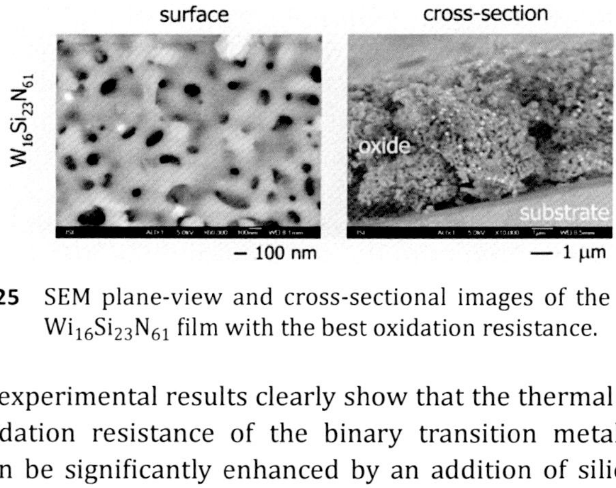

Figure 6.25 SEM plane-view and cross-sectional images of the annealed $Wi_{16}Si_{23}N_{61}$ film with the best oxidation resistance.

The experimental results clearly show that the thermal stability and oxidation resistance of the binary transition metal nitride films can be significantly enhanced by an addition of silicon. The amorphous glass-like microstructure without grain boundaries preventing inward oxygen diffusion is one of the decisive factors enabling to achieve the high-temperature oxidation resistance of protective coatings. Since a transport of species during oxidation is effectively hindered through a compact and dense amorphous silicon nitride tissue, fine-grained nanocomposite films exhibit much higher oxidation resistance compared to crystalline materials with well-developed diffusion paths. While this phenomenon is important mainly for the oxidation process at low temperatures, a structure of the oxide surface layer is decisive for high-temperature oxidation processes. This fact is crucial for better understanding the differences between thermal stability and resistance against oxidation of Me-Si-N films with a high (≥25 at.%) silicon content. Although all these films exhibit similarly high volume content of very thermally stable Si_3N_4 phase, they differ in the nature of the oxide layers formed on the sample surface during oxidation. Moreover, the difference in the oxidation behavior can be found in a different thermal stability of the second phase, in which the films are composed.

The oxide layer formed on the surface of the Zr-Si-N films exhibits very dense and compact structure, which acts as an efficient barrier for the transport of oxygen and thus prevents its penetration further into the film. Furthermore, the oxide layer is most probably not fully crystallized and its grain boundaries are filled with amorphous SiO_2 and SiO_xN_y phases. In this way, a high oxidation rate is effectively prevented. That ensures a high oxidation stability of the films since in addition both ZrN phase and formed surface oxide layer sustain extremely high temperatures exceeding 1000°C without disturbing their structure. On the contrary, the W-Si-N films oxidize already at 800°C. It is due to the fact that tungsten nitrides (WN, W_2N) and oxides (WO_3, WO_2) are unstable at high temperatures. The tungsten nitrides decompose around 700°C [23] while nitrogen subsequently releases from the film. Channels remaining after the nitrogen release serve for rapid inward oxygen diffusion. Oxygen diffusing through these paths readily reacts with free unbounded tungsten atoms to form WO_x. The evaporation of the tungsten oxide from the film leads to the destruction of the film microstructure where the amorphous Si_3N_4 matrix can no longer be efficient enough to hinder the oxidation. Thus, further oxidation of the bulk of the film is possible [72]. The oxidation process proceeds until the remaining film is fully oxidized. At temperatures higher than 1000°C the substrate may oxidize as well what typically leads to a damage of coated tools.

The stability of the film components and their oxides is the main reason why the Zr-Si-N films exhibit considerably better oxidation resistance than those of W-Si-N. A very thin passivating compact oxide layer without voids formed on the surface of the Zr-Si-N films makes these films also more stable than other high Si-containing Me-Si-N films, such as Ta-Si-N or Mo-Si-N [38,74]. All the facts clearly indicate that a high (≥50 vol.%) content of the Si_3N_4 phase and the presence of overstoichiometric $MeN_{x>1}$ nitride are not sufficient conditions to produce silicon based (Me-Si-N) nitride films with the oxidation resistance exceeding 1000°C. In fact, thermal stability and resistance against oxidation of all components, of which the films are composed, are the key prerequisite for exceptional performance of thin solid films in high-temperature applications.

References

1. F.C.T. So, E. Kolawa, X.-A. Zhao, M.-A. Nicolet (1987), WN_x: Properties and applications, *Thin Solid Films* **153**, 507–520.
2. J.S. Reid, E. Kolawa, R.P. Ruiz, M.-A. Nicolet (1993), Evaluation of amorphous (Mo, Ta, W)- Si-N diffusion barriers for (Si)|Cu metallizations, *Thin Solid Films* **236**, 319–324.
3. M. Hirano, K. Nishikawa, I. Toyoda, S. Aoyama (1997), Three-dimensional interconnect technology for ultra-compact MMICs, *Solid State Electron.* **41**, 1451–1455.
4. A. Cavaleiro, C. Louro, J.V. Fernandes, C.M.A. Brett (1999), Use of ultramicroindentation to evaluate the degradation of sputtered coatings, *Vacuum* **52**, 157–162.
5. B. Trindade, M.T. Vieira (1995), Influence of titanium on the structural stability of sputter-deposited W-Co-C films, *Surf. Coat. Technol.* **74–75**, 802–805.
6. A. Cavaleiro, M.T. Vieira, F. Ramos, J.P. Dias (1996), The ultimate vacuum pressure and the characteristics of sputtered coatings, *Thin Solid Films* **290–291**, 238–242.
7. C. Louro, A. Cavaleiro (1997), Thermal oxidation of tungsten-based sputtered coatings, *J. Electrochem. Soc.* **144**, 259–266.
8. M. Nose, M. Zhou, T. Nagae, T. Mae, M. Yokota, S. Saji (2000), Properties of Zr-Si-N coatings prepared by RF reactive sputtering, *Surf. Coat. Technol.* **132**, 163–168.
9. J.S. Reid, R.Y. Liu, P.M. Smith, R.P. Ruiz, M.-A. Nicolet (1995), W-B-N diffusion barriers for Si/Cu metallizations, *Thin Solid Films* **262**, 218–223.
10. J. Musil, R. Daniel, O. Takai (2005), Structure and mechanical properties of magnetron sputtered Zr-Si-N films with a high (≥25 at.%) Si content, *Thin Solid Films* **478**, 238–247.
11. M.-A. Nicolet (2000), Reactively sputtered ternary films of the type TM-Si-N and their properties (TM = early transition metal), *Vacuum* **59**, 716–720.

12. M. Zhou, M. Nose, Y. Deguchi, T. Mae, K. Nogi (2003), Influence of sputtering conditions on microstructure and mechanical properties of Zr-Si-N films prepared by radio-frequency-reactive sputtering, *J. Vac. Sci. Technol.* **A21**, 1791–1795.

13. M. Diserens, J. Patscheider, F. Lévy (1999), Mechanical properties and oxidation resistance of nanocomposite TiN-SiN_x physical-vapor-deposited thin films, *Surf. Coat. Technol.* **120–121**, 158–165.

14. A. Cavaleiro, C. Louro (2002), Nanocrystalline structure and hardness of thin films, *Vacuum* **64**, 211–218.

15. H.-D. Männling, D.S. Patil, K. Moto, M. Jilek, S. Vepřek (2001), Thermal stability of superhard nanocomposite coatings consisting of immiscible nitrides, *Surf. Coat. Technol.* **146–147**, 263–267.

16. C. Louro, A. Cavaleiro (2000), Mechanical behaviour of amorphous W-Si-N sputtered films after thermal annealing at increasing temperatures, *Surf. Coat. Technol.* **123**, 192–198.

17. K. Koski, J. Hölsä, P. Juliet (1998), Voltage controlled reactive sputtering process for aluminium oxide thin films, *Thin Solid Films* **326**, 189–193.

18. J. Musil, S. Kadlec, V. Valvoda, R. Kužel, Jr., R. Černý (1990), Ion-assisted sputtering of TiN films, *Surf. Coat. Technol.* **43–44**, 259–269.

19. J. Musil, H. Poláková, J. Šůna, J. Vlček (2003), Effect of ion bombardment on properties of hard reactively sputtered Ti(Fe)N_x films, *Surf. Coat. Technol.* **177–178**, 289–298.

20. J. Musil, R. Daniel, J. Soldán, P. Zeman, Properties of reactively sputtered W–Si–N films, *Surface and Coatings Technology*, **200**(12–13), 3886–3895.

21. T. Hurkmans, D.B. Lewis, H. Paritong, J.S. Brooks, W.D. Münz (1999), Influence of ion bombardment on structure and properties of unbalanced magnetron grown CrN_x coatings, *Surf. Coat. Technol.* **114**, 52–59.

22. J.A. Dean (1999), *Lange's Handbook of Chemistry*. New York: McGraw-Hill, Inc.

23. R. Kieffer, F. Benesovsky (1963), *Hartstoffe*. Wien: Springer-Verlag.

24. P. Rogl, J.C. Schuster (1992), *Phase Diagrams of Ternary Boron Nitride and Silicon Nitride Systems*. Materials Park, Ohio: ASM International.

25. J.-W. Lee, S.C.N. Cheng (2000), Development of SiN_x and AlN_x passivation layers, *Thin Solid Films* **358**, 215–222.

26. C. Louro, A. Cavaleiro, F. Montemor (2001), How is the chemical bonding of W-Si-N sputtered coatings? *Surf. Coat. Technol.* **142–144**, 964–970.

27. E.A. Brandes (1992), *Smithells Metals Reference Book*, Butterworth-Heinemann.

28. R. Daniel, J. Musil, P. Zeman, C. Mitterer (2006), Thermal stability of magnetron sputtered ZrSi-N films, *Surf. Coat. Technol.* **201**, 3368–3376.

29. T. Fu, Z.F. Zhou, K.Y. Li, Y.G. Shen (2005), Structure, stress and hardness of sputter deposited nanocomposite W-Si-N coatings, *Surf. Coat. Technol.* **200**, 2525–2530.

30. P.J. Martin, A. Bendavid, J.M. Cairney, M. Hoffman (2005), Nanocomposite Ti-Si-N, Zr-Si-N, Ti-Al-Si-N, Ti-Al-V-Si-N thin film coatings deposited by vacuum arc deposition, *Surf. Coat. Technol.* **200**, 2228–2235.

31. A. Winkelmann, J.M. Cairney, M.J. Hoffman, P.J Martin, A. Bendavid (2005), Zr-Si-N films fabricated using hybrid cathodic arc and chemical vapour deposition: Structure vs. properties, *Surf. Coat. Technol.* **200**, 4213–4219.

32. J. Patscheider, T. Zehnder, M. Diserens (2001), Structure-performance relations in nanocomposite coatings, *Surf. Coat. Technol.* **146–147**, 201–208.

33. F. Vaz, L. Rebouta, P. Goudeau, et al. (2001), Structural transitions in hard Si-based TiN coatings: The effect of bias voltage and temperature, *Surf. Coat. Technol.* **146–147**, 274–279.

34. C. Louro, A. Cavaleiro (1999), Hardness versus structure in W-Si-N sputtered coatings, *Surf. Coat. Technol.* **116–119**, 74–80.

35. A. Vomiero, E. B. Marchi, G. Mariotto, G. D. Mea, A. Scandurra, O. Puglisi (2005), The role of N in the re-sputtering inhibition of Si in W-Si-N reactively sputtered thin layers, *J. Appl. Phys.* **97**, 349051–349057.

36. J.I. Langford (1980). *Accuracy in powder diffraction.* In: S. Block, C.R. Hubbard (Eds.). Washington: National Bureau of Standards.

37. H. Zeman, J. Musil, P. Zeman (2004), Physical and mechanical properties of sputtered Ta-Si-N films with a high (≥40 at.%) content of Si, *J. Vac. Sci. Technol.* **A22**, 646–649.

38. J. Musil, P. Dohnal, P. Zeman (2005), Physical properties and high-temperature oxidation resistance of sputtered Si_3N_4/MoN_x nanocomposite coatings, *J. Vac. Sci. Technol.* **B23**, 1568–1575.

39. J.A. Thornton (1974), Influence of apparatus geometry and deposition conditions on the structure and topography of thick sputtered coatings, *J. Vac. Sci. Technol.* **A11**, 666–670.

40. R. Rosenberg, D.C. Edelstein, C.-K. Hu, K.P. Rodbell (2000), Copper metallization for high performance silicon technology, *Annu. Rev. Mater. Sci.* **30**, 229–262.

41. A.E. Kaloyeros, E. Eisenbraun (2000), Ultrathin diffusion barriers/liners for gigascale copper metallization, *Annu. Rev. Mater. Sci.* **30**, 363–385.

42. A.Y. Suh, A.A. Polycarpou, T.F. Conry (1993), Detailed surface roughness characterization of engineering surfaces undergoing tribological testing leading to scuffing, *Wear* **255**, 556–568.

43. J. Musil, P. Zeman, H. Hrubý, P.H. Mayrhofer (1999), ZrN/Cu nanocomposite film: A novel superhard material, *Surf. Coat. Technol.* **120–121**, 179–183.

44. J. Musil (2000), Hard and superhard nanocomposite coatings, *Surf. Coat. Technol.* **25**, 322–330.

45. J. Musil, P. Karvánková, J. Kasl (2001), Hard and superhard Zr-Ni-N nanocomposite films, *Surf. Coat. Technol.* **139**, 101–109.

46. J.G. Han, H.S. Myung, H.M. Lee, L.R. Shaginyan (2003), Microstructure and mechanical properties of Ti-Ag-N and Ti-Cr-N superhard nanostructured coatings, *Surf. Coat. Technol.* **174–175**, 738–743.

47. S.Vepřek (1999), The search for novel superhard materials, *J. Vac. Sci. Technol.* **A17**, 2401–2420.

48. S. Vepřek, S. Reiprich (1995), A concept for the design of novel superhard coatings, *Thin Solid Films* **268**, 64–71.

49. M. Diserens, J. Patscheider, F. Lévy (1998), Improving the properties of titanium nitride by incorporation of silicon, *Surf. Coat. Technol.* 108–109, 241–246.

50. J. Musil, R. Daniel (2003), Structure and mechanical properties of magnetron sputtered Zr-Ti-Cu-N films, *Surf. Coat. Technol.* **166**, 243–253.

51. H. Poláková, J. Musil, J. Vlček, J. Allaart, C. Mitterer (2003), Structure-hardness relations in sputtered Ti-Al-V-N films, *Thin Solid Films* **444**, 189–198.

52. E.O. Hall (1951), The deformation and ageing of mild steel: III Discussion of results, *Proc. Phys. Soc.* **B64**, 747–753.

53. S. Zhang, D. Sun, Y.Q. Fu, H.J. Du (2003), Recent advances of superhard nanocomposite coatings: A review, *Surf. Coat. Technol.* **167**, 113–119.

54. A. Leyland, A. Matthews (2000), On the significance of the H/E ratio in wear control: A nanocomposite coating approach to optimised tribological behaviour, *Wear* **246**, 1–11.

55. B. Jonsson, S. Hogmark (1984), Hardness measurements of thin films, *Thin Solid Films* **114**, 257–269.

56. S.V. Hainsworth, T. Bartlett, T.F. Page (1993), The nanoindentation response of systems with thin hard carbon coatings, *Thin Solid Films* **236**, 214–218.

57. T.F. Page, S.V. Hainsworth (1993), Using nanoindentation techniques for the characterization of coated systems: A critique, *Surf. Coat. Technol.* **61**, 201–208.

58. R. Saha, W.D. Nix (2002), Effects of the substrate on the determination of thin film mechanical properties by nanoindentation, *Acta Mater.* **50**, 23–38.

59. W.C. Oliver, G.M. Pharr (1992), An improved technique for determining hardness and elastic modulus using load and displacement sensing indentation experiments, *J. Mater. Res.* **7**(6) 1564–1583.

60. F. Regent, J. Musil (2001), Magnetron sputtered Cr-Ni-N and Ti-Mo-N films: Comparison of mechanical properties, *Surf. Coat. Technol.* **142–144**, 146–151.

61. A. Cavaleiro, B. Trindade, M.T. Vieira (1999), Deposition and characterization of fine-grained W-Ni-C/N ternary films, *Surf. Coat. Technol.* **116–119**, 944–948.

62. G.A. Malygin (1995), Break-down of the Hall-Petch law in micro- and nanocrystalline materials, *Phys. Solid. State* **37**, 1248–1253.

63. C. Rebholz, A. Leyland, J.M. Schneider, A.A. Voevodin, A. Matthews (1999), Structure, hardness and mechanical properties of magnetron-sputtered titanium-aluminium boride films, *Surf. Coat. Technol.* **120–121**, 412–417.

64. D. Pilloud, J.F. Pierson, A.P. Marques, A. Cavaleiro (2004), Structural changes in Zr-Si-N films vs. their silicon content, *Surf. Coat. Technol.* **180–181**, 352–356.

65. T. Mae, M. Nose, M. Zhou, T. Nagae, K. Shimamura (2001), The effects of Si addition on the structure and mechanical properties of ZrN thin films deposited by an r.f. reactive sputtering method, *Surf. Coat. Technol.* **142–144**, 954–958.

66. L. Hultman (2000), Thermal stability of nitride thin films, *Vacuum* **57**, 1–30.

67. R. Kullmer, C. Lugmair, A. Figueras, J. Bassas, M. Stoiber, C. Mitterer (2003), Microstructure, mechanical and tribological properties of PACVD Ti(B,N) and TiB_2 coatings, *Surf. Coat. Technol.* **174–175**, 1229–1233.

68. J.A. Thornton, D.W. Hoffman (1989), Stress-related effects in thin films, *Thin Solid Films* **171**, 5–31.

69. H. Holleck (1986), Material selection for hard coatings, *J. Vac. Sci. Technol.* **A4**, 2661–2669.

70. A.P. Marques, A. Cavaleiro (2003), Structural and mechanical properties of amorphous W-Si-N sputtered films after thermal annealing, *Thin Solid Films* **441**, 150–160.

71. C. Louro, A. Cavaleiro (1999), The oxidation behaviour of mixed tungsten silicon sputtered coatings, *Thin Solid Films* **343–344**, 51–56.

72. P. Zeman, J. Musil (2006), Difference in high-temperature oxidation resistance of amorphous Zr-Si-N and W-Si-N films with a high Si content, *Appl. Surf Sci.* **252**, 8319–8325.

73. L. Ferreira, C. Louro, A. Cavaleiro, B. Trindade (2002), Influence of heat treatment on the structure of W-Si-N sputtered films, *Key Eng. Mater.* **230–232**, 640–643.

74. P. Zeman, J. Musil, R. Daniel (2006), High-temperature oxidation resistance of Ta-Si-N films with a high Si content, *Surf. Coat. Technol.* **200**, 4091–4096.

Appendix

Characterization of Thin Films

Surface coating techniques are widely used to modify the surface properties of components in order to improve their performance and reliability. The effort to enhance the protecting capability of hard ceramic thin films used in industry applications is driven by high demands to prolong lifetime of working tools over the uncoated material and simultaneously reduce both costs and energy consumption. However, a development of a new generation of coatings with exceptional properties for industry applications requires understanding the relationships between the structure, elemental and phase composition of the films, and their properties and performance. It is necessary for optimization of coating processes and controlling the properties during the deposition. This is, however, not an easy task since different applications commonly require specific film properties and many of these properties are only useful in certain areas of applications. With the increasing importance of thin film technology, the need has arisen for information on the properties of the films. Fundamental film properties, such as composition, structure, mechanical properties, residual stress, adhesion, and thermal properties, are most relevant to a wide range of applications.

The dimensions of fabricated components and devices are steadily decreasing making the role of individual growth defects and formed interfaces more evident. They can even dominate

Novel Nanocomposite Coatings: Advances and Industrial Applications
Rostislav Daniel and Jindřich Musil

ISBN 978-981-4411-17-2 (Hardcover), 978-981-4411-18-9 (eBook)
www.panstanford.com

the physical properties of thin films. This has necessitated the development of methods of investigation suitable for thin films with thicknesses in the micro- and nanometer range. However, under such conditions the measurements are no longer trivial and obtaining correct experimental values is extremely difficult. A determination of materials characteristics or properties in a defined and reproducible way requires advanced theoretical background, which subsequently allows interpreting the obtained results based upon existing elaborated theoretical models. Considerable benefits would be gained if these basic film characteristics are determined with sufficiently high precision. Hence, it is important to develop new advanced analytical techniques for thin film characterization since it is a necessity for progress in the film development. Only through a complex knowledge on the coating/substrate system at the atomic level it is possible to improve, tailor, and control the performance of the coated tools in engineering applications. This is also why high-resolution analytical techniques are important to deploy. Especially in nanostructures where surface properties may differ from those of film bulk, commonly affected by the substrate, and where interfaces also play an important role, the possibility to satisfactory determine film characteristics is of a high interest. Consequently, the performance of a coated component can be improved by optimizing each of these regions based upon the known relationships which exist between them. For this purpose, various characterization techniques for the analysis of thin films should be combined to fully characterize a material.

A.1 Mechanical Properties

Knowledge of mechanical properties of the near-surface region of thin films is very important for their usage in various applications. However, not only properties of thin films determine performance of the coated systems. Especially in the case of thin films, their mechanical response under given load depends on the substrate on which the film is deposited. Complex information of the whole system is therefore required to fully understand the deformation behavior under the applied load. The underlying substrate plays an essential role in film adhesion and tribology as well as in protective film capability since failure can occur in the near-surface substrate material if the substrate surface is mechanically weak. Therefore, the mechanical properties, particularly hardness and elastic modulus of both the film and the substrate have to be determined.

Hardness is not a fundamental material property, since it is strongly dependent on how it is measured. The hardness of a material is stated as the resistance of material to plastic deformation. Various measurement techniques for assessment of the penetration resistance of materials have been developed while measurements of the permanent deformation of a surface by a specifically shaped indenter under a given load have become the most popular method. During indentation tests, an indenter of known geometry is pressed into the surface under predefined load where the measured depth of penetration or the area of the residual impression serves for calculation of the resistance of the material against plastic deformation. This is usually expressed in terms of the indentation hardness [1]. The hardness value is equivalent to the mean pressure which a material is able to support under the indenter load. Since the hardness of a material, and especially of thin films, may be influenced by a deformation mechanism involved in the course of indentation, and thus by the grain size, phase composition, microstructure, amount of defects, or density, all these effects have to be taken into account by interpreting the measured data.

A typical load–displacement indentation curve is shown in Fig. A.1. It represents the relationship between load P and displacement h, which is continuously monitored and recorded during indentation. During loading, both elastic and plastic deformations occur under the indenter as the contact area alters with increasing penetration depth. While the loading part corresponds mainly to the plastic deformation, the unloading part of the curve is dominated by elastic displacements. Hence, the indentation elastic modulus can be determined by analyzing the unloading part of the indentation curve. The hardness is usually found by calculating the mean pressure under the indenter typically at the point of the maximum load. To do so, the contact area at the maximum load has to be known. It is, however, not necessary to measure the residual impression to get the contact area. Several procedures have been developed in order to find the contact area from the load–displacement curve and the knowledge of the shape of the indenter. Thus, both hardness and elastic modulus can easily be determined from the maximum penetration depth compared to the residual depth of the indentation after the indenter is removed (see Fig. A.1).

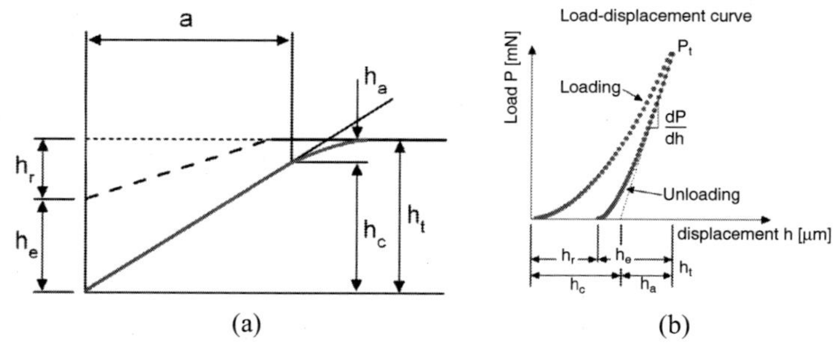

Figure A.1 Schematic diagram of (a) geometrical indentation parameters and (b) corresponding load–displacement curve.

The pattern of a real elastic-plastic deformation of a material during and after indentation can be fully described by following parameters: the depth measured during the indentation experiment h, which includes the contact depth h_c, representing the depth of indenter in contact with the material under load and the depression of the sample around the indentation $h_a = h - h_c$. That is caused by elastic displacements which must be subtracted from the data to obtain the actual depth of indentation. At maximum load, the load and displacement are P_t and h_t, respectively. The radius of the contact circle is a. Upon unloading, the elastic displacements in the contact region are recovered and when the indenter is fully removed, the final depth of the residual impression is h_r.

In order to remove the elastic contribution to the displacement and determine the projected area from the load–displacement curve using the calibrated indenter shape, several methods have been developed. Theoretical model of Doerner and Nix [2] and more complete and comprehensive model proposed by Oliver and Pharr [3] are the most widely used. By using the latter model, the reduced elastic modulus E^* can be calculated from

$$S = \frac{\mathrm{d}P}{\mathrm{d}h} = \frac{2}{\sqrt{\pi}} E^* \sqrt{A}, \tag{A.1}$$

where $S = \mathrm{d}P/\mathrm{d}h$ is the experimentally measured material stiffness derived from the upper portion of the unloading curve, and A is the projected area of the hardness impression. The reduced modulus is defined as

$$\frac{1}{E^*} = \frac{\left(1-\nu_f^2\right)}{E_f} + \frac{\left(1-\nu_i^2\right)}{E_i}. \tag{A.2}$$

Here, E_f and ν_f are the Young's modulus and the Poisson's ratio of the film, and E_i and ν_i are the same parameters for the indenter. Equation (A.1) has its origins in elastic contact theory. The projected area A of the residual impression is determined from the depth of contact and the geometry of the indenter, which, however, has to be known. Although the indenters are usually made with high precision and accuracy, the ideal geometry breaks down near the very tip, where tip rounding occurs either through use of the indenter or from the manufacturing process. Hence, the real area function $F(h_c)$ has to be experimentally determined for the used indenter tip before the measured data are analyzed. Under the assumption that the indenter does not itself deform significantly, the projected contact area at the maximum load can be computed from the relation

$$A = F(h_c) = C_1 h_c^2 + C_2 h_c + C_3 h_c^{\frac{1}{2}} + C_4 h_c^{\frac{1}{4}} + \ldots, \tag{A.3}$$

where C_1 is usually a number close to 26.43 in the case of the Vicker's diamond indenter and the remaining fitting constants account for the tip rounding and other departures from the ideal shape. In addition to the modulus, the indentation hardness can be computed from

$$H = \frac{P_t}{A}, \tag{A.4}$$

where A is the projected area of contact at the maximum load P_t evaluated from Eq. (A.3).

Material properties are easily to be characterized by comparing the load-displacement curves which characterize the response of the material to loading. It is schematically shown in Fig. A.2. A pure elastic material deforms elastically according to Young's modulus and the deformation is fully recovered during unloading. Hence, the unloading curve follows the loading one and no residual impression after unloading is observed. For a perfectly plastic material, no deformation occurs until yield stress is reached. Since plastic flow takes place during indentation, there is no recovery during unloading and the residual impression remains unchanged. On the contrary, an elastic-plastic material deforms first elastically according to Young's modulus and then it deforms plastically. There is a portion of recovery during unloading and thus the unloading curve more or less deviates from the loading curve.

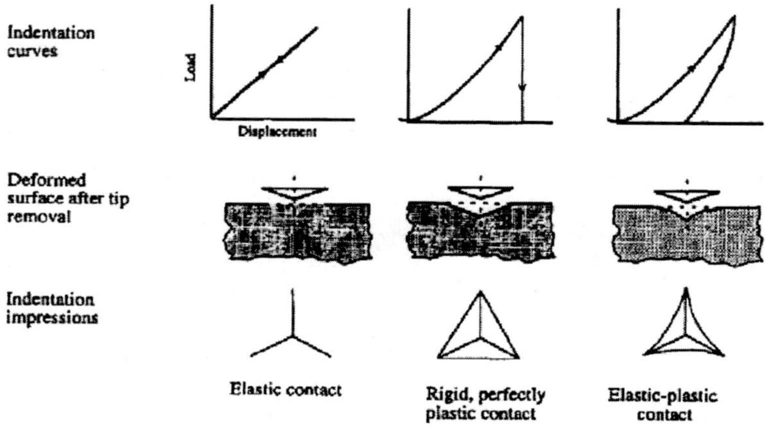

Figure A.2 Schematics of typical indentation curves, deformed surfaces after tip removal and residual impressions of indentation for ideal elastic, perfectly plastic, and real elastic-plastic solids.

In depth-sensing indentation hardness tests, the contact area is determined by measuring the indentation depth during the indentation measurement. However, the instrumented data are typically affected by various sources of error which lead to inaccurate values for modulus and hardness of the test specimen. All these errors need to be corrected before the measured indentation depth is used for calculation of both the hardness and modulus. Three of the most common and necessary corrections are the initial penetration, the instrument frame compliance and the indenter area function. The latter one is usually determined by performing a series of indentations into a material for which the elastic modulus or hardness is known. It is important to do so over the whole load range of the instrument that every measurement performed at any load can be consequently corrected for the real shape of the indenter tip. Another important procedure in nanoindentation testing is the correction of the initial contact of the indenter with the specimen. It is very important to know the initial contact since it determines the initial penetration depth of the load–displacement curve. In principal, the initial penetration correction has the effect of shifting the load–displacement curve so the first depth reading corresponds to the zero position. The instrument frame compliance is another contribution to the depth readings arising from deflections of the load frame instead of displacement into the specimen material. The deflection is usually linearly proportional to the applied load in the manner of a linear spring and needs to be corrected as well. The compliance correction is particularly important when testing very stiff materials. Since the contribution of soft materials is relatively low compared to the stiffer ones, the instrument compliance should be determined on a variety of standard specimens with different stiffness over the whole load range of the test. It is worthy to note that without the compliance correction, the calculated values for hardness and modulus can differ up to 50% from the accurate values.

In addition, there are some materials-related issues which are also important but more difficult to correct. They are the indentation size effect and piling-up and sinking-in effects around the indentation. Both raising up and sinking of the surface at the edge of the indentation significantly affect the calculated material properties. For instance, the piling-up supports the load applied to the indenter which consequently penetrates less than without this effect. Since it is difficult to be corrected, the tested material appears to be stiffer and harder. The only method to eliminate this error is a direct measurement of the residual impression after the indenter is removed. Both effects are illustrated in Fig. A.3.

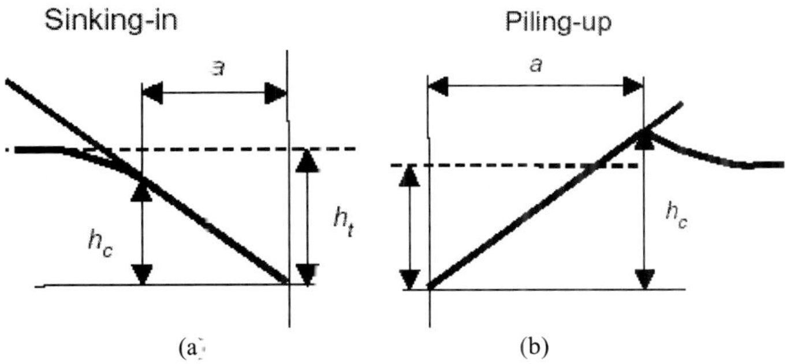

Figure A.3 Schematics of (a) sinking-in and (b) piling-up effects in indentation testing. Reprinted from Ref. [4], Copyright (2011), with permission from Elsevier.

Indentation in thin films which are deposited usually on various substrates is very difficult compared to the bulk materials since it is not easy to separate contributions of both the film and the substrate to the deformation occurred and to measure the properties of a thin film only. The thinner the film, the stronger the influence of the substrate. To obtain a bulk hardness value for a film, not affected by the presence of the underlying substrate, it is necessary to satisfy the requirement that the film thickness is minimally ten times greater than the penetration depth of the indenter under the maximum load [5,6]. This criterion for measuring the plastic properties of films ensures that the film yields before the substrate. It means that the maximum stresses induced by nanoindentation remain in the film even if the yield stress of the film is exceeded at maximum contact load [1]. The size of the contact in relation to the thickness of the film plays an essential role in the determination of the plastic hardness. The final results may be strongly affected by plastic deformation once it occurs not only in the film but also in the substrate. Especially for a combination of a hard film on a very soft substrate or a soft film on very hard substrate, this effect is greatly pronounced. What in general happens in both the film and the substrate during the indentation experiment is illustrated in Fig. A.4.

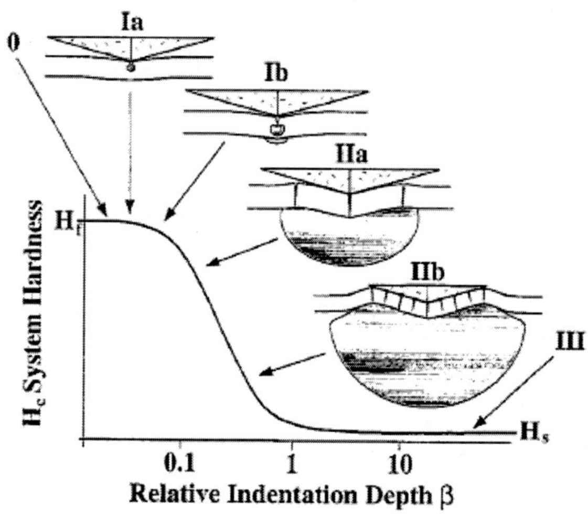

Figure A.4 Schematic representation of the principal deformation mechanisms occurring under the indenter load. Reprinted from Ref. [7], Copyright (1998), with permission from Elsevier.

While in region 0 only elastic response of the film-substrate system is typically observed, at higher applied loads initial highly localized elasto-plastic response in the film and fully elastic response in the substrate can be found. This transition to plastic behavior as the load applied increases is associated with the contact induced stresses exceeding the yield stress of the material being tested. The plastic zone under the indenter load in this range is, however, fully developed just if a sharp indenter penetrates into the measured sample. If the indenter tip is blunt or even a spherical indenter is used, pure elastic response mostly occurs and thus much higher load is necessary for onset of plastic deformation. Next region IIa is also characterized with plastic deformation in the substrate which results in through-thickness fracture which grows with increasing load. This region is characteristically dominated by the substrate but for a significant depth there still remains a contribution from the film as it bends and fractures around the indentation site. Region IIa is followed by region IIb where a complete through-thickness fracture occurs. The hardness value of the film-substrate system approaches slowly that one corresponding to the pure substrate. The true substrate-only response at high loads within region III reveals behavior where the energy absorbed by film, stretching, flexure, and fracture is insignificant compared to the amount dissipated by substrate plastic deformation. At this stage, the measured hardness values correspond to the hardness of the underlying substrate only [7].

As the substrate contribution becomes significant, the shape of the loading curves may alter as a consequence of the plastic deformation of the substrate or the onset of through-thickness fracture or interfacial detachment [8]. While a gradual change in the loading curve is attributed mainly to plastic deformation of the film, a sharper transition represents fracture in the film or in the substrate. A through-thickness fracture of the film, delamination, or even nucleation of plasticity in the underlying substrate is usually reflected by sharp displacement discontinuities on loading [9,10]. The effect of a hard substrate on the plastic behavior of a softer film is clearly visible in the load–displacement curve already at very low loads. The penetration depth is relatively great at a low load, however, the slope of the loading curve markedly increases with indenter penetration. It is caused by the transition between the film and substrate dominated behavior with increasing penetration depth. Although deformation of the coated system can be observed directly from the load-displacement curve, in order to clarify which deformation mechanism involved and where this occurred, microscopical analysis is highly recommended if features in the load–displacement curve are to be fully understood. Processes occurring in the film and the substrate in the course of the indentation testing are summarized in Fig. A.5.

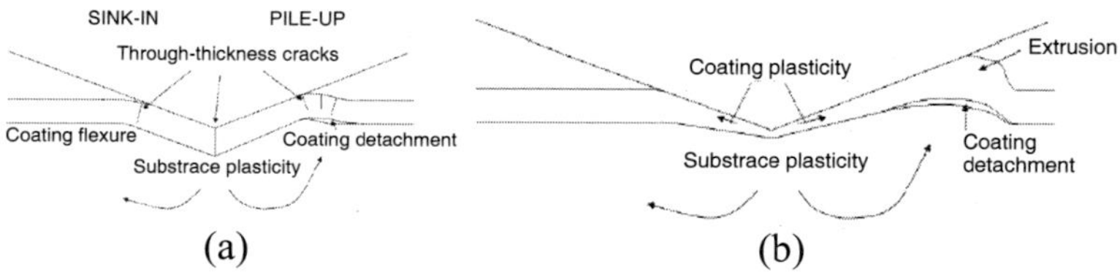

Figure A.5 Deformation in (a) a hard film on a soft substrate and (b) a soft film on a hard substrate. Reprinted from Ref. [1], Copyright (2005), with permission from IOP Publishing.

It has been widely observed that during indentation testing of a hard film deposited on a softer substrate, plastic deformation of both components may occur [11]. Radial and picture frame cracks are often observed at higher loads since the film is bent into the impression left by plastic deformation of the substrate. On the other side, just a little or no fracture occurs in the film but the film is dramatically thinned beneath the indenter by plastic deformation. This phenomenon is commonly associated with the extrusion of the film from the edge of the contact as it is trapped between a rigid indenter and a substrate. This can be clearly seen in real scanning electron microscopy (SEM) plane-view images shown in Fig. A.6.

A.1.1 Practical Implementation

In the last few years, a wide use of thin films in many applications initiated developing of new measuring devices that allow characterizing the mechanical properties of small volumes of material at the submicron scale. The main driving force in developing such instruments was to determine film properties without major contribution from substrate behavior. It is generally not an easy task since the deformed volumes of thin films are usually very small and ultrasensitive transducers for force and depth, fabrication of indenters of precise geometry and advances in theoretical analysis are highly required for such measurements. New generation of indentation instruments are designed so they continuously record the load and indenter displacement as it is pushed into and removed from the test surface. A schematic of a depth-sensing indentation tester is shown in Fig. A.7.

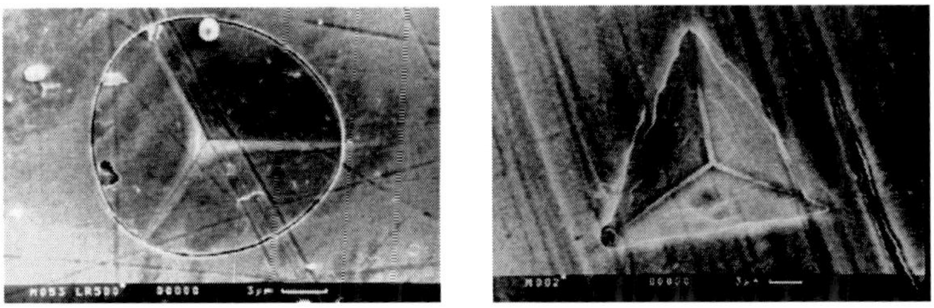

(a) (b)

Figure A.6 SEM micrographs of residual impressions after high load nanoindentation tests of hard NbN film. (a) A circular through-thickness crack at the edge of the plastic zone occur if the film is on a soft substrate, (b) picture frame cracking at the edge of the impression is observed when the substrate is hard. Reprinted from Ref. [1], Copyright (2005), with permission from IOP Publishing.

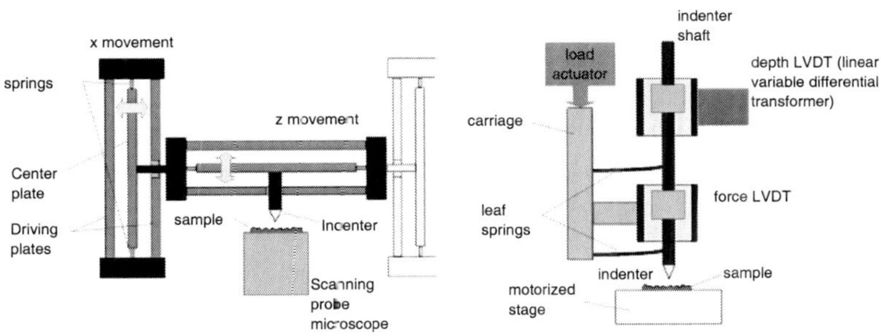

(a) (b)

Figure A.7 Schematics of the operating principle of depth-sensing nanoindentation instrument. (a) Capacitive and (b) inductive approach. (www.nanoindentation.cornell.edu)

Nanoindentation instruments are usually load-controlled machines. They typically measure depth of penetration using either inductance or capacitance displacement sensors. Load is applied by the expansion of the piezoelectric element, the movement of a coil in a magnetic field, or electrostatically. The high performance of the instruments attained by high sensitive electronics allows for the use of ultra-low loads which is necessary in determining the mechanical properties of tested materials at very shallow contact depths. The precise control of either the load or displacement during the measurement consequently allows the test to be performed with applied forces as low as a few micronewtons making indentations with depths in the nanometer range. Another very important feature of continuous depth and load recording is the fact that the film properties can be obtained directly from the measured data without the need to measure the residual impression. It is often very difficult to find, image, and precisely measure indentations made at such low loads. The depth-sensing indentation measurements significantly reduce errors associated with a determination of the residual impressions and serves for accurate values of both hardness and modulus of even ultra-thin films.

A.2 X-Ray Diffraction Analysis

X-ray diffraction (XRD) is a high-tech, nondestructive technique for analyzing a wide range of materials, including fluids, metals, thin films, ceramics, minerals, polymers, catalysts, plastics, and semiconductors. Throughout industry and research institutions, XRD has become an indispensable method for materials investigation, characterization, and quality control. Example areas of application comprise qualitative and quantitative phase analysis, crystallography, structure and relaxation determination, texture and residual stress investigations or micro-diffraction.

XRD analysis is a powerful method by which X-rays of a known wavelength are passed through a sample producing unique patterns by the diffraction of X-rays through the closely spaced lattice of atoms in a crystal. They are further analyzed in order to reveal the nature of the crystal structure. In principle, X-rays primarily interact with electrons in atoms. When X-ray photons collide with electrons, some photons from the incident beam will be deflected away from the direction where they original travel. If the X-ray photons did not lose any energy, the process is called elastic scattering, since only momentum has been transferred in the scattering process. By the elastic scattering, the wavelength of the scattered X-rays does not change. These scattered X-rays carry information about the electron distribution in materials and hence are measured in diffraction experiments. Diffracted waves from different atoms can interfere with each other and the resultant intensity distribution is strongly modulated by this interaction. If the atoms are arranged in a periodic fashion, as in crystals, the diffracted waves will consist of sharp interference maxima with the same symmetry as in the distribution of atoms. Measuring the diffraction pattern therefore allows us to deduce the distribution of atoms in a material. The peaks in an XRD pattern are directly related to the atomic distances. The interplanar distance d in the crystal lattice can be determined using Bragg's law, which represents the condition for a diffraction to occur. It is given by

$$n\lambda = 2d \sin \theta, \tag{A.5}$$

where λ is the wavelength of the X-ray, θ the angle of diffraction, and n an integer representing the order of the diffraction peak. In the course of the analysis, the X-ray detector moves around the sample and measures the intensity of the diffracted peaks and their position 2θ. The Bragg's law is one of most important laws used for interpreting XRD data since not just atoms but also any periodic distribution of electron density represents scattering centers and provides diffraction while the Bragg's law conditions for a diffraction is fulfilled. This means that any ordered structures, such as colloids, polymers, proteins, or virus particles, can be investigated by means of XRD.

A wide range of materials and their properties, which can be measured by XRD analysis, also make this method very popular in the field of thin film testing. Basic XRD measurements made on thin film samples include mainly precise lattice constants measurements derived from 2θ to θ scans, which provide information about lattice mismatch between the film and the substrate and therefore is indicative of strain and stress. By rocking curve measurements made by doing a θ scan at a fixed 2θ angle, the dislocation density in the film can be obtained from the width of the measured curve serving as a gauge of the quality of the film. Glancing incidence X-ray reflectivity measurements is often used to determine the thickness, roughness, and density of the film. XRD is also commonly used to determine the crystallite size and lattice strain by means of the analysis of the diffraction line broadening. Both these effects contribute to the change of the peak profile, which is characterized by the full width at half maximum height (FWHM) value of the diffracted peak [12]. If the peak broadening is corrected for the instrumental broadening, the crystallite size and lattice strain can be determined by the deconvolution of the measured profile. The broadening due to crystallite size β_c is given by the following equation:

$$\beta_c = \frac{K.\lambda}{d.\cos\theta}, \tag{A.6}$$

where d is the grain size, λ the X-ray wavelength, θ the Bragg angle, and K the shape factor for the crystallites which varies with reflection (hkl). Similarly, the broadening due to lattice strain, β_e, is given by the following equation:

$$\beta_e = 4e \tan\theta, \tag{A.7}$$

where e is the lattice strain. The total broadening is the sum of these contributions and if the constant K is set to unity we have the following equation:

$$\beta\frac{\cos\theta}{\lambda} = \frac{1}{L} + \frac{4e\sin\theta}{\lambda}, \tag{A.8}$$

where β is the measured broadening corrected for instrument effects. Analysis of the peak profile is usually performed by fitting the diffraction peak with a Pseudo-Voigt function, which is a linear combination of a Cauchy and a Gaussian function. A Cauchy function is related to the crystallite size and a Gaussian function arises from the lattice strain. Both components are obtained from the ratio of FWHM of the broadened diffraction peak to its integral breadth, which is calculated from the ratio of the peak area and the maximum intensity of the peak [12]. The Hall–Williamson plot, a plot of $\beta\cos\theta/\lambda$ vs. $2\sin\theta/\lambda$, represents another method for the separation of microstrain and crystallite size. In this plot, a straight line with a slope of $2e$ and an intercept $1/L$ is obtained. Since the lattice strain depends on the order of the reflection but the crystallite size is fully independent on it, two orders of the same set of crystallographic planes are usually used to separate L from e.

In addition, XRD enables residual stress analysis which relies on the study of the elastic deformations within a polycrystalline material. The deformations cause changes in the spacing of the lattice planes from their stress free state. While microstrain is associated to a broadening of the diffraction peak, the strain of a certain set of lattice planes in a certain direction is observed as a shift of the 2θ peak position with respect to the position of corresponding stress-free powder standard. By recording the peak shift as a function of the sample tilt angle ψ, the residual stress present can be determined using elasticity theory. The new lattice spacing will be the same in any similarly oriented planes, with respect to the applied stress and the crystal lattice therefore effectively acts as a very small strain gauge [13]. The residual stresses are related mainly to the nucleation and growth processes since particularly an amount of the generated defects contributes to the changes in the crystal lattice. Moreover, also the mismatch in the thermal expansion coefficients of the film and the substrate plays a significant role, as it is acting as a strong generator of the residual stress.

A.3 Stress Measurement

Controlling the stresses in thin films is very important for their successful application, since excessive tensile or compressive stress can lead to distortion of the substrate and can cause cracking or buckling of the film. High stress levels can even fracture substrates in some applications. Stress measurement is therefore needed for optimization of film properties through the film composition and deposition conditions as well. Residual stresses of thin films can be composed of stresses that arise from the growing process itself due to the adatoms not being located in their most energetically favorable positions. Thermal stress originates from different thermal expansion coefficients between the substrate and the film when the deposition is performed at elevated temperatures. Extrinsic stress can then be caused by phase transformation and precipitation processes, chemical reactions, and recrystallization processes after deposition namely at room or elevated temperatures. The total residual macrostress in the film is determined by summing up all the individual stress components. The residual stress can vary through the film thickness as a consequence of the growth mode and differing thermal histories of the various layers of the film. Measurements of residual stress gradients within the film thickness reveal a strong relationship between the grain size, amount of defects, and the stress level showing higher stresses near the interface than near the surface [14]. The effect of the film morphology on the stress buildup throughout the film thickness significantly influences the adhesion and stability of films and thus sets high demands for the film optimization. Moreover, the anisotropic stress distribution in thin films may result in compressive stress in one direction and tensile in another. The residual stresses also cause a lattice strain. It represents a stored energy in the lattice that plays an important role in the resistance of the film to plastic deformation. The lattice strain can be measured by XRD analysis by measuring the lattice parameters and translated into a stress by assuming the mechanical properties of the material. However, it is not summing over all the stress components and characterizes the stress just over small regions [15].

An illustrative model for the residual stress generation during the film deposition is depicted in Fig. A.8. If the film initially tends to shrink relative to the substrate due to the surface tension forces or the coalescence phenomenon, the tensile forces developed in the film are balanced by inverse compressive forces in the substrate. That is driven by the compatibility requirement of both the substrate and the film to have the same length. However, the combination is still not in mechanical equilibrium because of the uncompensated end moments. If the coated system is not restrained from moving, it will elastically bend to counteract the unbalanced moments. Hence, films containing tensile residual stresses bend the substrate concavely upward. Since the film with the tensile stress try to contract, bowing the substrate so the film is on the concave side, the stresses relieve typically by microcracking the film once the critical values are exceeded. On the other hand, films under compression will try to expand relative to the substrate which is compensated by tensile forces in the substrate. The compressive stresses in the films tend to bend the substrate convexly outward so the film is on the convex side which results typically in buckling or a wavy pattern after the stress relief. In an entirely similar manner, bending of thin substrates occur due to the mismatch in the thermal expansion coefficients between the film α_f and the substrate α_s after cooling the coated system to room temperature showing that this behavior is generally regardless of the specific mechanism that cause films to stretch or shrink relative to substrates. Films prepared at elevated temperatures are residually compressed when measured at room temperature if $\alpha_s > \alpha_f$. Thus, the substrate shrinks more than the film. Since the substrate is not allowed to contract fully, it is placed in tension and the film hindered from shrinking is consequently forced into compression [17].

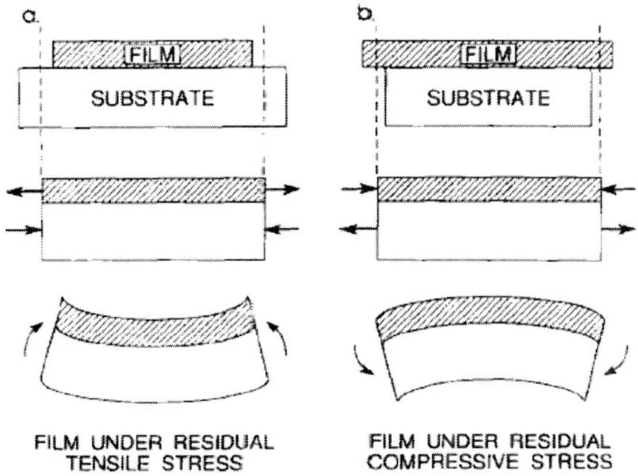

Figure A.8 Model for the generation of internal stress during the deposition of films. Reprinted from Ref. [16], Copyright (2011), with permission from Elsevier.

Determination of deflection of one side coated thin substrates due to lattice misfit, differential thermal expansion coefficients, or film growth effects is the most general technique for measuring the film stress generated by these mechanisms. By knowing the mechanical properties of the substrate and film material, the film thickness, and the substrate deflection, the film stress can be calculated. A major uncertainty in measuring film stress is the elastic modulus and Poisson's ratio of the film material which has to be assumed in most cases. However, if the thicknesses of the substrate h_s and the film h_f are chosen such that the h_s/h_f ratio is very large, the magnitude of the residual stresses can be determined by means of the modified Stoney's formula without knowing or assuming E_f. The substrate to film thickness ratio plays an essential role, since too thin film deposited on a thick substrate cannot induce forces high enough to bend the substrate which leads to an expressive error in calculated stress values. The Stoney's equation

$$\sigma = \frac{E_s}{1-\nu_s}\frac{h_s^2}{6h_f}\frac{1}{r} \tag{A.9}$$

represents the extent of the film-substrate curvature caused by the generated residual stress [18]. Here, E_s and ν_s are the Young's modulus and the Poisson's ratio of a substrate, respectively, and r is the radius of a film-substrate curvature. Since the force on a substrate induced by the film stress is proportional to the film thickness, the thickness of the film h_f and substrate h_s have to be known. In order to ensure a development of the biaxial stress in the films and homogeneous film properties along the substrate, the right geometric setup of the substrate has to be chosen. It is commonly accepted that the width of the substrate should be less than a half the length.

The curvature of the substrate is used to measure by optical methods using parallel laser beams [16], interferometers [19], or long-working-distance microscopes [20]. XRD analysis can be used as an alternative technique for determining the substrate bending. The radius of curvature of Si substrate is, for instance, determined by the measurement of rocking curves of Si 400 reflections at different sample positions. XRD also serves to determine stresses in thin film from changes of the lattice parameter d with respect to that of unstressed film. Here, positions of strong X-ray reflections are analyzed by $\sin^2\psi$ method providing information of the stresses in films from the $\sin^2\psi$-d plot.

Important contributions to film stress represent thermal effects. Films prepared at elevated temperature are thermally stressed after cooling to room temperature. If the temperature of the coated system is reduced, it tends it to shrink in length by an amount attributed to the coefficient of linear expansion. By knowing the thermal expansion coefficients of the substrate and the film, thermal stresses can be calculated according to the following equation:

$$\sigma_{th} = \frac{E_f}{1-\nu_f}(\alpha_s - \alpha_f)(T - T_s) \tag{A.10}$$

Here, E_f and ν_f are the Young's modulus and Poisson's ratio of the film, respectively, T_s is the substrate temperature during the film deposition, and T is the temperature at which σ is measured. This modification of Stoney's equation represents the extent of bowing when differential thermal expansion effects cause the stress.

The coefficients of thermal expansion of thin films can be determined by performing the film stress measurement at various temperatures. During post-deposition annealing, σ in the film linearly increases with increasing annealing temperature T_a and follows the thermoelastic line defined by Eq. (A.10). The thermal expansion coefficient of the film α_f can be calculated from the slope of this thermoelastic line corresponding to the heating or cooling cycle. Using Eq. (A.10), the following formula for calculation of α_f can be derived:

$$\alpha_f = \left(\frac{1-\nu_f}{E_f} \frac{\Delta\sigma}{\Delta T} \right) + \alpha_s \tag{A.11}$$

where $\Delta\sigma$ is the increase in σ in the interval ΔT. In order to calculate the thermal expansion coefficient of the film, the thermal expansion coefficient of the substrate has to be known.

A.4 Film Thickness Measurement

The thickness of a film is among the first quoted attributes of its nature. The reason is that thin film properties very often markedly depend on thickness. In some applications, such as decorative or metallurgical, the actual film thickness is not crucial to the film performance within broad limits. However, in microelectronic or optical applications, the maintenance of precise and reproducible film thicknesses is required. The film thickness can be determined by several contact and noncontact techniques, while the most generally used technique is the surface stylus profilometry. This contact thickness measuring technique measures a defined step height from substrate to film surface, which is typically generated by masking the substrate during deposition or by post-deposition etching. The stylus method is based upon measuring the mechanical movement of a stylus as it is made to trace the topography of a film-substrate step. For this purpose, a diamond needle stylus with a tip radius of several microns is commonly used, serving as the electromagnetic pickup. The stylus is adjustable from 1 to 30 mg and can scan a length of several centimeters with a horizontal position resolution of <0.2 microns and give height measurements of <10 Å. The film thickness is directly read out as the height of the resulting step-contour trace. In order to get high accuracy and precision with thickness measurements, it is necessary to ensure constant temperature during the measurement and to avoid the influence of the data readings through the background vibrations. The accuracy of the stylus method is limited by several factors. By measuring very soft films, stylus can penetrate to the film and scratch the surface. Furthermore, uncertainties in the position of the step can be caused by the substrate roughness introducing an excessive noise into the measurement.

A.5 Scanning Electron Microscopy

Scanning electron microscopy (SEM) is the most useful surface imaging technique. This electron optical system uses electron-beam-generated secondary electrons for imaging of solids. The electron gun produces an electron beam which is a demagnified image of the source. The beam is imaged by the probe forming lens onto specimen, where it is deflected by scanning coils over a rectangular raster. Its size, relative to the display screen, determines the magnification. The emitted electron signal is collected by detectors and used to modulate the intensity of the beam of the display video screen. It is typically rastered in synchronism with the probe to image the sample surface. When an electron is incident on the surface of the sample, several signals can be detected. The primary electron results in ionization of atoms along its path in the solid, which consequently results in generation of secondary electrons very close to the incident beam position. As the high energy primary electrons penetrate the solid, they undergo scattering which increases the interaction volume. The primary electrons can be backscattered toward the surface with little or no loss in energy and in turn collected by detectors. Moreover, energetic primary electrons ionize atoms in the solid producing X-rays. They are characteristic of the elements that are present in the sample and with suitable detectors may be detected to provide elemental analysis [21].

The secondary electron yield per primary electron is relatively high and hence, secondary electrons (SE) are mostly utilized to image the surface. In order to efficiently collect the secondary electrons, they are attracted to a positively charged detector which is positioned in proximity to the sample (Fig. A.9). The signal is converted to light and fed out through a light pipe to a photomultiplier tube. Compared to the optical microscopes, scanning electron microscopes provide surface images with higher magnification and with greater depth resolution [22].

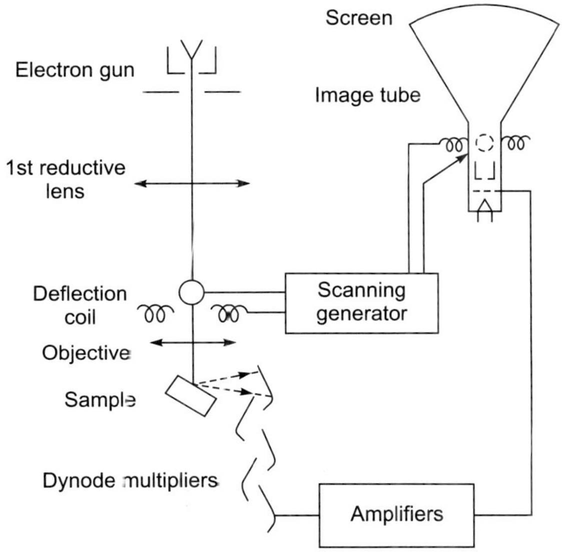

Figure A.9 Schematic diagram of a SEM microscope showing the principle of sample imaging (copyright CAMECA Archives).

The primary electrons can be deflected through large angles which commonly cause their backscattering from a region surrounding the incident probe. The backscattered electrons (BSE) are emerged from the surface with high energy, simultaneously generating secondary electrons. Since the backscattering yield of primary electrons within an elemental sample of atomic number Z increases with Z, BSE provides contrast which is composition dependent. Furthermore, the high energy BSE sample much greater depths than low energy secondary electrons [23].

A.6 Energy Dispersive X-Ray Spectrometry

One of the most common analytical attachments to the SEM is the energy dispersive X-ray spectrometer (EDX). Since the primary electron beam impacts the sample surface with very high energy, it causes emission of a core electron, which leaves the atom in an excited state. In the course of de-excitation process, the characteristic X-ray is emitted corresponding to the element from which it originates. Peaks at energies characteristic of the elements within that volume are detected by a solid state detector positioned in the vicinity of the sample and consequently identified. X-rays interacting with the detector create electron-hole pairs converted to a voltage pulse by a charge-sensitive preamplifier. The analyzed depth of the sample depends on the path of the X-rays; however, the EDX signals mostly originate from depths around a half of micrometer or more. Within this bulk region, the concentrations of the elements can be determined. The characteristic X-ray spectrum can be detected for all elements above F in the periodic table. In order to obtain an elemental composition of the investigated sample, a wavelength dispersive X-ray analysis (WDX) is also often used, however, in a much longer time [21].

A.7 Differential Scanning Calorimetry

Thermal analysis and calorimetry represent measuring methods investigating changes in the sample, which is subjected to elevated temperatures. These methods are based upon monitoring changes in variables of state of the sample, such as temperature, mass or volume, to determine processes involved at varying sample temperature. In this way, heat of transition, heat capacity, temperatures and enthalpies of phase transformation, thermal hysteresis data, and melting temperatures can be obtained. The generation or consumption of heat connected with these processes is picked up by sensors and transformed into electric signals serving to obtain quantitative information about occurred changes [24].

Differential scanning calorimetry (DSC) is a thermoanalytical technique in which the change of the difference in heat flow to the sample and to a reference sample is analyzed while they both are exposed to a temperature alteration. Generally, the temperature program for a DSC analysis is designed such that the temperature increases linearly as a function of time. Both the sample and reference are maintained at the same temperature throughout the experiment.

The reference sample should have a well-defined heat capacity over the range of temperatures to be scanned. The temperature difference between sample and reference sample is recorded using a differential thermocouple as a direct measure of the difference in the heat flow rates to these samples. When the sample undergoes a physical transformation, such as phase transitions, more or less heat will need to flow to it than the reference to maintain both at the same temperature. Whether more or less heat must flow to the sample depends on whether the process is exothermic or endothermic. While a solid sample melts to a liquid it will require more heat flowing to the sample to increase its temperature at the same rate as the reference. This is due to the absorption of heat by the sample as it undergoes the endothermic phase transition from solid to liquid. Similarly, as the sample undergoes exothermic processes, such as crystallization, less heat is required to raise the sample temperature. By observing the difference in heat flow between the sample and reference, differential scanning calorimeters are able to measure the amount of energy absorbed or released during such transitions [25].

A typical differential scanning calorimeter consists of two sealed crucibles, as it is depicted in Fig. A.10. The first one is a sample crucible and the second a reference crucible, which is generally empty. These crucibles are composed of alumina, which acts as a radiation shield. The two crucibles are heated, or cooled, uniformly while the heat flow difference between the two is monitored. This can be done at a constant temperature, but is more commonly done by changing the temperature at a constant rate. This mode of operation is called temperature scanning. During the experiment, the instrument detects differences in the heat flow between the sample and reference through the differential thermocouple sensors. This signal is sent to a computer and results in a plot of the differential heat flow between the reference and sample cell as a function of temperature. When there are no thermodynamic chemical processes occurring the heat flow, difference between the sample and reference varies only slightly with temperature and shows up as a flat or very shallow base line on the plot. However, an exothermic or endothermic process within the sample results in a significant deviation in the difference between the two heat flows. The result is a peak in the DSC curve. Generally, the differential heat flow is calculated by subtracting the sample heat flow from the reference heat flow. When following this convention, exothermic processes will show up as positive peaks above the baseline while peaks resulting from endothermic processes are negative and displayed below the baseline. In general, DSC analysis requires very small amount of the measured material of only a few milligrams in mass. The dynamical DSC experiments are typically carried out in a flow of inert gas to avoid oxidation of the investigated sample during the measurement.

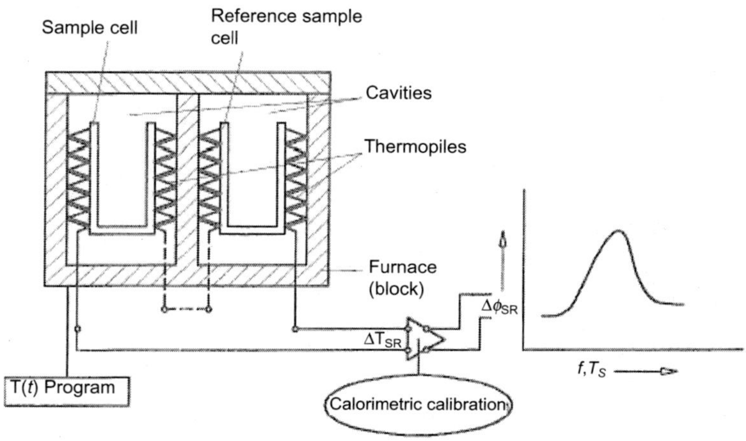

Figure A.10 Cylinder-type DSC measuring device with two sealed crucibles. Reprinted from Ref. [24], Copyright (2011), with permission from Elsevier.

All thermoanalytical methods are nonequilibrium methods, since the temperature in the course of the experiments does not change slowly enough that there is thermal equilibrium both between the sample and its surroundings and inside the sample. Therefore, the changes in the sample are heating-rate-dependent.

A.8 Thermogravimetric Analysis

Thermogravimetric analysis (TGA) is a measuring technique in which the change in the sample weight is analyzed while the sample is subjected to a controlled temperature alteration in a defined atmosphere. Thermogravimetry thermal analysis provides testing for a wide range of materials including metals, thin films, polymers, composites, laminates, adhesives, organic materials, chemicals, or biological samples to determine their characteristics, degradation temperatures, absorbed moisture content of materials, or decomposition points. It is also often used to estimate the corrosion kinetics in high temperature oxidation [26].

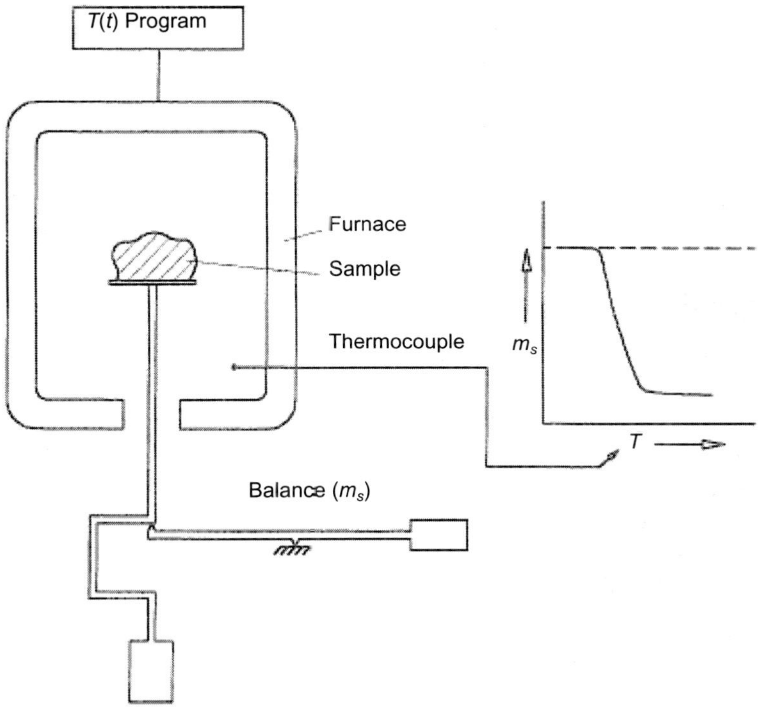

Figure A.11 Schematic of TGA measuring system showing the principle of thermogravimetric analysis. Reprinted from Ref. [27], Copyright (2011), with permission from Elsevier.

In principle, TGA uses heat to force reactions and physical changes in materials and by monitoring of mass change in materials provides quantitative measurement of changes associated with transition and thermal degradation, as schematically depicted in Fig. A.11. Changes in mass from dehydration, decomposition, and oxidation of a sample with time and temperature are recorded and displayed as thermogravimetric curves. These curves are characteristic, since physicochemical reactions in individual materials and chemical compounds occur over specific temperature ranges and heating rates. In TGA, a high degree of precision in the measurement of weight and temperature is required to exactly determine at which temperature a weight loss or gain occurs.

References

1. S.J. Bull, Nanoindentation of coatings, *J. Phys.* 38 (2005) 393–413. doi:10.1088/0022-3727/38/24/R01
2. M.F. Doerner, W.D. Nix, A method for interpreting the data from depth-sensing indentation instruments, *J. Mater. Res.* 1 (1986) 601–609.
3. W.C. Oliver, G.M. Pharr, An improved technique for determining hardness and elastic modulus using load and displacement sensing indentation experiments, *J. Mater. Res.* (1992) 1564–1583.
4. A.C. Fischer-Cripps, Critical review of analysis and interpretation of nanoindentation test data, Surf. *Coat. Technol.* 200 (2006) 4153–4165.
5. S.V. Hainsworth, M.R. McGurk, T.F. Page, The effect of coating cracking on the indentation response of thin hard-coated systems, *Surf. Coat. Technol.* 102 (1998) 97–107.
6. E.G. Berasetegui, T.F. Page, The contact response of thin SiC-coated silicon systems-characterisation by nanoindentation, *Surf. Coat. Technol.* 163–164 (2003) 491–498.
7. A.M Korsunsky, M.R. McGurk, S.J. Bull, T.F. Page, On the hardness of coated systems, *Surf. Coat. Technol.* 99 (1998) 171–183.
8. W.C. Oliver, C.J. McHargue, Characterizing the hardness and modulus of thin films using a mechanical properties probe, *Thin Solid Films* 161 (1988) 117–122.
9. S.V. Hainsworth, T. Bartlett, T.F. Page, The nanoindentation response of systems with thin hard carbon coatings, *Thin Solid Films* 236 (1993) 214–218.
10. T. Wright, T.F. Page, Nanoindentation and microindentation studies of hard carbon on 304 stainless steel, *Surf. Coat. Technol.* 54–55 (1992) 557–562.
11. C. Feldman, F. Ordway, J. Bernstein, Distinguishing thin film and substrate contributions in microindentation hardness measurements, *J. Vac. Sci. Technol.* A8 (1990) 117–122.
12. H.P. Klug, L.E. Alexander, *X-Ray Diffraction Procedures for Polycrystalline and Amorphous Materials*, Wiley Interscience, New York, 1974.

13. D.S. Rickerby, A.M. Jones, B.A. Bellamy, X-ray diffraction studies of physically vapour-deposited coatings, *Surf. Coat. Technol.* 37 (1989) 111–137.

14. H. Oettel, R. Wiedemann, Residual stresses in PVD hard coatings, *Surf. Coat. Technol.* 76–77 (1995) 265–273.

15. P.S. Prevey, X-ray diffraction residual stress techniques, in R.E. Whan (Ed.), *Materials Characterization*, ASM Metals Handbook, 1986.

16. M. Ohring, *The Material Science of Thin Films*, Academic Press, San Diego, 1992.

17. H. Oettel, R. Wiedemann, S. Preißler, Residual stresses in nitride hard coatings prepared by magnetron sputtering and arc evaporation, *Surf. Coat. Technol.* 74–75 (1995) 273–278.

18. J.D. Wilcock, D.S. Campbell, A sensitive bending beam apparatus for measuring the stress in evaporated thin films, *Thin Solid Films* 3 (1969) 3–12.

19. A.E. Ennos, Stresses developed in optical film coatings, *Appl. Opt.* 5 (1966) 51–62.

20. H.P. Murbach, H. Wilman, The origin of stress in metal layers condensed from the vapour in high vacuum, *Proc. Phys. Soc.* B66 (1953) 905–910.

21. K.F.J. Heinrich, in R. Castaing, P. Deschamps, J. Philbert (Eds.), *X-Ray Optics and Microanalysis*, Paris, 1966.

22. D.C. Joy, A.D. Romig, Jr., J.I. Goldstein, *Principles of Analytical Electron Microscopy*, Plenum Press, New York, 1986.

23. D.E. Newbury, H. Yakowitz, in S.I. Goldstein, H. Yakowitz (Eds.), *Practical Scanning Electron Microscopy*, Plenum Press, New York,1975.

24. P.J. Haines, M. Reading, F.W. Wilburn, Differential thermal analysis and differential scanning calorimetry, in P.K. Gallagher, M.E. Brown (Eds.), *Handbook of Thermal Analysis and Calorimetry*, Elsevier, Amsterdam, 1998.

25. G. Hühne, W. Hemminger, H.J. Flammersheim, *Differential Scanning Calorimetry*, Springer-Verlag, Berlin, 1996.

26. P.H. Mayrhofer, H. Willmann, C. Mitterer, Oxidation kinetics of sputtered Cr-N hard coatings, *Surf. Coat. Technol.* 146–147 (2001) 222–228.

27. P.K. Gallagher, Thermogravimetry and thermomagnetometry, in P.K. Gallagher, M.E. Brown (Eds.), *Handbook of Thermal Analysis and Calorimetry*, Elsevier, Amsterdam, 1998.

Index